BARRON'S

THE TRUSTED NAME IN TEST PREP

AP® Biology
PREMIUM

Mary Wuerth, M.S.

AP® is a registered trademark of the College Board, which is not affiliated with Barron's and was not involved in the production of, and does not endorse, this product.

Acknowledgments

I wish to thank:
— My husband, Austin, for his love and support
— My children, Jamie and Didi, for making every day a great one
— My editor, Samantha Karasik, for her guidance and valuable input
— The evaluators of this book, who made great recommendations
— The students and teachers I have worked with during my career—I have learned so much from all of you

AP® is a registered trademark of the College Board, which is not affiliated with Barron's and was not involved in the production of, and does not endorse, this product.

© Copyright 2025, 2024, 2023, 2022 by Kaplan North America, LLC, d/b/a Barron's Educational Series

All rights reserved.
No part of this publication may be reproduced in any form, or by any means, without the written permission of the copyright owner.

Published by Kaplan North America, LLC, d/b/a Barron's Educational Series
1515 West Cypress Creek Road
Fort Lauderdale, Florida 33309
www.barronseduc.com

ISBN: 978-1-5062-9670-8

10 9 8 7 6 5 4 3 2 1

Kaplan North America, LLC, d/b/a Barron's Educational Series print books are available at special quantity discounts to use for sales promotions, employee premiums, or educational purposes. For more information or to purchase books, please call the Simon & Schuster special sales department at 866-506-1949.

About the Author

Mary Wuerth has taught AP Biology for more than 20 years at Tamalpais High School in Mill Valley, California. She earned her B.S. degree in Biochemistry at UCLA and her M.S. degree in Biological Sciences at Clemson University. In addition to having taught Biology at the College of Marin, she has taught teens and teachers in a variety of environments, under various circumstances, including as a visiting teacher and a remote teacher prior to, and during, the COVID-19 pandemic.

Mary has written items for the AP Biology exam, served as a Table Leader for the scoring of the free-response questions on the AP Biology exam, and served as chair for the Test Development Committee for the SAT Subject Test in Biology. She has also been presenting AP Biology workshops to new and experienced teachers around the world since 1999.

Mary is a winner of the Presidential Award for Excellence in Math and Science Teaching and has received national awards for the use of technology in the classroom. She was selected as one of 25 Lead Teachers for WGBH-TV's Evolution Project and is currently serving as an HHMI Biointeractive Ambassador.

Table of Contents

How to Use This Book xi
Barron's Essential 5 xii

ABOUT THE EXAM

1 Introduction 3
    Exam Format 3
    Tips for Section I: Multiple-Choice Questions 4
    Tips for Section II: Free-Response Questions 5
    Scoring of the AP Biology Exam 7
    Suggested Study Plans 7

2 Statistics in AP Biology 9
    Overview 9
    What Is a Null Hypothesis? 9
    Chi-Square Test 9
    Descriptive Statistics 11
    Practice Questions 16
    Answer Explanations 20

UNIT 1: CHEMISTRY OF LIFE

3 Water 25
    Overview 25
    Water and the Importance of Hydrogen Bonds 25
    pH 27
    Practice Questions 29
    Answer Explanations 33

4 Macromolecules 37
    Overview 37
    Biological Macromolecules 37
    Protein Structure 40
    Nucleic Acids 41
    Practice Questions 43
    Answer Explanations 49

UNIT 2: CELL STRUCTURE AND FUNCTION

5 Cell Organelles, Membranes, and Transport 55
    Overview 55
    Cell Organelles and Their Functions 55
    Endosymbiosis Hypothesis 61
    The Advantages of Compartmentalization 61

The Importance of Surface Area to Volume Ratios 62
Structure of the Plasma Membrane 62
What Can (and Cannot) Cross the Plasma Membrane 63
Passive Transport 64
Active Transport 64
Practice Questions 66
Answer Explanations 70

6 Movement of Water in Cells 73
Overview 73
Water Potential 73
Osmolarity and Its Regulation 75
Practice Questions 77
Answer Explanations 82

UNIT 3: CELLULAR ENERGETICS

7 Enzymes 89
Overview 89
Enzyme Structure and Function 89
Environmental Factors that Affect Enzyme Function 90
Activation Energy in Chemical Reactions 91
Energy and Metabolism/Coupled Reactions 92
Practice Questions 94
Answer Explanations 98

8 Photosynthesis 101
Overview 101
Light-Dependent Reactions 102
Light-Independent Reactions (The Calvin Cycle) 103
Practice Questions 105
Answer Explanations 110

9 Cellular Respiration 113
Overview 113
Glycolysis 114
Oxidation of Pyruvate 115
Krebs Cycle (Citric Acid Cycle) 115
Oxidative Phosphorylation 116
Fermentation 118
Practice Questions 119
Answer Explanations 122

UNIT 4: CELL COMMUNICATION AND CELL CYCLE

10 Cell Communication and Signaling ... 127
- Overview ... 127
- Types of Cell Signaling ... 127
- Signal Transduction ... 129
- Disruptions in Signal Transduction Pathways ... 129
- Feedback Mechanisms ... 130
- Practice Questions ... 131
- Answer Explanations ... 136

11 The Cell Cycle ... 139
- Overview ... 139
- Phases of the Cell Cycle ... 139
- Regulation of the Cell Cycle, Cancer, and Apoptosis ... 140
- Practice Questions ... 143
- Answer Explanations ... 147

UNIT 5: HEREDITY

12 Meiosis and Genetic Diversity ... 151
- Overview ... 151
- How Meiosis Works ... 151
- How Meiosis Generates Genetic Diversity ... 153
- Practice Questions ... 156
- Answer Explanations ... 160

13 Mendelian Genetics and Probability ... 165
- Overview ... 165
- Mendelian Genetics ... 165
- Probability in Genetics Problems ... 166
- Practice Questions ... 169
- Answer Explanations ... 173

14 Non-Mendelian Genetics ... 177
- Overview ... 177
- Linked Genes ... 177
- Codominance, Incomplete Dominance, and Pleiotropy ... 178
- Nonnuclear Inheritance ... 179
- Phenotype = Genotype + Environment ... 179
- Practice Questions ... 180
- Answer Explanations ... 184

UNIT 6: GENE EXPRESSION AND REGULATION

15 DNA, RNA, and DNA Replication 191
- Overview 191
- Structure of DNA and RNA 191
- DNA Replication 193
- Practice Questions 195
- Answer Explanations 198

16 Transcription and Translation 201
- Overview 201
- Transcription in Prokaryotes vs. Transcription in Eukaryotes 201
- Translation 203
- Flow of Information from the Nucleus to the Cell Membrane 205
- Practice Questions 207
- Answer Explanations 211

17 Regulation and Mutations 213
- Overview 213
- Regulation of Gene Expression in Prokaryotes 213
- Regulation of Gene Expression in Eukaryotes 216
- Gene Expression Helps Cells Specialize 216
- Mutations 217
- Practice Questions 218
- Answer Explanations 222

18 Biotechnology 225
- Overview 225
- Bacterial Transformation 225
- Gel Electrophoresis 226
- Polymerase Chain Reaction (PCR) 226
- CRISPR-Cas9 226
- Practice Questions 228
- Answer Explanations 232

UNIT 7: NATURAL SELECTION

19 Types of Selection 237
- Overview 237
- Evidence of Evolution 237
- Natural Selection 238
- Artificial Selection 242
- Sexual Selection 243
- Practice Questions 244
- Answer Explanations 249

20 Population Genetics 253
- Overview 253
- Population Genetics and Genetic Drift 253
- Hardy-Weinberg Equilibrium 255
- Practice Questions 258
- Answer Explanations 262

21 Phylogeny and Speciation 265
- Overview 265
- Phylogeny and Common Ancestry 265
- Speciation 267
- Modern-Day Examples of Continuing Evolution 269
- Practice Questions 270
- Answer Explanations 274

UNIT 8: ECOLOGY

22 The Basics of Ecology 279
- Overview 279
- How Organisms Respond to Changes in the Environment 279
- Energy Flow Through Ecosystems 280
- Practice Questions 284
- Answer Explanations 289

23 Population Ecology, Community Ecology, and Biodiversity 293
- Overview 293
- Population Ecology 293
- *K*-Selected vs. *r*-Selected Populations 294
- Community Ecology and Simpson's Diversity Index 295
- Biodiversity 297
- Practice Questions 298
- Answer Explanations 302

LAB REVIEW

24 Labs 307
- Overview 307
- Lab 1: Artificial Selection 308
- Lab 2: Hardy-Weinberg 309
- Lab 3: BLAST 310
- Lab 4: Diffusion and Osmosis 310
- Lab 5: Photosynthesis 313
- Lab 6: Cellular Respiration 314
- Lab 7: Mitosis and Meiosis 315
- Lab 8: Bacterial Transformation 317

Lab 9: Restriction Enzyme Analysis of DNA 318
Lab 10: Energy Dynamics 320
Lab 11: Transpiration 321
Lab 12: Fruit Fly Behavior 322
Lab 13: Enzyme Activity 323
Practice Questions 324
Answer Explanations 329

PRACTICE TESTS

Practice Test 1 335
Section I: Multiple-Choice 337
Section II: Free-Response 349
Answer Explanations 354

Practice Test 2 367
Section I: Multiple-Choice 369
Section II: Free-Response 382
Answer Explanations 388

APPENDICES

A Frequently Used Formulas and Equations 401

B Choosing the Right Graph in AP Biology 405

C Biogeochemical Cycles 409

Index 411

Visit Barron's Online Learning Hub for more full-length practice tests.

How to Use This Book

This book provides comprehensive review and extensive practice for the latest AP Biology course and exam.

About the Exam

Start with Chapter 1, which summarizes the Big Ideas and the eight units of this course, provides an exam overview, and discusses scoring. Review the tips for answering each question type. Consult the suggested study plans to map out your test prep.

Next, review Chapter 2, which covers the statistical tests and descriptive statistics you need to be familiar with for this exam. Don't worry if math isn't your strongest skill—this chapter will teach you simple calculations needed for test day.

Review and Practice

Study Chapters 3 through 23, which are organized according to the eight units of AP Biology. The most complex material is divided into multiple chapters to provide you with more manageable chunks of content. Every chapter includes Learning Objectives that will be covered, a review of each topic, dozens of figures that illustrate key concepts, and a set of multiple-choice and short and long free-response practice questions (with detailed answer explanations) to check your progress.

Then, review Chapter 24, which focuses on labs. While no specific labs are required for this course, the curriculum emphasizes inquiry-based labs that require you to evaluate data, make predictions, and justify your conclusions with evidence. This chapter covers 13 common labs that allow you to refine your scientific thinking skills.

Practice Tests

This book contains two full-length practice tests that mirror the actual exam in format, content, and level of difficulty. Each test is followed by detailed answers and explanations for all questions.

Appendices

Be sure to familiarize yourself with the added information in the Appendices. Review the formulas and equations that will be provided to you on test day. Learn about the most commonly used graphs in AP Biology, including their key features and the types of data they can represent. Study the four major biogeochemical cycles that are necessary for life on Earth and may be tested on the exam.

Online Practice

There are also four additional full-length practice tests online. You may take these tests in practice (untimed) mode or in timed mode. All questions are answered and explained.

For Students

Whether you are using this book at the start of the school year or in the weeks leading up to the exam, this book provides the support you need to maximize your score. Try to answer as many questions as you can before checking the explanations to determine which topics you know well and which you need to review further. After studying the test-taking tips, reviewing each topic, and completing every practice question and test, you can (and will) achieve success on the AP Biology exam.

For Teachers

This book is fully aligned with the latest AP Biology Course and Exam Description. You can use this book as a resource in the classroom, or you can assign chapters as supplemental reading or practice questions as homework or test material.

BARRON'S ESSENTIAL 5

As you review the content in this book to work toward earning that **5** on your AP Biology exam, here are five things that you **MUST** know above everything else:

Everything in AP Biology connects to the four Big Ideas. The AP Biology course and exam are structured around four Big Ideas:

1. Evolution
2. Energetics
3. Information Storage and Transmission
4. Systems Interactions

As you review the eight units of AP Biology, notice how the Big Ideas are interwoven throughout multiple units. For example, Big Idea 3 (Information Storage and Transmission) is covered in five units. This idea is first introduced in Unit 1 during the discussion of nucleic acids. This idea reappears in many topics of Unit 4 (such as cell communication, signal transduction, and the cell cycle) and in several topics of Unit 5 (including Mendelian genetics and non-Mendelian genetics). Big Idea 3 is also included in *every* topic of Unit 6 (gene expression and regulation). Finally, this concept is covered in Unit 8, when discussing ecological responses to the environment. As seen from this example, the Big Ideas are the foundation of this course and exam, and it is important to recognize how each of these four concepts connects with multiple units and topics.

Learn how to explain concepts and how to analyze visual representations (Science Practices 1 and 2). You must be able to:

- Describe and explain biological concepts in applied situations
- Construct a visual model (a diagram or a graph) of the characteristics of a biological system
- Given a visual representation of a biological system, describe the interactions and relationships between the components of the system

Know how to ask testable questions and how to design experiments and methods to test those questions (Science Practice 3). You must be able to:

- State the null hypothesis and design an experiment to test it
- Identify the components of an experiment, including appropriate controls, the independent and dependent variables, and experimental constants

- Use experimental data to evaluate a hypothesis
- Design a follow-up experiment

Show that you can represent data in an appropriate graph and that you can accurately describe that data (Science Practice 4). You must be familiar with the uses of the following types of graphs:

- Line graph
- Bar graph
- Histogram
- Pie chart
- Scatterplot
- Box and whisker plot
- Graphs with two *y*-axes (dual *Y*)

Demonstrate that you can use statistical tests to analyze data and that you know how to use the results of these statistical tests to support or reject claims and hypotheses (Science Practices 5 and 6). You should be familiar with the following calculations and statistical tests:

- Means
- Rate calculations
- Ratios and percentages
- 95% confidence intervals (error bars)
- Chi-square

About the Exam

Introduction

Before beginning your review, it is important to understand the guiding principles and units that make up the AP Biology course and exam. The AP Biology curriculum focuses on four Big Ideas:

- **Big Idea 1: Evolution**—The process of evolution drives the diversity and unity of life.
- **Big Idea 2: Energetics**—Biological systems use energy and molecular building blocks to grow, reproduce, and maintain dynamic homeostasis.
- **Big Idea 3: Information Storage and Transmission**—Living systems store, retrieve, transmit, and respond to information essential to life processes.
- **Big Idea 4: Systems Interactions**—Biological systems interact, and these systems and their interactions exhibit complex properties.

These Big Ideas are the overarching themes covered in the eight units of content that make up the AP Biology course and exam. The review chapters that follow this introduction are all grouped according to these eight units, so you can test which units you are strongest in and which you may want to study more closely. Table 1.1 lists each of these eight units and the approximate percentage of questions that will be devoted to each unit on the AP Biology exam.

Table 1.1 AP Biology Units

Unit	% of Questions
1—Chemistry of Life	8–11%
2—Cell Structure and Function	10–13%
3—Cellular Energetics	12–16%
4—Cell Communication and Cell Cycle	10–15%
5—Heredity	8–11%
6—Gene Expression and Regulation	12–16%
7—Natural Selection	13–20%
8—Ecology	10–15%

Exam Format

You will have three hours total to complete the AP Biology exam, which consists of the two sections outlined in Table 1.2.

Table 1.2 AP Biology Exam Format

	Section I	Section II
Question Type	Multiple-Choice	Free-Response
Number of Questions	60	2 Long Free-Response 4 Short Free-Response
Time	90 minutes	90 minutes
% of Overall Score	50%	50%

Section I: Multiple-Choice

You will have 90 minutes to complete 60 multiple-choice questions, which will make up 50% of your overall score. Each question will have four possible answer choices, and you need to select the choice that best answers the question. Some questions (in both Section I and Section II of the exam) may require you to use your math skills and the AP Biology Equations and Formulas sheet, which will be provided to you on test day.

Many of the multiple-choice questions will require you to evaluate data in tables, graphs, or diagrams. Thus, to prepare yourself for Section I, practice analyzing and interpreting as many different types of tables and graphs as possible. A wide variety of tables and graphs are incorporated throughout this book to provide you with as much practice with them as possible.

> **TIP**
> Get familiar with the AP Biology Equations and Formulas sheet. You do NOT need to memorize any of those formulas, but you DO need to know when to use them and how to apply them.

Section II: Free-Response

You will have 90 minutes to complete six free-response questions, two of which are long free-response questions and four of which are short free-response questions. Each of the six free-response questions will consist of four parts. All six free-response questions combined will make up 50% of your overall exam score.

Long Free-Response Questions

Questions 1 and 2 of Section II will be the long free-response questions. Each will be worth 9 points, for a total of 18 points between both questions. Both will likely involve **interpreting** and **evaluating** experimental results. Question 1 may ask you to **evaluate** data presented in a table or graph, while question 2 may ask you to **construct** a graph using the appropriate confidence intervals.

Short Free-Response Questions

Questions 3, 4, 5, and 6 of Section II will be the short free-response questions. Each of these questions will be worth 4 points.

Question 3 may describe an experimental scenario. You may be asked to **identify** the parts of the experiment (such as any controls, the independent variable, and the dependent variable), **predict** results, and **justify** your predictions. You may also be asked to **describe** the biological processes covered in the experiment.

Question 4 is typically a conceptual analysis question. You may be asked to **describe** and **explain** a biological process. Given a disruption in the process, you must **predict** how that disruption will affect the process and **justify** your prediction with evidence.

Question 5 may ask you to **analyze** a model or visual representation of a biological concept. You may be presented with a diagram and asked to **describe** the characteristics of the process represented in this model. Then, you may be asked to **explain** the relationships between the different parts of the model and relate or apply the model to a larger biological concept.

Question 6 may ask you to **analyze** data. You may see data in a graph or table, and you might be asked to **describe** the data and to use the data presented to **evaluate** a claim. Finally, you may be asked to **explain** how the data presented relates to a larger biological concept.

Tips for Section I: Multiple-Choice Questions

- **Do NOT skip over the scenarios and/or diagrams presented in the stem of the question.** A stem that contains a description of a scenario and/or a diagram or graph will precede many of the multiple-choice questions. In a testing situation where time is limited, students are sometimes tempted to save time by skipping over the stem and proceeding directly to the question. *Don't do this!* Often, taking just 30 seconds to read over the data or scenario presented will make it easier to answer the question or questions that follow it. The scenario presented in the stem of the question often will have important background information that will *help* you answer the question. If you are presented with a graph, note the *variables* shown on each axis and their *units*, and try to detect any *patterns* in the data. In data tables or charts, note the *column headings* and their *units*, and observe any *trends* or patterns in the data.

- **Do NOT be afraid of organisms or genes you may not have heard of before.** There are so many great examples of organisms, genes, and ecosystems that apply to the content of the AP Biology course, and no teacher or textbook can mention all of them. Any example that is not explicitly included in the AP Biology Course and Exam Description will be described in enough detail in the question so that you will have enough background information to answer the question. Therefore, don't worry if you see a question about the CYP6M2 gene in *Anopheles gambiae* and you've never heard of either before! The stem of the question will tell you what you need to know about that gene and organism (for example, that the CYP6M2 gene confers insecticide resistance to *Anopheles* mosquitoes), so all you need to do is apply your knowledge and skills to that background information to find the correct answer.

- **Do NOT be tempted by the "distractors."** Incorrect answer choices are called distractors. As you read each question, cover the answer choices with your hand. Before you reveal the answer choices, think of the characteristics that a good answer to the question at hand will contain. Then, reveal the answer choices and choose the answer that best fits the characteristics you know a good answer will have. It is often easier to focus your brain on finding the best answer rather than trying to eliminate each of the distractors.

- **DO pace yourself.** You will have 90 minutes to answer 60 multiple-choice questions. If it is taking you more than two minutes to answer a question, move on to the next question and go back to that question later. Just make sure to go back and answer the question(s) you skipped.

- **DO answer every question.** There is no guessing penalty on the AP Biology exam. If you leave a question blank, you are guaranteed to not earn points for that question, so answer every question, even if you have to guess. Never leave a question blank on the AP Biology exam! Reserve the last two or three minutes of the time allotted for Section I to check that you have answered all of the questions and have not left any questions blank.

Tips for Section II: Free-Response Questions

- **Do NOT leave any questions blank.** Even if you think you don't know how to answer the question, reread the question to see what terms in the question you do know something about. Then, use those terms as the basis for your answer, keeping in mind the task verbs in the question. As in Section I, if you leave a question blank, you are guaranteed to not earn points on that question, but if you write something, you may earn some points that could make the difference between a score of a 3, 4, or 5. Never give up—remember, you CAN do this!

- **Do NOT make any contradictory statements.** For example, if you state that the function of the mitochondria is to generate energy for the cell (a correct statement) but then later in your response state that the function of the mitochondria is also to perform photosynthesis (an incorrect statement), you have made two contradictory statements. Thus, you will not earn any points for either of those statements.

- **DO plan your approach to Section II.** Take the first 5–10 minutes allotted for Section II to "read and rank." Read all six free-response questions, and then place the number 1 next to the question you think will be the easiest for you, the number 2 next to the next easiest question, and so on. You do not have to answer the questions in the order they appear in the test. Sometimes the easiest free-response questions are at the end of this section, and if you get hung up on a more challenging question that appears earlier, you may never get to the easier questions you are likely to earn points on.

- **DO read each question carefully.** Read each question carefully at least two times. Each time you read the question, pay attention to key words, especially any **bolded** words (which are the action or task verbs), any numbers, or any words like *and* or *or* (which indicate whether all or some of the items mentioned need to be addressed).

- **DO pace yourself.** You will have 90 minutes to complete all six free-response questions. Some of the free-response questions will require less time; others will require more time. Here is a suggested time plan for Section II:
  - First 5–10 minutes for "read and rank" (planning the order in which you will answer the questions)
  - 20 minutes for each of the two long free-response questions (Questions 1 and 2) for a total of 40 minutes
  - 5–10 minutes for each of the short free-response questions (Questions 3, 4, 5, and 6) for a total of 20–40 minutes
- **DO write legibly.** This may seem obvious, but if your answer is unclear or unreadable, the AP reader cannot award you points for it. Use a black ballpoint pen to write your answer. If you make a mistake, just cross it out with a single strikethrough—any more than that is unnecessary.
- **DO label your graphs completely with units.** If a question asks you to construct a graph, always make sure the axes are labeled clearly with the appropriate units. A unitless graph will not earn points. Use consistent scaling on your axes, and give your graph a legend, if necessary. Also, remember to include 95% confidence intervals (error bars).
- **DO label the parts of your answer appropriately.** This makes it easier for the reader who scores your exam to award you points. However, if you happen to answer Part A of a question in the section you labeled Part B, the reader will still award you points for it.
- **DO use complete sentences.** As per the instructions for Section II, use complete sentences in your answers. You will not be awarded points for bulleted lists. If you use a drawing in your answer, make sure to also describe it in complete sentences.
- **DO ATP (Answer the Prompt).** Do not waste time writing an introductory paragraph, a thesis statement, or a concluding paragraph. Do not restate the question—the reader knows what the question is! While you need to be clear in your writing, you are not being evaluated on your ability to write a well-constructed essay, as you might be in an AP English course. You ARE being evaluated on your knowledge of biology. Make sure you understand the question prompt and what it is asking you to do. Then, reread your answer to make sure you addressed all of the task verbs in the question and did not make any contradictory statements.
- **DO pay attention to the task verbs!** Pay attention to these action verbs, which are typically bolded in the long and short free-response questions, as these words indicate what the question requires you to provide in your response. Some of the most frequently used task verbs are the following:
  - **Predict**—state what you think will happen if a change is made in a system or process
  - **Justify**—give evidence to support your prediction
  - **Make a claim**—make a statement based on the available data or evidence
  - **Support a claim**—give evidence to defend a claim
  - **Describe**—note the characteristics of something
  - **Explain**—state "why" or "how" something happens (Note: This is *more* demanding than describing.)
  - **Identify**—provide the information that is asked for (Note: This is *less* demanding than describing.)
  - **Calculate**—perform the requested calculation, and ALWAYS show your work and your units!
  - **Construct**—make a graph (show units!) or a diagram that illustrates data or a relationship
  - **Determine**—make a conclusion based on evidence
  - **State**—give a null hypothesis or an alternative hypothesis that is supported by data/evidence
  - **Evaluate**—assess the validity or accuracy of a claim or hypothesis

Scoring of the AP Biology Exam

The AP Biology exam is scored on a scale from 1 to 5, with 5 being the highest possible score. Table 1.3 describes each score.

Table 1.3 AP Biology Exam Scores

AP Exam Score	Recommendation
5	Extremely Well Qualified
4	Well Qualified
3	Qualified
2	Possibly Qualified
1	No Recommendation

Scores of 3 or above may earn you college credit or allow you to skip introductory courses and take more advanced courses earlier in your college career. Policies regarding credit for AP exam scores vary widely between schools and may even vary between majors at the same school. Always check with the college or university you plan to attend to find out the latest information.

Fifty percent of your total score is based on your performance on Section I (the multiple-choice section), and the other 50% of your total score is based on your performance on Section II (the free-response section). For this reason, it is very important to practice answering all question types (multiple-choice, short free-response, and long free-response)—that is why you will see all of these types of questions at the end of every chapter of this book.

Suggested Study Plans

The following are suggested study plans depending on how much time is left until test day. If there's a lot of time left before the exam, read through all of the chapters in this book, answer all of the practice questions, and complete all the practice tests. If time is limited, refer to these study plans to skip to the areas that you may want to study further. Follow what works best for you and your schedule. Remember, by reviewing and practicing with this book, you are already taking the first step toward achieving success on the AP Biology exam!

Six Weeks Until the Exam

- Start by taking all of Practice Test 1.
- Once you've completed Practice Test 1, review the answer explanations and use the Self-Analysis Chart for Section I to determine what your strengths are and to diagnose the four units where you need the most improvement.
- Read through the chapters that cover those four units, and answer all the practice questions in those chapters.
- Reread the preceding tips for Section I and Section II.
- Review Chapter 2, which focuses on statistics (since 95% confidence intervals, the null hypothesis, and the chi-square test are key tools used to evaluate experimental results in AP Biology).
- Take all of Practice Test 2.
- Once you've completed Practice Test 2, review the answer explanations and use the Self-Analysis Chart for Section I to determine where you've improved and what two units you're still having trouble with.
- Reread the chapters related to those two units, and answer all the practice questions in those chapters.
- Review Chapter 24, which focuses on the lab component of the course. Answer all the practice questions at the end of this chapter, and review the answer explanations for any questions you may have answered incorrectly.
- Revisit the preceding tips for Section I and Section II one last time so that those reminders are fresh in your mind for test day.

Two Weeks Until the Exam

- Complete all of Practice Test 1.
- Once you've completed Practice Test 1, review the answer explanations and use the Self-Analysis Chart for Section I to determine what your strengths are and to diagnose the three units where you need the most improvement.
- Read through the chapters that cover those three units, and answer all the practice questions in those chapters.
- Reread the preceding tips for Section I and Section II.
- Review Chapter 2, which focuses on statistics.
- Complete all of Practice Test 2.
- Once you've completed Practice Test 2, review the answer explanations and use the Self-Analysis Chart for Section I to determine where you've improved and the one unit you're still having trouble with.
- Reread the chapters related to that unit, and answer all the practice questions in those chapters.
- Review Chapter 24, which focuses on the lab component of the course. Answer all the practice questions at the end of this chapter, and review the answer explanations for any questions you may have answered incorrectly.

One Week Until the Exam

- Complete all of Practice Test 1.
- Once you've completed Practice Test 1, review the answer explanations and use the Self-Analysis Chart for Section I to determine what your strengths are and to diagnose the two units where you need the most improvement.
- Read the chapters related to those two units, and answer all the practice questions in those chapters.
- Reread the preceding tips for Section I and Section II.
- Review Chapter 2, which focuses on statistics.
- Complete all of Practice Test 2.
- Once you've completed Practice Test 2, review the answer explanations and use the Self-Analysis Chart for Section I to determine where you've improved and the one unit you're still having trouble with.
- Reread the chapters related to that unit, and answer all the practice questions in those chapters.
- Review Chapter 24, which focuses on the lab component of the course. Answer all the practice questions at the end of this chapter, and review the answer explanations for any questions you may have answered incorrectly.

The Day Before the Exam

- Complete just Section I of one of the two practice tests in this book.
- Once you've finished Section I, review the answer explanations and use the Self-Analysis Chart to diagnose the one unit where you need the most improvement.
- Skim through the chapters related to that unit, and answer all the practice questions in those chapters.
- Reread the preceding tips for Section I and Section II.
- Review Chapter 2, which focuses on statistics.
- Review Chapter 24, which focuses on the lab component of the course.

2

Statistics in AP Biology

> **Learning Objectives**
>
> In this chapter, you will learn:
> → What Is a Null Hypothesis?
> → Chi-Square Test
> → Descriptive Statistics

Overview

Scientists make hypotheses and then design experiments to test these hypotheses. Data are gathered during these experiments and then analyzed. Scientists use these analyses to draw conclusions about the data. An important tool in data analysis is statistics. Statistical tests are used to evaluate hypotheses. Descriptive statistics describe data sets. This chapter will review some of the statistical tests and descriptive statistics you need to understand for the AP Biology course and exam.

What Is a Null Hypothesis?

The **null hypothesis** (H_0) states that there is no statistically significant difference between two groups in an experiment. For example, a student designs an experiment to see if plants watered with bottled water will exhibit more growth than plants watered with tap water. The null hypothesis for this experiment would be that there will be no statistically significant difference in plant growth between the plants watered with bottled water and the plants watered with tap water.

Here's another example: You want to test if dogs prefer dog food brand A over dog food brand B. The null hypothesis would be that there will be no statistically significant difference between the number of dogs choosing dog food brand A and the number of dogs choosing dog food brand B. If 100 dogs are presented with both dog food brand A and dog food brand B, the null hypothesis would predict that 50 dogs would choose brand A and 50 dogs would choose brand B.

Chi-Square Test

The **chi-square test** is a statistical test that is used to compare the observed results to the expected results in the experiment. In AP Biology, the chi-square test is used to evaluate the null hypothesis, and it is often used in genetics problems, in the lab on mitosis (Lab 7), and in the lab on animal behavior (Lab 12).

> It is important to note that the chi-square test is used to compare primary or raw data, such as the number of items in each category of data. The chi-square test should not be used to compare processed data, such as percentages or means. For example, it *would* be appropriate to use the chi-square test if you were comparing the number of purple flowers and the number of white flowers that resulted from a genetic cross. It would *not* be appropriate to use the chi-square test to compare the percentage of purple flowers to the percentage of white flowers.

Consider the experiment that measured whether dogs prefer dog food brand A or dog food brand B. If there are 100 dogs, the null hypothesis would predict that 50 dogs would choose brand A and 50 dogs would choose brand B. These are the *expected* results. If the experiment was carried out and 45 dogs chose brand A and 55 dogs chose brand B, those are the *observed* results. The chi-square test could be used to evaluate the null hypothesis that there will be no statistically significant difference between the expected results and the observed results of the experiment. The steps of the chi-square test are as follows:

1. Calculate the chi-square value. The formula for chi-square is:

$$\chi^2 = \sum \frac{(\text{observed} - \text{expected})^2}{\text{expected}}$$

The symbol $\sum$ means "summation." This means you need to do this calculation for each category of data (brand A and brand B) and then add the values.

Using the observed and expected values (45 observed and 50 expected for brand A; 55 observed and 50 expected for brand B):

$$\chi^2 = \frac{(45-50)^2}{50} + \frac{(55-50)^2}{50} = \frac{25}{50} + \frac{25}{50} = 1$$

2. Determine the number of **degrees of freedom (df)** in the experiment. The number of degrees of freedom in an experiment is defined as the number of possible outcomes in the experiment minus 1. In this experiment, there are two possible outcomes, brand A or brand B, so the df is $2 - 1 = 1$.

3. Using the degrees of freedom and the *p*-value, find the critical value in the chi-square table.

The **p-value** is defined as the probability that the observed data would be produced by random chance alone. For biology, the typical *p*-value used is 0.05. Using the df of 1 calculated above and a *p*-value of 0.05, the critical value is 3.84 according to the chi-square table that follows.

> **TIP**
> You do NOT need to memorize any of the formulas reviewed in this chapter—they are included on the AP Biology Equations and Formulas sheet, which will be supplied to you on test day. For quick reference, you can review those formulas in Appendix A of this book.

Chi-Square Table

p-value	\multicolumn{8}{c}{Degrees of Freedom (df)}
	1	2	3	4	5	6	7	8
0.05	3.84	5.99	7.81	9.49	11.07	12.59	14.07	15.51
0.01	6.63	9.21	11.34	13.28	15.09	16.81	18.48	20.09

4. Compare your calculated chi-square value to the critical value. The calculated chi-square value from the first step (1) is less than the critical value (3.84).

5. Based on the comparison, decide whether to reject the null hypothesis or whether you fail to reject the null hypothesis. If the calculated chi-square value is *less than or equal to* the critical value, *fail to reject* the null hypothesis. If the calculated chi-square value is *greater than* the critical value, *reject* the null hypothesis.

In this example, since the calculated chi-square value is less than the critical value, you would fail to reject the null hypothesis that there is no statistically significant difference between the observed and expected data. This does not mean the null hypothesis is *proven*—it just means you cannot reject the null hypothesis. In other words, the null hypothesis cannot be ruled out.

Here is another example: A coin has heads on one side and tails on the other side. If the coin is flipped 40 times, you would expect the coin to come up heads 20 times and come up tails 20 times. These are the expected results. The coin is flipped 40 times, and the coin comes up heads 12 times and it comes up tails 28 times. Those are the observed results. The null hypothesis is that there is no statistically significant difference between the observed and expected numbers of heads and tails. To evaluate this null hypothesis, here are the steps involved:

1. Calculate the chi-square value.

$$\chi^2 = \sum \frac{(\text{observed} - \text{expected})^2}{\text{expected}} = \frac{(12-20)^2}{20} + \frac{(28-20)^2}{20} = \frac{64}{20} + \frac{64}{20} = 6.40$$

2. Determine the number of degrees of freedom (df) in the experiment. There are two possible outcomes in this experiment (heads or tails), so the df is $2 - 1 = 1$.
3. Using a df of 1 and a *p*-value of 0.05, the critical value is 3.84 according to the chi-square table.
4. Compare your calculated chi-square value to the critical value. The calculated chi-square value (6.40) is greater than the critical value (3.84).
5. Decide whether to reject the null hypothesis or whether you fail to reject the null hypothesis. In this example, you should reject the null hypothesis because the calculated chi-square value is greater than the critical value. Since the null hypothesis is rejected, you can then come up with an alternative hypothesis—for example, perhaps this is a "trick coin"!

Descriptive Statistics

Descriptive statistics are used to describe data sets. There are descriptive statistics that describe the *center* of a data set, and other descriptive statistics describe the amount of variability or *spread* of a data set.

The mean and median can be used to characterize the center of a data set. Thus, the mean and median are sometimes referred to as measures of the central tendency of a data set.

The **mean**, or average, of a data set is calculated with the following formula:

$$\bar{x} = \frac{1}{n}\sum_{i=1}^{n} x_i$$

In the above formula:

n = sample size (the number of data points in the data set)

$\sum$ = summation (add all members of the data set, starting with the first member of the set, and continue to the nth member of the data set)

x_i = the ith member of the data set

To calculate the mean, add the values of all the data points in the data set, and then divide that by the sample size. (Multiplying by $\frac{1}{n}$ gives the same result as dividing by n.) Here's how you would calculate the means for data set A and data set B:

Data Set A: 1, 2, 3, 4, 5; $n = 5$

$$\bar{x} = \frac{(1+2+3+4+5)}{5} = 3$$

Data Set B: 3, 3, 3, 3, 3; $n = 5$

$$\bar{x} = \frac{(3+3+3+3+3)}{5} = 3$$

Notice that data set A looks very different from data set B, but they both have the same mean. Means can be distorted or skewed by extreme values in the data set. These two data sets illustrate how looking at just the mean is sometimes not enough to accurately describe the data set.

The **median** is the midpoint of the data set. To find the median, place the members of the data set in numerical order from lowest to highest value. The middle of the data set is the median. If there are an even number of data points in the set, calculate the mean of the two numbers in the middle of the data set to find the median. Here's how you would find the median for data set A and data set B:

Data Set A: 1, 2, 3, 4, 5

Since the members of the data set are already arranged from lowest to highest, look for the middle value, which in this case is 3.

Data Set B: 3, 3, 3, 3, 3

Since the members of this data set are all the same, the median is 3.

Again, even though data sets A and B are very different, they have the same median.

Extremes in a data set do not affect the median, but they do affect the mean of the data set. For example, add the numbers 0 and 100 to data set A and data set B to form new data sets C and D, respectively. This changes the means greatly from the original calculations, but the medians do not change:

Data Set C: 0, 1, 2, 3, 4, 5, 100

$$\bar{x} = 16.43; \text{median} = 3$$

Data Set D: 0, 3, 3, 3, 3, 3, 100

$$\bar{x} = 16.43; \text{median} = 3$$

Using the mean and median alone are often not enough to accurately describe a data set. It is also important to use descriptive statistics that describe the spread of a data set. Standard deviation and standard error of the mean can be used to describe how spread out the data points are.

> You will *not* be required to calculate standard deviation or standard error of the mean on the AP Biology exam. However, it is important to understand what standard deviation and standard error of the mean can tell you about a data set. You also must be able to use standard error of the mean to construct 95% confidence intervals.

Standard deviation (s) averages how far each data point is from the mean of the data set. The formula for standard deviation is:

$$s = \sqrt{\frac{\sum (x_i - \bar{x})^2}{n - 1}}$$

Here's how you would calculate the standard deviation for data set A and data set B:

Data Set A: 1, 2, 3, 4, 5; $\bar{x} = 3$ and $n = 5$

$$s = \sqrt{\frac{(1-3)^2 + (2-3)^2 + (3-3)^2 + (4-3)^2 + (5-3)^2}{5-1}} = 1.58$$

Data Set B: 3, 3, 3, 3, 3; $\bar{x} = 3$ and $n = 5$

$$s = \sqrt{\frac{(3-3)^2 + (3-3)^2 + (3-3)^2 + (3-3)^2 + (3-3)^2}{5-1}} = 0$$

If a data set is more spread out, the standard deviation will be larger. The less spread out the data points, the smaller the standard deviation.

The **standard error of the mean** ($SE_{\bar{x}}$) is another measure of how spread out a data set is. Each time you repeat an experiment, random chance will lead to slightly different means. If an experiment is repeated multiple times and a mean is calculated for each experiment, the standard error of the mean predicts the distribution of the means of those repeated experiments. The prediction would be that 95% of the means would fall within two standard errors of the mean (above or below the mean of the original experiment). If the standard error of the mean is large, repeating the experiment will result in a larger range of means than if the standard error of the mean was small. Data sets with smaller standard errors of the mean are considered more accurate.

Standard error of the mean is calculated with the following formula:

$$SE_{\bar{x}} = \frac{s}{\sqrt{n}}$$

In the above formula:

s = standard deviation

n = sample size

Notice that as the sample size (n) increases, the standard error of the mean decreases. You may already understand that an experiment performed with a larger sample size is considered more reliable than an experiment performed with a smaller sample size. Standard error of the mean gives a mathematical reason for why experiments with larger sample sizes are typically more reliable. Here's how you would calculate the standard error of the mean for data set A and data set B:

Data Set A: 1, 2, 3, 4, 5; $s = 1.58$ and $n = 5$

$$SE_{\bar{x}} = \frac{1.58}{\sqrt{5}} = \frac{1.58}{2.24} = 0.705$$

Data Set B: 3, 3, 3, 3, 3; $s = 0$ and $n = 5$

$$SE_{\bar{x}} = \frac{0}{\sqrt{5}} = 0$$

Standard error of the mean can be used to create a type of error bar on a graph called a **95% confidence interval** (95% CI). A 95% confidence interval does NOT mean you are 95% confident in your data! What a 95% confidence interval does mean is that if you repeated an experiment 100 times and calculated the mean of the data you collected each time, the mean would fall within the 95% confidence interval 95 of those 100 times. In other words, you would expect your mean to fall within the 95% confidence interval 95% of the time, but 5% of the time you would expect the mean to be outside of the 95% confidence interval.

To construct a 95% confidence interval, you need to know the upper limit of the interval and the lower limit of the interval. The upper limit of the 95% confidence interval is found by starting with the mean and then *adding* two times the standard error of the mean:

$$\text{Upper limit of 95\% CI} = \bar{x} + 2(SE_{\bar{x}})$$

To find the lower limit of the 95% confidence interval, start with the mean and then *subtract* two times the standard error of the mean:

$$\text{Lower limit of 95\% CI} = \bar{x} - 2(SE_{\bar{x}})$$

Here's an example to practice working with 95% confidence intervals.

An AP Biology class grows "Fast Plants" (*Brassica rapa*) and records the height of each plant on day 28 of growth. The mean plant height and standard error of the mean are calculated. The 10 tallest plants are cross-pollinated and produce seeds. These seeds are harvested and planted to form a second generation of plants. The height of each plant in this second generation is measured, and again the mean plant height and standard error of the mean are calculated. The data are shown in the following table:

	Mean Plant Height on Day 28 (mm)	$SE_{\bar{x}}$
Generation 1	158.4	8.8
Generation 2	203.1	9.6

Construct a graph of the mean plant height for each generation, showing 95% confidence intervals. First, find the upper and lower limits for the 95% confidence intervals for each generation.

For Generation 1:
Upper limit of 95% CI = 158.4 + 2(8.8) = 176.0 mm
Lower limit of 95% CI = 158.4 − 2(8.8) = 140.8 mm

For Generation 2:
Upper limit of 95% CI = 203.1 + 2(9.6) = 222.3 mm
Lower limit of 95% CI = 203.1 − 2(9.6) = 183.9 mm

Graphing the means for each generation with the 95% confidence intervals leads to Figure 2.1.

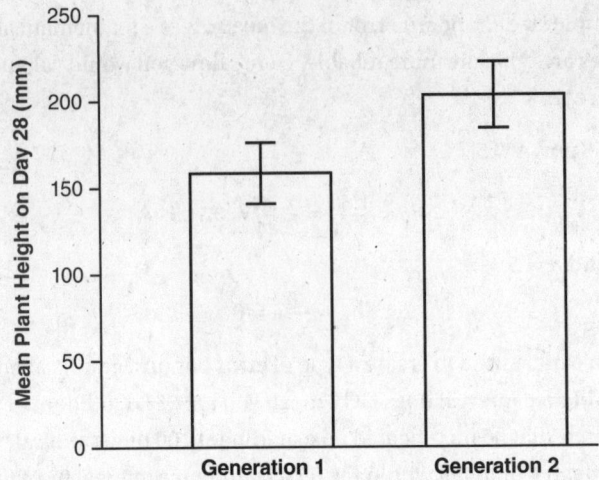

Figure 2.1 Mean Plant Height on Day 28 for Generations 1 and 2

Now that you know how to construct graphs with 95% confidence intervals, it is important to understand what they can tell us when comparing data sets. In the previous example, the 95% confidence intervals for generations 1 and 2 do not overlap. If the 95% confidence intervals for two sets of data do not overlap, it is likely there is a statistically significant difference between the two groups. In that example, there is likely a statistically significant difference between the mean plant heights on day 28 for generations 1 and 2.

However, if the 95% confidence intervals do overlap, the data are inconclusive, and it is not possible to say whether or not there is a significant difference between the groups. The experiment would need to be repeated. Here is another example that illustrates this point.

On field trips to a salt marsh, an AP Biology class counted the number of crustaceans in a 100 cm^2 quadrant. Three different locations in the marsh were visited, and multiple quadrants were counted at each location. The mean number of crustaceans per quadrant and the standard error of the mean were calculated from the data at each location. The data are shown in the following table.

Location	Mean Number of Crustaceans per Quadrant	Standard Error of the Mean
A	8	1.5
B	11	0.9
C	14	1.1

The data from each location were plotted with 95% confidence intervals (calculated as follows). Figure 2.2 shows the graph of this data.

Location A: Upper limit of 95% CI = 8 + 2(1.5) = 11
                    Lower limit of 95% CI = 8 − 2(1.5) = 5

Location B: Upper limit of 95% CI = 11 + 2(0.9) = 12.8
                    Lower limit of 95% CI = 11 − 2(0.9) = 9.2

Location C: Upper limit of 95% CI = 14 + 2(1.1) = 16.2
                    Lower limit of 95% CI = 14 − 2(1.1) = 11.8

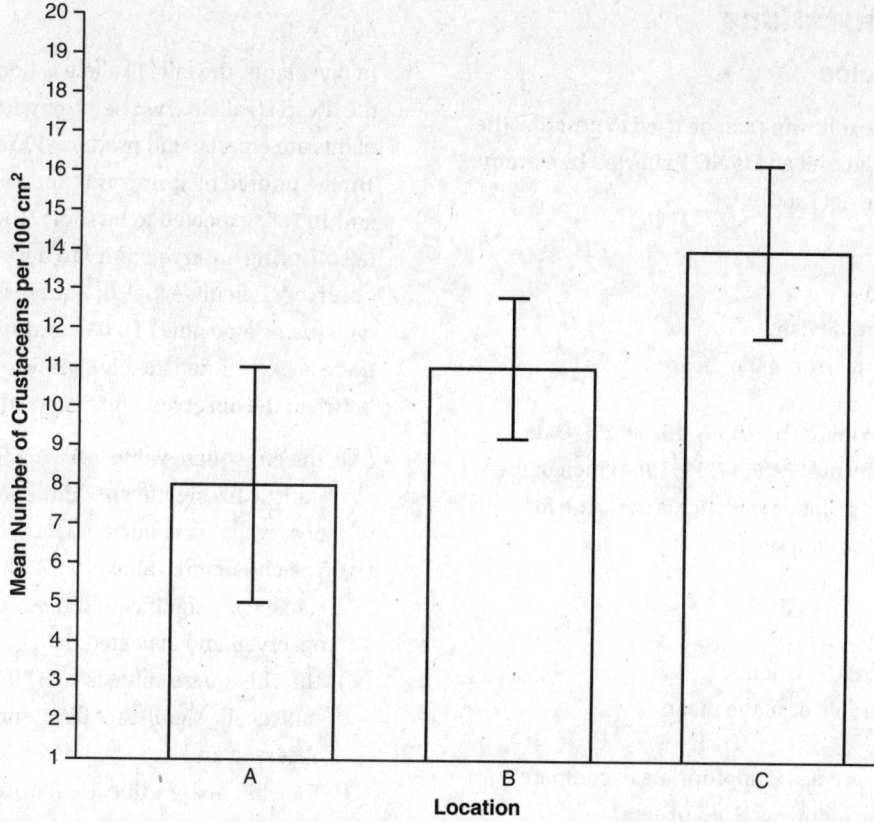

Figure 2.2 Mean Number of Crustaceans per 100 cm^2 in Three Locations in the Salt Marsh

Which of the two locations are most likely to have a statistically significant difference in the number of crustaceans found per 100 cm^2? The answer is that locations A and C would most likely have statistically significant differences because the 95% confidence intervals for locations A and C do not overlap. The 95% confidence intervals for locations A and B do overlap, so it is not possible to make any conclusions about statistically significant differences between those two locations. Similarly, the 95% confidence intervals for locations B and C overlap, so it is also not possible to make any conclusions about statistically significant differences between locations B and C.

Practice Questions

Multiple-Choice

1. Which of the following can be used to describe the center of a data set and is NOT affected by extreme values in the data set?

    (A) mean
    (B) median
    (C) standard deviation
    (D) standard error of the mean

2. Data Set A consists of {10, 15, 15, 20, 25}. Data Set B consists of {15, 15, 17, 19, 19}. Which of the following descriptive statistics is the same for both data set A and data set B?

    (A) mean
    (B) median
    (C) standard deviation
    (D) standard error of the mean

3. The chi-square test is appropriate to compare which of the following types of data?

    (A) means
    (B) percentages
    (C) processed data
    (D) raw data

4. In a dihybrid cross of two organisms (*AaBb* × *AaBb*), four different phenotypes of offspring can be produced. How many degrees of freedom would there be?

    (A) 1
    (B) 2
    (C) 3
    (D) 4

5. In pea plants, the tall (*T*) allele is dominant to the dwarf (*t*) allele. Two heterozygous (*Tt*) pea plants are crossed and produce 400 offspring. Three hundred offspring are expected to be tall, and 100 are expected to be short. There are 290 tall offspring observed and 110 dwarf offspring observed. Calculate the chi-square value from this data (using a *p*-value of 0.05), and state whether there is likely a statistically significant difference between the observed and expected data.

    (A) The chi-square value is 0.67. There is a statistically significant difference between the observed and expected data.
    (B) The chi-square value is 0.67. There is NOT a statistically significant difference between the observed and expected data.
    (C) The chi-square value is 1.33. There is a statistically significant difference between the observed and expected data.
    (D) The chi-square value is 1.33. There is NOT a statistically significant difference between the observed and expected data.

6. A student wanted to see if isopods preferred banana or watermelon as a food source. Twenty isopods were placed in the center of a choice chamber with two compartments. Banana was placed in one compartment, and watermelon was placed in the second compartment. After 15 minutes, the number of isopods in each compartment was counted. Data are shown in the table.

Type of Food in Compartment	Number of Isopods in Compartment After 15 Minutes
Banana	6
Watermelon	14

The null hypothesis was that the isopods would have no food preference and would be found in equal numbers in both compartments. Calculate the chi-square value (using a *p*-value of 0.05), and determine if there was a statistically significant difference between the observed and expected data in this experiment.

(A) The chi-square value is 0.80. There is a statistically significant difference between the observed and expected data.
(B) The chi-square value is 0.80. There is NOT a statistically significant difference between the observed and expected data.
(C) The chi-square value is 3.20. There is a statistically significant difference between the observed and expected data.
(D) The chi-square value is 3.20. There is NOT a statistically significant difference between the observed and expected data.

7. Data Set X and Data Set Y have the same standard deviations. Data Set X has a sample size of 10, while Data Set Y has a sample size of 50. How will their standard errors of the mean compare?

(A) Data Set X will have a larger standard error of the mean than Data Set Y.
(B) Data Set Y will have a larger standard error of the mean than Data Set X.
(C) Both data sets will have the same standard error of the mean.
(D) The difference in the standard errors of the mean cannot be determined from the information given.

Questions 8 and 9

The heights of oak trees in four different locations were measured. The mean height and the standard error of the means were calculated for each location and are shown in the table.

Location	Mean Height of Oak Trees (m)	Standard Error of the Mean
Oakville	20.1	2.5
Sacramento	16.4	1.6
San Rafael	28.7	4.3
Oak Valley	34.1	2.0

8. Which location showed the greatest variability in the heights of the oak trees?

(A) Oakville
(B) Sacramento
(C) San Rafael
(D) Oak Valley

9. Which of the following two locations are likely to have a statistically significant difference between the heights of their oak trees?

(A) Oakville and Sacramento
(B) Oakville and Oak Valley
(C) San Rafael and Oak Valley
(D) San Rafael and Oakville

10. Which of the following correctly describes how to calculate a 95% confidence interval?

(A) Mean ± 1(Standard Deviation)
(B) Mean ± 1(Standard Error of the Mean)
(C) Mean ± 2(Standard Deviation)
(D) Mean ± 2(Standard Error of the Mean)

Short Free-Response

11. Some chemicals are known to increase the frequency of chromosome breakage in dividing cells. Thirty Petri dishes with dividing cells were treated with either lead (10 Petri dishes), cadmium (10 Petri dishes), or control solutions (10 Petri dishes). Forty-eight hours after treatment, each Petri dish was examined, and the percentage of cells that showed chromosome breakage in each Petri dish was calculated. The mean percentage of cells that showed chromosome breakage and the standard error of the mean for each treatment were calculated and are shown in the table.

	Mean Percentage of Cells That Showed Chromosome Breakage	Standard Error of the Mean
Control	5.5	0.05
Lead	24.3	5.2
Cadmium	46.1	9.1

Part A

(i) **Describe** which treatment produced the greatest variability in the data.

Part B

(i) **Describe** which treatment was least likely to result in chromosome breakage.

Part C

A student claims that exposure to cadmium is more likely to result in chromosome breakage than is exposure to lead.

(i) **Evaluate** this claim using the data provided.

Part D

(i) **Identify** the independent variable in this experiment.

(ii) **Identify** the dependent variable in this experiment.

12. In tobacco plants, green (*G*) is dominant to albino (*g*). A heterozygous green (*Gg*) tobacco plant is crossed with an albino (*gg*) tobacco plant, and 100 offspring are produced. Fifty offspring are expected to be green (*Gg*), and 50 offspring are expected to be albino (*gg*). However, when the offspring are counted, 60 green plants and 40 albino plants are observed.

Part A

(i) **Identify** the degrees of freedom in this experiment.

Part B

(i) **Calculate** the chi-square value for this data.

Part C

(i) **Make a claim** about whether or not the observed data are likely significantly different statistically from the expected data. Use a *p*-value of 0.05.

Part D

(i) **Justify** your claim from Part C with your knowledge of the chi-square test.

Long Free-Response

13. Organic gardeners will sometimes use ladybugs as a method of reducing aphid populations in their gardens. A garden infested with aphids is divided into three sections. In the first section, no treatment (to reduce the aphid populations) is applied. In the second section, ladybugs are introduced to reduce the aphid populations. In the third section, a chemical pesticide is used to reduce the aphid populations. One week after the treatments are applied, 10 measurements of the density of the aphid populations (in aphids per square decimeter) are taken in each section of the garden. The means and the 95% confidence intervals for each section of the garden are shown in the graph.

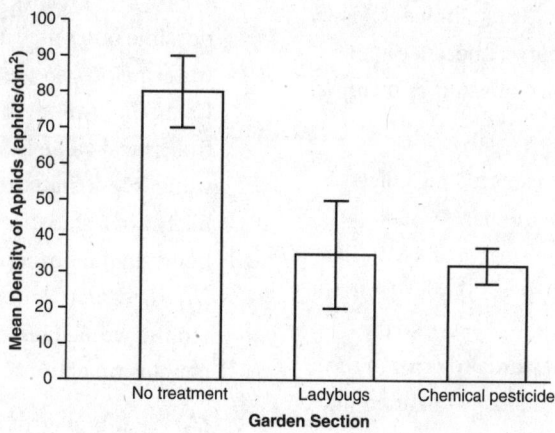

Part A

(i) Based on the graph, **identify** the section of the garden with the least variability in aphid density one week after treatment.

(ii) Based on the graph, **identify** the section of the garden with the most variability in aphid density one week after treatment.

Part B

(i) **Analyze** the data shown in the graph to determine which sections of the garden are most likely to have a statistically significant difference in aphid density.

Part C

In this experiment, 10 measurements ($n = 10$) of aphid density were taken in each section of the garden.

(i) If 50 measurements ($n = 50$) were taken, **make a prediction** about the effect, if any, this would have on the 95% confidence intervals.

Part D

(i) Using your knowledge of how 95% confidence intervals are calculated, **justify** your prediction from Part C.

Answer Explanations

Multiple-Choice

1. **(B)** The median is the midpoint of a data set and is not as affected by extreme values in a data set in the same way that the mean is. Extreme values in a data set skew the mean, so choice (A) is incorrect. The mean is used to calculate standard deviation, so standard deviation would be affected by extreme values in a data set. Thus, choice (C) is incorrect. Choice (D) is incorrect because standard error of the mean is also affected by extreme values in a data set.

2. **(A)** The mean for both data sets is 17. Choice (B) is incorrect because the median for data set A is 15, while the median for data set B is 17. The spread for data set A ($25 - 10 = 15$) is larger than the spread for data set B ($19 - 15 = 4$), so the standard deviation and the standard error of the means for those data sets will likely be different. Therefore, choices (C) and (D) are incorrect.

3. **(D)** The chi-square test is most appropriately used on raw data, such as the numbers of individuals in each category. The values used in the chi-square table have been calculated by statisticians and are designed to be used with raw data. Processed data, such as means and percentages, cannot be used with the chi-square test, making choices (A), (B), and (C) incorrect.

4. **(C)** Since there are four different phenotypes that could be produced, there are 3 degrees of freedom. Degrees of freedom = number of possible outcomes in the experiment $- 1 = 4 - 1 = 3$.

5. **(D)** $\chi^2 = \sum \frac{(\text{observed} - \text{expected})^2}{\text{expected}}$. Using the observed and expected values in the problem:

   $\chi^2 = \frac{(290 - 300)^2}{300} + \frac{(110 - 100)^2}{100} = 1.33$. There are two possible outcomes in the experiment (tall or dwarf), so the degrees of freedom $= 2 - 1 = 1$. Using the p-value of 0.05 and the chi-square table, the critical value is 3.84. The calculated chi-square value from the data is less than the critical value, so there is likely no statistically significant difference between the observed data and the expected data. Choices (A) and (B) are incorrect because they have incorrect calculations of the chi-square data. Choice (C) is incorrect because even though the chi-square value from the data is calculated correctly, the wrong conclusion is drawn from the result.

6. **(D)** The chi-square value = $\frac{(6 - 10)^2}{10} + \frac{(14 - 10)^2}{10} = 3.20$. There are two possible outcomes in the experiment (banana or watermelon), so there is one degree of freedom. Using the p-value of 0.05 and the chi-square table, the critical value is 3.84. The calculated chi-square value is less than the critical value, so there is likely not a statistically significant difference between the observed and expected data. Choices (A), (B), and (C) are incorrect because they present the wrong chi-square value and/or the wrong conclusion about the data.

7. **(A)** Standard error of the mean is calculated by dividing the standard deviation by the square root of the sample size ($SE_{\bar{x}} = \frac{s}{\sqrt{n}}$). If the standard deviation is the same for both data sets, the standard error of the mean will decrease when the sample size increases. Since both data sets have the same standard deviation, the data set with the smaller sample size (Data Set X, which has $n = 10$) will have a larger standard error of the mean than Data Set Y, which has a larger sample size ($n = 50$).

8. **(C)** The standard error of the mean is a way to measure the variability or spread of a data set. The larger the standard error of the mean, the greater the variability in the data set. Since data collected in San Rafael has the highest standard error of the mean, San Rafael has the greatest variability in the data.

9. **(B)** The upper limit of the 95% confidence interval for data from Oakville (25.1) is less than the lower limit of the 95% confidence interval for data from Oak Valley (30.1). So their 95% confidence intervals do not overlap and there is likely a significant difference between the oak tree heights in those two locations. Choice (A) is incorrect because the 95% confidence interval

for the data from Sacramento overlaps with the 95% confidence interval for the data from Oakville (the upper limit for Sacramento of 19.6 is greater than the lower limit for Oakville of 15.1). Similarly, the 95% confidence intervals for data from San Rafael and Oak Valley overlap (the upper limit for San Rafael of 37.3 is greater than the lower limit for Oak Valley of 30.1), so choice (C) is incorrect. Choice (D) is incorrect because the 95% confidence intervals for data from San Rafael and Oakville overlap (the upper limit for Oakville of 25.1 is greater than the lower limit for San Rafael of 20.1).

10. **(D)** The upper limit of a 95% confidence interval is determined by adding two times the standard error of the mean to the mean of the data set, which is represented by choice (D). The lower limit of the 95% confidence interval is determined by subtracting two times the standard error of the mean from the mean of the data set. Choices (A), (B), and (C) are incorrect because they do not represent the upper limit or the lower limit of a 95% confidence interval.

Short Free-Response

11. (A-i) The treatment with cadmium resulted in the greatest variability in the data because it has the greatest standard error of the mean.

    (B-i) The control treatment was the least likely to result in chromosome breakage. Its mean percentage of cells that showed chromosome breakage was far lower than the mean for the other two treatments.

    (C-i) The upper limit of the 95% confidence interval for lead $(24.3 + 2(5.2) = 34.7)$ is greater than the lower limit of the 95% confidence interval for cadmium $(46.1 - 2(9.1) = 27.9)$, so the 95% confidence intervals for those two treatments overlap. When the 95% confidence intervals overlap, the data are inconclusive and it is not possible to claim that there is a statistically significant difference between the groups. The data provided do not support the student's claim that exposure to cadmium is more likely to result in chromosome breakage than exposure to lead.

    (D-i) The independent variable is the type of treatment (control, lead, or cadmium).

    (D-ii) The dependent variable is the mean percentage of cells that showed chromosome breakage.

12. (A-i) There are two possible outcomes in this experiment (green or albino), so the number of degrees of freedom $(df) = 2 - 1 = 1$.

    (B-i) $\chi^2 = \sum \dfrac{(\text{observed} - \text{expected})^2}{\text{expected}}$
    $= \dfrac{(60 - 50)^2}{50} + \dfrac{(40 - 50)^2}{50}$
    $= \dfrac{100}{50} + \dfrac{100}{50} = 4$

    (C-i) The observed data are likely statistically significantly different from the expected data.

    (D-i) With one degree of freedom and using a p-value of 0.05, the critical value from the chi-square table is 3.84. Since the calculated chi-square value (4) is greater than the critical value, there is likely a statistically significant difference between the observed and expected values.

Long Free-Response

13. (A-i) The section of the garden with the least variability in aphid density is the section treated with chemical pesticide because its 95% confidence interval is the smallest.

    (A-ii) The section with the most variability is the section treated with ladybugs because its 95% confidence interval is the largest.

    (B-i) The 95% confidence interval for the section that received no treatment does not overlap with the 95% confidence intervals for the ladybug section or the chemical pesticide section. Thus, the section that received no treatment is least like the others and is most likely to have a statistically significant difference in aphid density from the ladybug- and chemical pesticide–treated sections.

    (C-i) If 50 measurements were taken in each section, the 95% confidence intervals would be smaller.

    (D-i) To calculate 95% confidence intervals, add or subtract two times the standard error of the mean from the mean of the data set. The standard error of the mean is inversely proportional to the sample size (if the sample size increases, the standard error of the mean decreases). So if the sample size, n, increased to 50 for all three sections, the standard errors of the mean for each section would likely decrease, and the 95% confidence intervals would be smaller.

UNIT 1
Chemistry of Life

3
Water

> **Learning Objectives**
>
> In this chapter, you will learn:
> → Water and the Importance of Hydrogen Bonds
> → pH

Overview

Living organisms contain more water than any other compound. The environment of most living organisms is dominated by water. Understanding water and its properties is key for the study of life on Earth. This chapter will review water's unique properties and how these properties affect living organisms.

Water and the Importance of Hydrogen Bonds

Water is a polar molecule. Its polarity allows it to form hydrogen bonds. These hydrogen bonds give water properties that are essential to life on Earth.

Water contains covalent bonds (shared electrons) between the oxygen and hydrogen atoms. The element oxygen has a high electronegativity (ability to attract electrons), while the element hydrogen has a lower electronegativity. Because of this electronegativity difference, the electrons in the covalent bond between oxygen and hydrogen are unequally shared, with the electrons spending more time around the oxygen atom. This results in a **polar covalent bond**, with a partial negative charge around the oxygen atom and a partial positive charge around the hydrogen atoms, as shown in Figure 3.1.

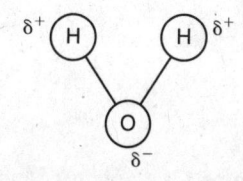

Figure 3.1 Polarity of Water

As a result, the partial negative charge on an oxygen atom in one water molecule is attracted to the partial positive charge on a hydrogen atom in another water molecule, resulting in a hydrogen bond. This causes water molecules to be attracted to one another, as shown in Figure 3.2.

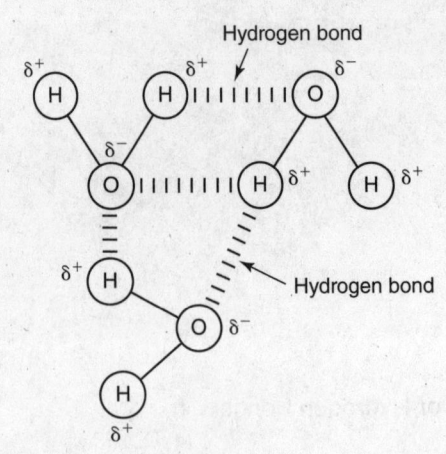

Figure 3.2 Hydrogen Bonds Between Water Molecules

Since water molecules can form hydrogen bonds, they have properties that help sustain life on Earth, including:

- **Exhibiting cohesive and adhesive behavior:** Water molecules are "sticky." They are attracted to other water molecules and to other polar molecules. This is what gives water its unique properties like water's high surface tension and its ability to climb up the xylem in plants through capillary action. (See Figure 3.3.)

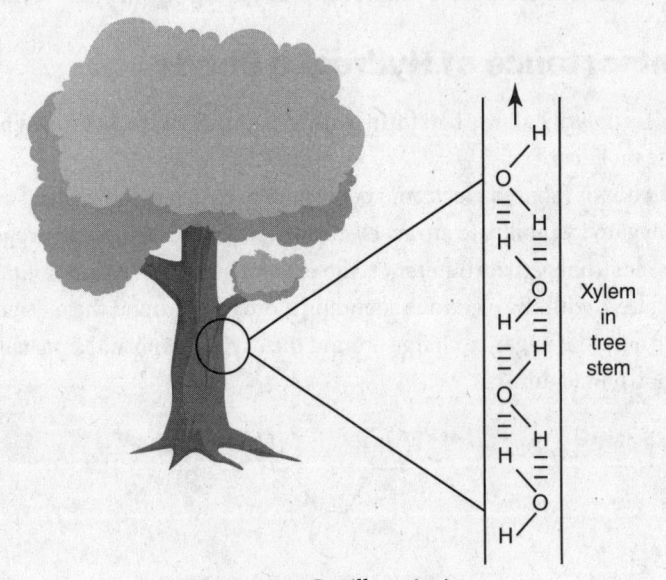

Figure 3.3 Capillary Action

- **Having a high specific heat:** As a result of water's ability to form hydrogen bonds, more energy is required to separate water molecules during phase changes, giving water both a high specific heat and a high heat of vaporization. When a person sweats, the water in the sweat on the skin absorbs heat from the person's body as the water/sweat evaporates, having a cooling effect on the person's body temperature.
- **Moderating climate:** Since water has a high heat capacity, it can absorb and release large amounts of energy. This stabilizes climates in locations near large bodies of water. (See Figure 3.4.)

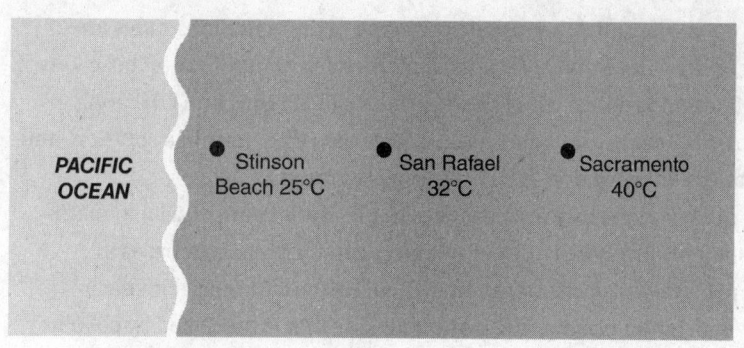

Figure 3.4 Effects of a Large Body of Water on Climate

- **Expanding upon freezing:** Since water has the ability to form hydrogen bonds, there is more space between water molecules in the solid state than in the liquid state. As a result, ice has a lower density than that of liquid water, and thus ice floats on liquid water. This has profound consequences for organisms in ponds or lakes that freeze in the winter. The layer of ice on the surface of the lake helps protect the organisms below from temperature extremes in the atmosphere, increasing their chances of surviving the cold winter. If ice were denser than liquid water, the ice would sink, leaving the remaining water in the lake exposed and vulnerable to more freezing and increasing the likelihood that the lake would freeze solid during the winter. This would result in fewer organisms in the lake surviving the winter.
- **Acting as a great solvent for other polar molecules and for ions:** Water has a partially positive end and a partially negative end. Thus, water can readily dissolve ionic compounds (see Figure 3.5) and other polar molecules. This makes water an excellent solvent for many biological molecules.

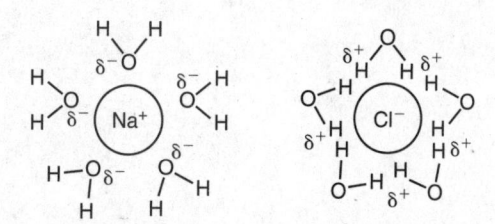

Figure 3.5 Water Interacting with Sodium Chloride

pH

pH (or "power of hydrogen") measures the concentration of H^+ ions in a solution. The formula for pH is as follows:

$$pH = -\log[H^+]$$

> While you will not be required to calculate pH from [H^+] on the AP Biology exam, you should know the following:
> A pH less than 7 is acidic.
> A pH greater than 7 is basic.
> A pH of 7 is neutral.

Due to the negative sign in the formula for pH, a higher [H^+] leads to a lower pH value and a lower [H^+] leads to a higher pH value. Thus, a solution with a pH of 3 would have a higher [H^+] than that of a solution with a pH of 5. Also, because the pH scale is a logarithmic scale, a pH change of one unit corresponds to a tenfold difference in H^+ concentration. For example, a pH of 3 would have 10 times the H^+ concentration of a pH of 4 and 100 times the concentration of a pH of 5.

The pH of a water-based solution depends on how many of the water molecules are dissociated (separated into H⁺ ions and OH⁻ ions) and the relative numbers of these ions. Pure water will dissociate and produce equal concentrations of H⁺ ions and OH⁻ ions, resulting in a pH of 7. Acids increase the relative concentration of H⁺ ions in a solution, and bases increase the relative concentration of OH⁻ ions in a solution.

Biological systems can be very sensitive to changes in pH. **Buffers** are crucial in maintaining relatively constant pH levels in living cells. Buffers can form acids or bases in response to changing pH levels in a cell. An example of this is the carbonic acid–bicarbonate buffering system in blood plasma. The carbon dioxide that is produced by cellular respiration reacts with water in blood plasma to produce carbonic acid (H_2CO_3). Carbonic acid can dissociate into bicarbonate ions (HCO_3^-) and hydrogen ions, as shown in the following equation:

$$H_2CO_3 \rightleftharpoons HCO_3^- + H^+$$

> **TIP**
> Remember, the concentration of hydrogen ions is *inversely* proportional to pH. If the pH *decreased*, that means the [H⁺] *increased*.

If the pH of a cell becomes too low (excess H⁺), the reaction will shift to the left, allowing the basic bicarbonate ions to neutralize the excess H⁺ ions, returning the pH to normal levels. When the pH of a cell becomes too high (excess OH⁻), the reaction shifts to the right, adding more H⁺ ions to the cell that can neutralize the excess OH⁻, which lowers the pH back to normal levels.

Chapter 4 ("Macromolecules") will review how changes in pH can denature proteins, changing their functions. Chapter 7 ("Enzymes") will discuss how most enzymes have a pH at which they function optimally.

Practice Questions

Multiple-Choice

1. In a water molecule, hydrogen atoms are attached to oxygen atoms through which type of bond?

    (A) hydrogen bond
    (B) nonpolar covalent bond
    (C) polar covalent bond
    (D) ionic bond

2. The attraction between the partially positive charge on a hydrogen atom on one water molecule and the partially negative charge on an oxygen atom on another water molecule is called a(n)

    (A) hydrogen bond.
    (B) nonpolar covalent bond.
    (C) polar covalent bond.
    (D) ionic bond.

3. Water's high specific heat is due to

    (A) the lower density of solid ice compared to that of liquid water.
    (B) the amount of energy required to break the covalent bonds within a water molecule.
    (C) the amount of energy required to break the hydrogen bonds between water molecules.
    (D) the low electronegativity of oxygen atoms compared to that of hydrogen atoms.

4. Which of the following solutions has the greatest concentration of H^+?

    (A) stomach acid with a pH of 2
    (B) acetic acid with a pH of 3
    (C) coffee with a pH of 5
    (D) bleach with a pH of 12

5. Solution A has a pH of 4; solution B has a pH of 7. How do the [H^+] in these solutions compare?

    (A) Solution A has 3 times the [H^+] concentration of solution B.
    (B) Solution A has 30 times the [H^+] concentration of solution B.
    (C) Solution A has 1,000 times the [H^+] concentration of solution B.
    (D) Solution A has 3,000 times the [H^+] concentration of solution B.

6. Coastal areas near large bodies of water tend to have more stable climates than inland areas at the same latitude. Which of the following is the property of water that best explains this difference in climate?

    (A) high surface tension
    (B) high specific heat
    (C) capillary action
    (D) density of ice

7. Small, lightweight insects can walk on the surface of water, as seen in the following figure:

    Which of the following is the property of water that best explains this phenomenon?

    (A) high surface tension
    (B) high specific heat
    (C) capillary action
    (D) density of ice

8. Arctic seals and walruses rely on ice floes for survival. Which of the following best explains why these ice floes exist?

    (A) high surface tension
    (B) high specific heat
    (C) capillary action
    (D) density of ice

9. Redwood trees, which are over 200 feet tall, can move water upward from their roots to other parts of the tree, despite the downward pull of gravity. Which of the following properties of water best explains this?

(A) high surface tension
(B) high specific heat
(C) capillary action
(D) density of ice

10. In hot weather, humans can cool their body temperature by sweating. Which of the following properties of water makes this possible?

(A) high surface tension
(B) high specific heat
(C) capillary action
(D) density of ice

Short Free-Response

11. On hot summer days, misters will sometimes be used to cool participants at outdoor events.

    Part A

    (i) **Describe** the property of water that allows misters to have an effective cooling effect.

    Part B

    (i) **Explain** why the evaporation of water makes the participants in these events more comfortable.

    Part C

    Instead of water, nonpolar oil is spread on the skin.

    (i) **Predict** whether this would have a less effective cooling effect, a more effective cooling effect, or the exact same cooling effect as water on the skin.

    Part D

    (i) Using what you know about the comparative properties of water and nonpolar substances, **justify** your prediction from Part C.

12. Refer to the following figure, which depicts water and methane.

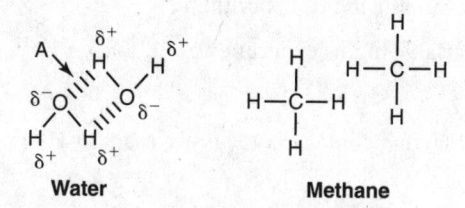

    Part A

    (i) **Describe** the type of bond indicated by arrow A.

    Part B

    (i) **Explain** why the bond indicated by arrow A forms between water molecules.

    Part C

    (i) Would an ionic salt dissolve more readily in water or methane? **Explain** your reasoning.

    Part D

    Plants in arid climates often need to conserve water loss due to evaporation through the leaves of the plant. Some plant species have a waxy, nonpolar cuticle on the outer surface of their leaves.

    (i) A student claims that this waxy cuticle reduces water loss from the leaves. **Support the student's claim** with reasoning.

Long Free-Response

13. Aquatic animals produce carbon dioxide as a product of cellular respiration. Carbon dioxide combines with water to form carbonic acid (H_2CO_3), which releases hydrogen ions (H^+) into solution. Four test tubes (containing 10 mL of water each and different numbers of aquatic snails) are prepared. pH levels were measured in each tube at the beginning of the experiment and after 20 minutes. The results are shown in the following table.

Tube	Number of Aquatic Snails in Tube	Initial pH	pH After 20 Minutes
A	0	7.0	7.0
B	1	7.0	6.0
C	2	7.0	5.0
D	3	7.0	4.0

Part A

(i) **Explain** why tubes B, C, and D all had lower pH levels after 20 minutes.

Part B

(i) **Identify** the independent variable in this experiment.

(ii) **Identify** the dependent variable in this experiment.

Part C

(i) **Identify** the tube (B, C, or D) that contained 100 times as many H^+ ions as that of tube A after 20 minutes.

(ii) **Explain** your reasoning.

Part D

Aquatic plants, such as *Elodea*, perform cellular respiration, but they also perform photosynthesis. Photosynthesis removes carbon dioxide from the water, reducing the amount of carbonic acid.

(i) **Predict** the effect of adding *Elodea* to all four tubes at the start of the experiment.

(ii) **Justify** your prediction.

Answer Explanations
Multiple-Choice

1. **(C)** Oxygen atoms and hydrogen atoms are joined in covalent bonds within a water molecule. Because oxygen has a higher electronegativity than hydrogen does, the electrons in this bond are unequally shared, and this results in a polar covalent bond. Choice (A) is incorrect because hydrogen bonds occur between different water molecules, not within a given water molecule. Choice (B) is incorrect because nonpolar covalent bonds form between atoms with similar electronegativities. Choice (D) is incorrect because the electrons in the bonds within a water molecule are shared; they are not transferred as is the case in an ionic bond.

2. **(A)** Hydrogen bonds occur between the hydrogen atom of one water molecule and the oxygen atom of another water molecule. Choices (B) and (C) are incorrect because both describe *intra*molecular bonds, not *inter*molecular bonds between different molecules. Choice (D) is incorrect because there is no transfer of electrons between water molecules.

3. **(C)** Because of the attraction between water molecules that is the result of hydrogen bonds, more energy is required to separate water molecules. Choice (A) is incorrect because although solid ice does have a lower density than that of liquid water, this does not affect water's specific heat. Choice (B) is incorrect because specific heat does not depend on the energy needed to break the *intra*molecular covalent bonds *within* a water molecule; rather, specific heat depends on the energy needed to break the *inter*molecular bonds *between* molecules. Choice (D) is incorrect because oxygen atoms have a higher electronegativity than that of hydrogen atoms.

4. **(A)** Of the choices presented, a stomach acid with a pH of 2 has the greatest concentration of H^+. All the other answer choices have a higher pH value than that of choice (A) and therefore a lower $[H^+]$.

5. **(C)** pH is a logarithmic scale, so a difference of three pH units would result in a difference of 10^3 times the $[H^+]$ concentration. Choice (A) is incorrect because pH is not a linear scale. Choice (B) is incorrect because 10^3 is not 30 times. Choice (D) is incorrect because 10^3 is not 3,000 times.

6. **(B)** Water's ability to form hydrogen bonds allows it to absorb and release large amounts of heat, giving water a higher specific heat than that of many other liquids. Locations near large bodies of water have more stable climates because the nearby bodies of water can absorb atmospheric heat during the day and then release it at night, leading to more stable climates. Choice (A) is incorrect because surface tension only affects the surface of the body of water and does not affect climate. Choice (C) is incorrect because capillary action describes water's ability to climb up narrow tubes. While ice may cool limited areas, the density of ice does not explain why coastal areas have more stable climates than inland areas, so choice (D) is incorrect.

7. **(A)** Surface tension describes the attraction of molecules to each other on the surface of a liquid. Since water has strong hydrogen bonds, it has a higher surface tension that is sufficient enough to support the mass of very lightweight insects on its surface. Choice (B) is incorrect because specific heat is not involved in supporting the mass of an insect on the surface of water. Choice (C) is incorrect because capillary action describes water's ability to climb up narrow tubes. While the density of ice is less than the density of liquid water, this does not explain why lightweight insects can walk on the surface of water, so choice (D) is incorrect.

8. **(D)** Ice has a lower density than that of liquid water, which allows ice floes to float on the ocean surface. Choice (A) is incorrect because surface tension describes the attraction between molecules of water at the surface of a liquid. Specific heat does not contribute to the creation of ice floes, so choice (B) is incorrect. Choice (C) is incorrect because capillary action describes water's ability to climb up narrow tubes.

9. **(C)** Capillary action describes water's ability to climb up narrow tubes. The trunks of redwood trees contain many narrow tubes made of xylem cells, which allow water to travel from the roots to the rest of the tree. Choice (A) is incorrect because surface tension describes the attraction between molecules of water at the surface of a liquid. Specific heat is not involved in capillary action, so choice (B) is incorrect. Choice (D) is incorrect because the density of ice has nothing to do with water's ability to move upward from the roots to the rest of the tree.

10. **(B)** Due to water's high specific heat, more energy is required to evaporate water molecules than other liquids. Sweat on the skin absorbs heat from the body as it evaporates, lowering body temperature. Choice (A) is incorrect because surface tension describes the attraction between molecules of water at the surface of a liquid. Choice (C) is incorrect because capillary action describes water's ability to climb up narrow tubes. Choice (D) is incorrect since ice is not involved in sweating.

Short Free-Response

11. (A-i) The polarity of water molecules leads to the formation of hydrogen bonds between water molecules, so water requires more energy to evaporate than molecules that do not form hydrogen bonds.

    (B-i) Energy is required to break the hydrogen bonds between water molecules before those molecules can evaporate. As the hydrogen bonds are broken, heat energy is absorbed from the participant's body, which has a cooling effect.

    (C-i) A nonpolar molecule would have a less effective cooling effect.

    (D-i) Nonpolar molecules do not form hydrogen bonds between them, so they would require less energy to evaporate and therefore have a less effective cooling effect.

12. (A-i) Arrow A indicates a hydrogen bond between water molecules.

    (B-i) Oxygen has a much greater electronegativity than hydrogen does. So the electrons in the covalent bond between oxygen and hydrogen in a water molecule are not shared equally and form a polar covalent bond. This gives the oxygen atom a partially negative charge and the hydrogen atom a partially positive charge.

    (C-i) An ionic salt would dissolve more readily in water because the polar water molecules could form hydration shells around the ions, as shown in the following figure.

    Since methane is nonpolar, it could not form hydration shells around the ions.

    (D-i) Polar water molecules cannot cross a waxy, nonpolar cuticle layer, so less water can evaporate from leaves surrounded by a waxy, nonpolar cuticle.

Long Free-Response

13. (A-i) The aquatic snails in tubes B, C, and D all produced carbon dioxide. The carbon dioxide combined with the water in the tubes to form carbonic acid, which released H^+ ions into the solution and lowered the pH. The more H^+ ions there are, the lower the pH is.

    (B-i) The independent variable in this experiment is the number of aquatic snails in each tube.

    (B-ii) The dependent variable is the pH.

    (C-i and ii) The equation for pH ($pH = -\log[H^+]$) is a logarithmic scale, which means each pH change of 1 unit will increase the number of H^+ ions by a factor of 10. A decrease of 2 pH units would indicate a 100-fold increase in H^+ concentration. Since tube A has a pH of 7.0 after 20 minutes, it makes sense that tube C, with a pH of 5.0, would have 100 times the H^+ concentration as that found in tube A.

    (D-i and ii) In tube A, the pH will increase after 20 minutes. The *Elodea* in tube A will remove carbon dioxide by the process of photosynthesis and therefore increase the pH. In tubes B, C, and D, which contain aquatic animals as well as *Elodea*, the pH will still decrease after 20 minutes since the animals are performing cellular respiration. However, since the *Elodea* will absorb some of the carbon dioxide produced, the decrease in pH will not be as large as it was when the tubes only contained aquatic snails.

4

Macromolecules

Learning Objectives

In this chapter, you will learn:
- → Biological Macromolecules
- → Protein Structure
- → Nucleic Acids

Overview

Biological macromolecules form the basis for the structure and function of living organisms. This chapter reviews the basic reactions that form and break down these molecules. Then, the structure and function of carbohydrates, lipids, nucleic acids, and proteins are reviewed, with special attention given to the four levels of structure found in proteins.

Biological Macromolecules

The macromolecules necessary for life are primarily made of six elements: nitrogen, carbon, hydrogen, oxygen, phosphorus, and sulfur.

Carbon is the "backbone" of these molecules. Carbon has four valence electrons and is extremely versatile in the way it can bond to other atoms. It can form single, double, or even triple bonds. Carbon can also form linear, branched, or ring-type structures. Carbon is found in *all* types of macromolecules.

Oxygen and sulfur each have six valence electrons, and each of these elements typically forms two bonds. Oxygen is found in all types of macromolecules. Sulfur is typically found in proteins.

Nitrogen and phosphorus each have five valence electrons, and each of these elements typically forms three bonds. Nitrogen is found in nucleic acids and proteins. Phosphorus is found in nucleic acids and some lipids.

Hydrogen has one valence electron and forms a single bond. Hydrogen is found in all types of macromolecules. In fact, hydrogen atoms are so ubiquitous that often they are not drawn in molecular structures.

The structure and function of a macromolecule is determined by the types of monomers from which it is made and how the monomers are linked.

Making and Breaking Biological Macromolecules

Biological macromolecules are formed from building blocks (i.e., monomers) that are linked by **dehydration synthesis** to form larger molecules. These biological macromolecules are broken down by **hydrolysis** reactions. (See Figure 4.1.)

> **TIP**
> An easy mnemonic device to help you remember these elements is **NCHOPS** for Nitrogen, Carbon, Hydrogen, Oxygen, Phosphorus, Sulfur.

> **TIP**
> While you do not need to know the structures of *specific* molecules for the AP Biology exam, you do need to be familiar with the different types of structures and functions of macromolecules.

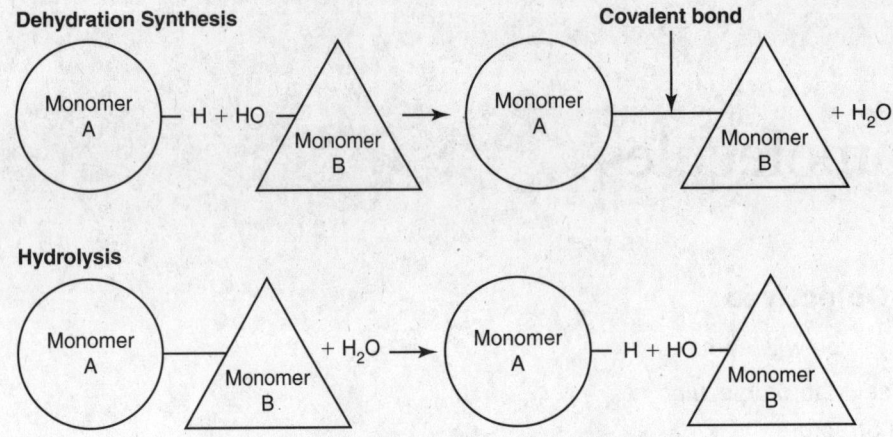

Figure 4.1 Dehydration Synthesis and Hydrolysis

Carbohydrates

Carbohydrates are polymers of sugar monomers. The types of sugars used to make the carbohydrate and how the sugars are linked determines the structure and function of the carbohydrate. The sugars may be joined in linear structures or in branched chains. Carbohydrates can be used to store energy (such as in starch or glycogen) and can also have structural functions (such as in cellulose). The type of linkages between the sugars in carbohydrates that store energy is different from the type of linkages found in carbohydrates that have a structural function. (See Figure 4.2.)

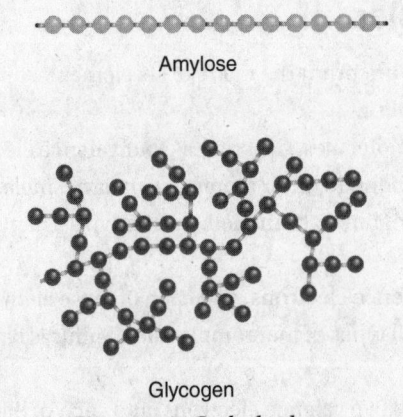

Figure 4.2 Carbohydrates

Lipids

Lipids are nonpolar (a.k.a. "hydrophobic") macromolecules that function in energy storage, cell membranes, and insulation. (See Figure 4.3.) One of the building blocks of lipids is fatty acids. Fatty acids with the maximum number of C—H single bonds are called saturated, are solid at room temperature, and usually originate in animals. Fatty acids with at least one C=C double bond are called unsaturated, are liquid at room temperature, and usually originate in plants. How a lipid functions in a cell is dependent on the lipid's saturation level.

Phospholipids are extremely important in cell membranes. They are built from a glycerol molecule, two fatty acids, and a phosphate group. Because the fatty acids are nonpolar and the phosphate is polar, phospholipids are amphipathic, meaning they have both hydrophobic and hydrophilic regions.

Figure 4.3 Lipids

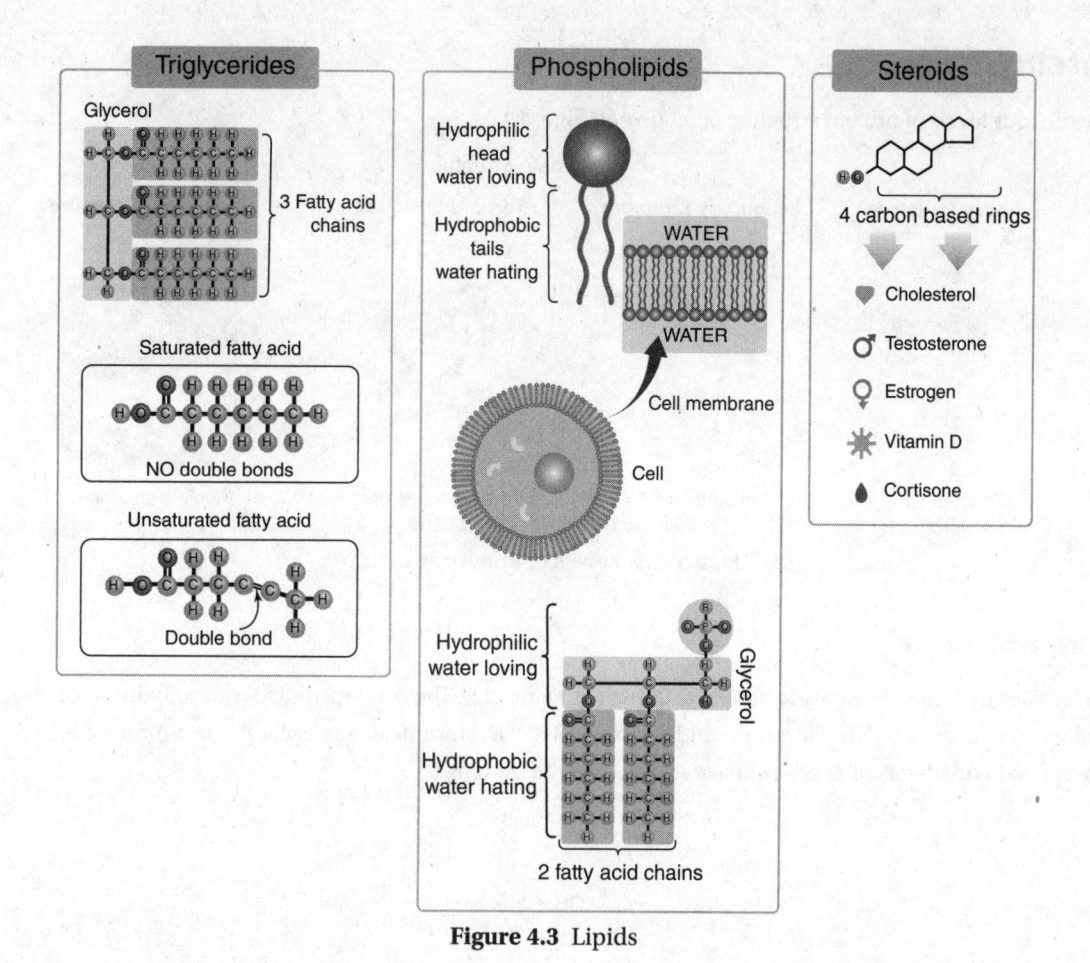

Another class of lipids are steroids. Steroids are relatively flat, nonpolar molecules. Many steroids are formed by modifying cholesterol molecules. Examples of steroids include estradiol, testosterone, and cortisol. Cholesterol can also help provide stability to cell membranes.

Nucleic Acids

Nucleic acids are polymers of nucleotides and function as the carriers of genetic information. Nucleotides will be discussed in more detail later in this chapter.

Proteins

Proteins are polymers of amino acids. Amino acids have an amino group, a carboxylic acid group, a hydrogen atom, and a side chain (R-group) attached to a central carbon, as shown in Figure 4.4. The R-group is unique for each amino acid; it determines the amino acid's identity and whether the amino acid will be nonpolar, polar, acidic, or basic. Proteins function in enzyme catalysis, maintaining cell structures, cell signaling, cell recognition, and more.

Figure 4.4 Structure of an Amino Acid

Protein Structure

There are four levels of protein structure, as shown in Figure 4.5.

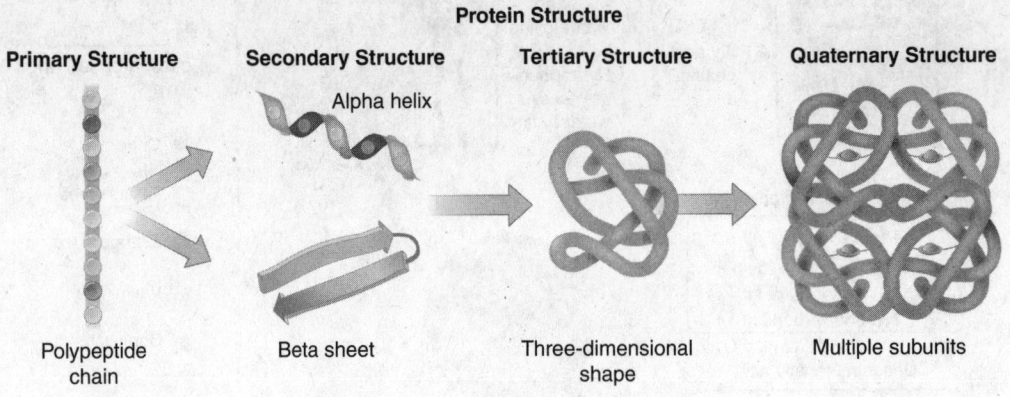

Figure 4.5 Levels of Protein Structure

Primary Structure

Amino acids are joined by **peptide bonds**, as shown in Figure 4.6. The resulting polypeptide chains have directionality, with an amino (NH_2) terminus and a carboxyl (COOH) terminus. The order of the amino acids in the polypeptide chain determines the primary structure of the protein.

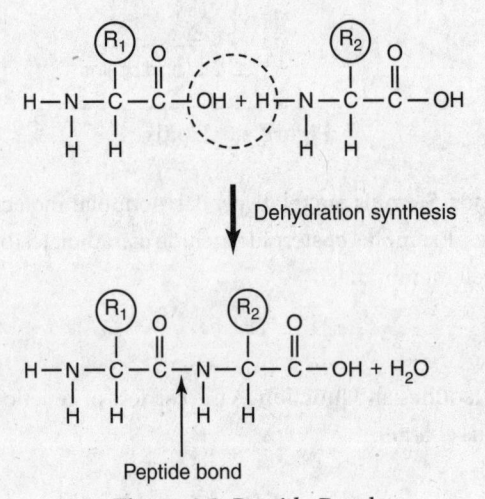

Figure 4.6 Peptide Bond

A change in the primary structure of a protein may have severe effects on the function of the protein, as seen in sickle cell disease. Just one amino acid is changed out of over 140 amino acids in sickle cell hemoglobin.

Secondary Structure

Once the primary structure is formed, hydrogen bonds may form between adjacent amino acids in the polypeptide chain. This drives the formation of the secondary structure of the protein. These secondary structures include alpha helixes and beta-pleated sheets.

> **NOTE**
> Changes in pH can disrupt hydrogen bonding and ionic interactions between amino acids in a protein, causing denaturation of the protein. This can result in a change in shape, and, consequently, reduced function of the protein.

Tertiary Structure

Tertiary structure is the three-dimensional folded shape of the protein, often determined by the hydrophobic/hydrophilic interactions between R-groups in the polypeptide. The most stable tertiary structures will have hydrophilic R-groups on the surface of the protein (in contact with the watery environment of the cell's cytosol), while the amino acids with hydrophobic R-groups will be found in the interior of the protein (away from the watery cytosol). Tertiary structures may also include disulfide bridges between sulfur atoms. Special proteins called chaperonins often help fold a polypeptide into its three-dimensional structure.

Quaternary Structure

Some proteins consist of multiple polypeptide chains (subunits), which are joined together to form the complete protein and function as a unit. For example, hemoglobin has four subunits in its quaternary structure, and collagen has three subunits in its quaternary structure.

Nucleic Acids

Nucleic acids (DNA and RNA) are polymers of nucleotides, as shown in Figure 4.7. The genetic information is stored and communicated through the order of these nucleotides. Nucleotides consist of a five-carbon sugar (deoxyribose or ribose), a nitrogenous base (adenine, thymine, cytosine, guanine, or uracil), and a phosphate group. Nucleotides have directionality in that the phosphate group is always attached to the 5′ carbon in the sugar, and the 3′ carbon always has a hydroxyl group to which new nucleotides may be added.

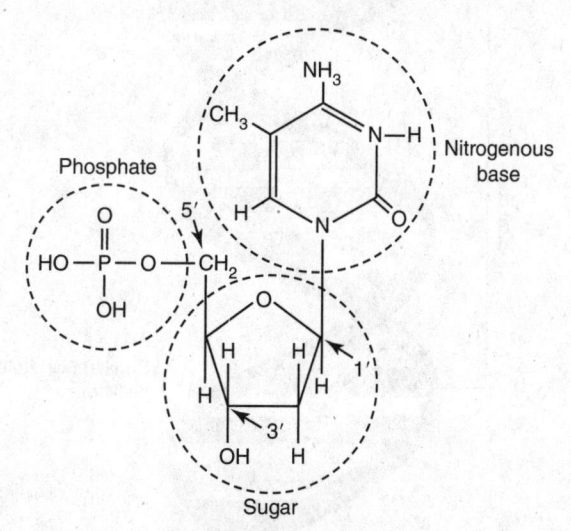

Figure 4.7 Nucleotides

While DNA and RNA are both made of nucleotides, their structures differ. Unit 6 of this book will review how these structural differences relate to their different functions. Table 4.1 summarizes some quick initial facts to know about DNA and RNA.

Table 4.1 DNA vs. RNA

	DNA	RNA
Five-Carbon Sugar	Deoxyribose	Ribose
Nitrogenous Bases	Adenine, thymine, cytosine, and guanine	Adenine, uracil, cytosine, and guanine
Strands	Double-stranded helix	Usually single-stranded but can form three-dimensional structures when folded
Function	Holds genetic information	Transcribes and regulates the expression of genetic information
Location	Usually found in the nucleus	Found in both the nucleus and the cytoplasm

The two strands in the double helix of DNA are antiparallel; the 5′ phosphate group of one of the strands in the double helix is at the opposite end from the 5′ phosphate group of the other strand, as shown in Figure 4.8.

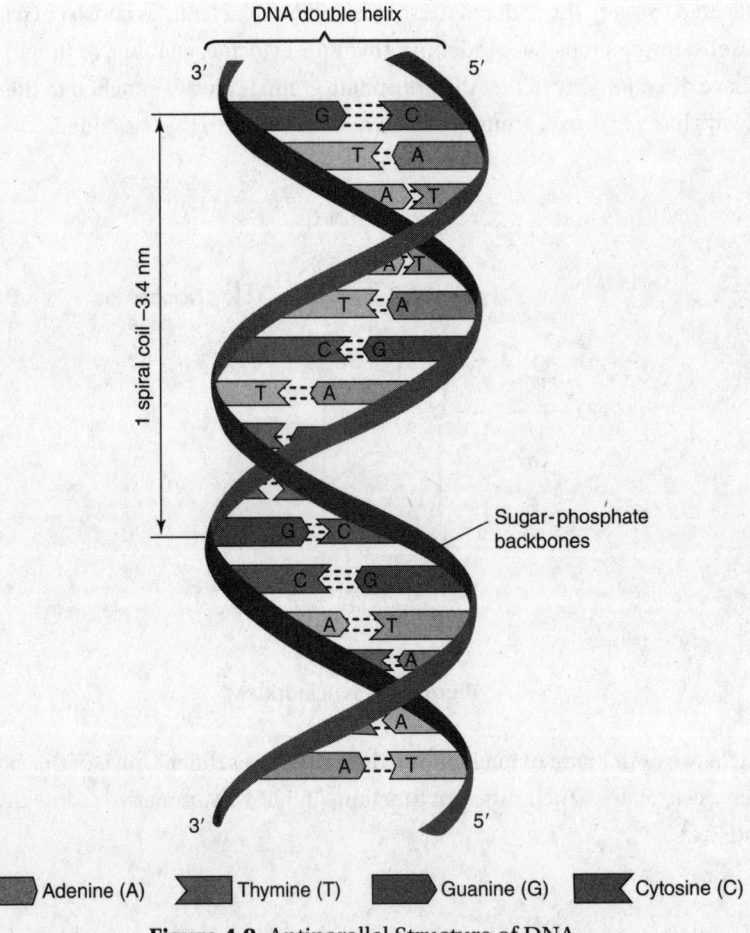

Figure 4.8 Antiparallel Structure of DNA

Thymine, uracil, and cytosine are all in a class of macromolecules called **pyrimidines**. Adenine and guanine are **purines**. When DNA strands form, a purine complements a pyrimidine and forms a series of hydrogen bonds, making the "rungs" of the ladder in the double helix. Thymine and adenine form two hydrogen bonds with one another, whereas guanine and cytosine form three hydrogen bonds with each other. Chapter 15 will discuss the profound implications this has for how DNA is replicated.

Practice Questions

Multiple-Choice

1. COVID-19 is a single-stranded RNA virus. Which molecules would most likely be found in a single-stranded RNA virus, such as COVID-19?

    (A) adenine, cytosine, deoxyribose, guanine, thymine
    (B) adenine, cytosine, deoxyribose, guanine, uracil
    (C) adenine, cytosine, ribose, guanine, thymine
    (D) adenine, cytosine, ribose, guanine, uracil

2. A scientist conducted an experiment to find out what type of macromolecule a virus injects into a cell. Using radiolabeled atoms, the scientist found that phosphorus from the virus entered the cell but sulfur did not. Which of the following molecules would most likely be injected from this virus into the cell?

    (A) carbohydrate
    (B) nucleic acid
    (C) protein
    (D) steroid

3. Which of the following best describes the formation of the primary structure of a protein?

    (A) A dehydration reaction forms an ionic bond between the carboxyl group of one amino acid and the amino group of another amino acid.
    (B) A dehydration reaction forms a covalent bond between the carboxyl group of one amino acid and the amino group of another amino acid.
    (C) A hydrolysis reaction forms an ionic bond between the carboxyl group of one amino acid and the amino group of another amino acid.
    (D) A hydrolysis reaction forms a covalent bond between the carboxyl group of one amino acid and the amino group of another amino acid.

4. Which of the following best describes the differences between saturated and unsaturated lipids?

    (A) Saturated lipids have at least one C=C double bond and tend to be solid at room temperature. Unsaturated lipids have no double bonds and tend to be liquid at room temperature.
    (B) Saturated lipids have at least one C=C double bond, which makes them amphipathic. Unsaturated lipids have no double bonds and are hydrophilic.
    (C) Saturated lipids have no C=C double bonds and tend to be solid at room temperature. Unsaturated lipids have at least one C=C double bond and tend to be liquid at room temperature.
    (D) Saturated and unsaturated lipids both have C=C double bonds. Saturated lipids are hydrophobic, and unsaturated lipids are hydrophilic.

5. The molecular formula for glucose is $C_6H_{12}O_6$. The molecule maltose is formed by a dehydration reaction that links two glucose molecules together. What is the molecular formula for maltose?

    (A) $C_2H_4O_2$
    (B) $C_6H_{10}O_5$
    (C) $C_{12}H_{22}O_{11}$
    (D) $C_{12}H_{24}O_{12}$

6. In an aqueous environment like the cytosol, the most stable tertiary protein structures would have hydrophilic amino acids in which part of the protein's structure?

    (A) in the interior of the protein, interacting with water in the cytosol
    (B) in the interior of the protein, avoiding water in the cytosol
    (C) on the surface of the protein, interacting with water in the cytosol
    (D) on the surface of the protein, avoiding water in the cytosol

7. Which level of protein structure is formed by peptide bonds between amino acids?

   (A) primary
   (B) secondary
   (C) tertiary
   (D) quaternary

8. Which of the following drives the formation of a protein's secondary structure?

   (A) Hydrophobic interactions form between R-groups of amino acids.
   (B) Hydrogen bonds form between amino acids in a polypeptide chain.
   (C) Disulfide bridges form between amino acids in a polypeptide chain.
   (D) Multiple subunits/domains of a protein are connected by covalent bonds.

9. Which of the following is common to both DNA and RNA?

   (A) the nitrogenous bases: adenine, cytosine, guanine, and thymine
   (B) a double-stranded antiparallel helix
   (C) a phosphate group attached to the 5′ carbon
   (D) the five-carbon sugar ribose

10. Which of the following correctly describes DNA but not RNA?

    (A) It contains adenine, cytosine, guanine, and uracil.
    (B) A hydroxyl group is attached to the 5′ carbon.
    (C) Nucleotides are linked by phosphodiester bonds.
    (D) It contains the five-carbon sugar deoxyribose.

Short Free-Response

11. Cell membranes are made of phospholipid bilayers, which may contain both unsaturated fatty acids and saturated fatty acids.

    Part A

    (i) **Describe** the structural differences between an unsaturated fatty acid and a saturated fatty acid.

    Part B

    A student conducts an experiment to compare the melting points of an unsaturated fatty acid and a saturated fatty acid. The data are shown in the table below.

Fatty Acid	Melting Point (°C)
Oleic acid (unsaturated)	16.3
Stearic acid (saturated)	69.3

    (i) **Identify** the independent variable in this experiment.

    (ii) **Identify** the dependent variable in this experiment.

    Part C

    One type of cell (cell A) contains a much higher percentage of unsaturated fatty acids in its membrane than that of another type of cell (cell B).

    (i) **Predict** which cell would have a more flexible membrane.

    Part D

    (i) **Justify** your prediction from Part C.

12. The melting temperature (Tm) is defined as the temperature at which 50% of double-stranded DNA is separated into single-stranded DNA. The greater the guanine-cytosine (poly-GC) of the DNA, the higher the Tm compared to DNA with more adenine-thymine (poly-AT) content. The following graph shows the Tm of a poly-AT DNA strand, a poly-GC DNA strand, and DNA from two different organisms (A and B).

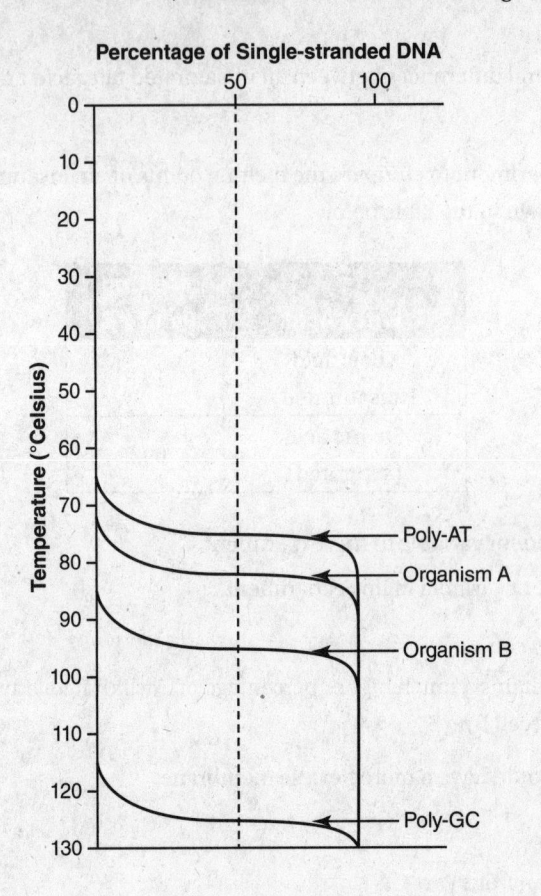

Part A

(i) **Describe** how the Tm of DNA from organism A compares to the Tm of DNA from organism B.

Part B

(i) **Describe** the differences between the Tm of poly-AT and the Tm of poly-GC DNA.

Part C

DNA sequencing finds that DNA from organism A has a GC content of 39% and that DNA from organism B has a GC content of 48%. A student claims that DNA from organism C (with a GC content of 55%) would have a Tm greater than 95° Celsius.

(i) Using the data from the graph, **evaluate** the student's claim.

Part D

(i) **Explain** how these experimental results relate to the structure of DNA.

Long Free-Response

13. Genomes with a higher GC content are more likely to resist denaturation.

    Part A

    (i) Using what you know about DNA structure, **explain** why genomes with a higher GC content might be more stable and more resistant to denaturation than genomes with a lower GC content.

    Part B

    A plant biologist hypothesized that plants with a higher GC content in their genomes are more likely to be found in colder climates. Four species of plants were examined for the percentage of GC content in their genomes. *Brachypodium pinnatum* and *Dioscorea caucasica* are native to areas with cold winters. *Juncus inflexus* and *Carex acutiformis* are native to areas with warmer winters. DNA was examined from 50 plants of each species. Results are shown in the table.

Plant Species	Mean Percent of GC Content	Standard Error of the Mean
Brachypodium pinnatum	46.0	1.9
Dioscorea caucasica	42.5	2.0
Juncus inflexus	33.7	2.0
Carex acutiformis	35.6	0.8

    (i) On the axes provided, **construct** an appropriately labeled graph of the mean percent of GC content in each species' genome. Include 95% confidence intervals.

Part C

(i) Do the data support the claim that plants in colder climates have a higher GC content in their genomes? Use the data provided to **support your answer**.

Part D

A plant species has a genome with a GC content of 43.4%.

(i) **Predict** whether that plant is more likely to be native to an area with cold winters or an area with warmer winters.

(ii) **Justify** your prediction.

Answer Explanations
Multiple-Choice

1. **(D)** RNA contains ribose and uracil. Choice (A) is incorrect because deoxyribose and thymine are found in DNA, not RNA. Deoxyribose is found in DNA, not RNA, so choice (B) is incorrect. Choice (C) is incorrect because thymine is not found in RNA; it is found in DNA.

2. **(B)** Nucleic acids contain phosphorus but not sulfur. So if phosphorus entered the cell but sulfur did not, a nucleic acid was most likely injected. Choice (A) is incorrect because carbohydrates typically do not contain phosphorus. Proteins typically contain sulfur but not phosphorus, so choice (C) is incorrect. Choice (D) is incorrect because steroids are a type of lipid that do not contain phosphorus.

3. **(B)** The primary structure of a protein is formed by linking amino acids, and dehydration reactions link amino acids in covalent bonds between the carboxyl group of one amino acid and the amino group of another amino acid. Choice (A) is incorrect because dehydration reactions do not form ionic bonds. Choices (C) and (D) are incorrect because hydrolysis reactions break, not form, bonds between monomers.

4. **(C)** Saturated lipids have carbon atoms linked through single bonds and form straight chains; this makes saturated lipids easier to pack tightly and more likely to be solid at room temperature. Choice (A) is incorrect because saturated lipids do not have C=C double bonds. Choice (B) is incorrect because saturated lipids are not amphipathic (phospholipids are amphipathic). Choice (D) is incorrect because while unsaturated lipids do have C=C double bonds, saturated lipids do not. Also, both saturated and unsaturated lipids are hydrophobic.

5. **(C)** A dehydration reaction removes a water molecule. Two glucose molecules would contain 12 carbon atoms, 24 hydrogen atoms, and 12 oxygen atoms, but the dehydration reaction (linking the two glucose molecules to form maltose) would remove two hydrogen atoms and one oxygen atom to form the water molecule removed in a dehydration reaction. This leaves 12 carbon atoms but only 22 hydrogen atoms and 11 oxygen atoms. Choice (A) is incorrect because it has fewer atoms than a single glucose molecule, so it could not be maltose (which is made from two glucose molecules). Choice (B) is incorrect because it is what would remain if a water molecule was removed from a single glucose molecule, not from two glucose molecules. Choice (D) is incorrect because it shows the number of atoms present in two glucose molecules without taking into account the atoms lost from the water molecule that was removed in the dehydration synthesis reaction.

6. **(C)** Tertiary protein structures in the cell are most stable when their hydrophilic amino acids are on the surface of the protein, in contact with the watery cytosol of the cell. Choices (A) and (B) are incorrect because *hydrophobic* amino acids are more likely to be found in the interior of a protein, away from water in the cytosol. Choice (D) is incorrect because hydrophilic amino acids would not avoid water; they would be more stable when interacting with water.

7. **(A)** The primary structure of a protein is the sequence of amino acids held together by peptide bonds. Choice (B) is incorrect because secondary structure is formed by the hydrogen bonds between amino acids in a polypeptide chain. Tertiary structure is the globular shape formed by a polypeptide chain, so choice (C) is incorrect. Choice (D) is incorrect because quaternary structure is formed when multiple subunits come together to form the functional protein.

8. **(B)** Secondary structure is driven by the formation of hydrogen bonds between the carboxyl groups and amino groups in a polypeptide chain. Choice (A) is incorrect because hydrophobic interactions between R-groups influence tertiary and quaternary structures but not secondary structures. Choice (C) is incorrect because disulfide bridges are formed in tertiary and quaternary structures, not secondary structures. Choice (D) is incorrect because multiple subunits/domains are only present in quaternary structures.

9. **(C)** Both DNA and RNA have a phosphate group attached to the 5′ carbon. Choice (A) is incorrect because RNA does not contain thymine. A double-stranded antiparallel helix is only present in DNA, not RNA, so choice (B) is incorrect. Choice (D) is incorrect because ribose is present in RNA but not in DNA.

10. **(D)** Deoxyribose is found in DNA; ribose is found in RNA. Choice (A) is incorrect because DNA does not contain uracil. Choice (B) is incorrect because the hydroxyl group is attached to the 3′ carbon, not the 5′ carbon. Phosphodiester bonds are found in both DNA and RNA, so choice (C) is incorrect.

Short Free-Response

11. (A-i) Saturated fatty acids have a chain of carbon atoms connected by single bonds. Unsaturated fatty acids have carbon chains that contain at least one double bond between the carbon atoms.

    (B-i) The independent variable is the type of fatty acid.

    (B-ii) The dependent variable is the melting point in degrees Celsius.

    (C-i) Cell A would have a more flexible cell membrane.

    (D-i) Cell A would have a more flexible cell membrane because it contains more unsaturated fatty acids. Unsaturated fats contain double bonds in random locations, causing bends in the long chains. These bends in the fatty acid chains prevent unsaturated fats from forming tightly packed sheets, which leaves some spaces between the fatty acid chains. Saturated fats, on the other hand, are straight chains; they form tightly packed sheets and have less space between their fatty acid chains. Since unsaturated fats cannot pack tightly like saturated fats, this allows unsaturated fats to be more flexible and thus be liquid at room temperature.

12. (A-i) The Tm of DNA from organism A is lower than the Tm of DNA from organism B.

    (B-i) The Tm of poly-GC DNA is much higher than the Tm of poly-AT DNA.

    (C-i) DNA from organism B has a GC content of 48% and a Tm of approximately 95° Celsius. Since increased GC content leads to a higher Tm and the DNA from organism C has a GC content of 55%, it would follow the pattern of the data that organism C's Tm would be higher than 95° Celsius.

    (D-i) In DNA, GC pairs form three hydrogen bonds while AT pairs form only two hydrogen bonds. More hydrogen bonds would require more energy to separate, so it makes sense that DNA with a higher GC content would have a higher Tm.

Long Free-Response

13. (A-i) The two strands of a DNA double helix are held together by hydrogen bonds between the base pairs. AT pairs form two hydrogen bonds, and the GC pairs form three hydrogen bonds. Since GC pairs form more hydrogen bonds, that may make DNA with a higher GC content more stable across a wider range of temperatures than DNA with a lower GC content.

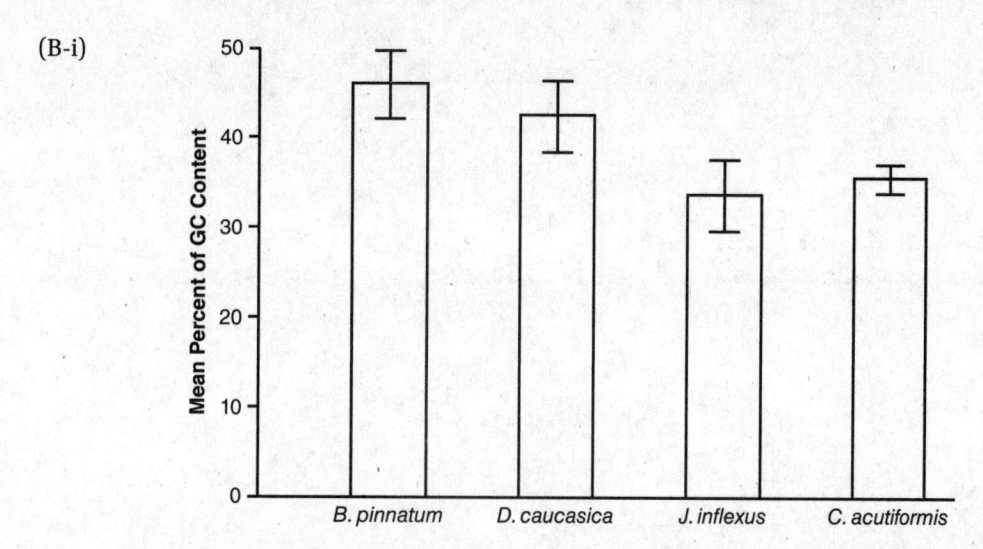

(B-i)

[Bar graph showing Mean Percent of GC Content for four species: B. pinnatum (~46%), D. caucasica (~42%), J. inflexus (~33%), C. acutiformis (~35%), with error bars.]

(C-i) Yes, the claim that plants in colder climates have a higher GC content in their genomes is supported by the data. *B. pinnatum* and *D. caucasica* are native to colder climates. Both also have statistically significantly higher GC content in their genomes as compared to the species found in warmer winter locations (*J. inflexus* and *C. acutiformis*), as shown in the graph for Part B.

(D-i and ii) The plant with a GC content of 43.4% is more likely be native to an area with cold winters since its GC content is closer to the GC contents found in *B. pinnatum* and *D. caucasica*.

UNIT 2
Cell Structure and Function

5

Cell Organelles, Membranes, and Transport

Learning Objectives

In this chapter, you will learn:

- → Cell Organelles and Their Functions
- → Endosymbiosis Hypothesis
- → The Advantages of Compartmentalization
- → The Importance of Surface Area to Volume Ratios
- → Structure of the Plasma Membrane
- → What Can (and Cannot) Cross the Plasma Membrane
- → Passive Transport
- → Active Transport

Overview

In prokaryotic and eukaryotic organisms (no matter the size), the basic unit of life is the cell. This chapter reviews the functions of different parts of the cell and how cells maintain an internal environment (**homeostasis**) that supports life.

Cell Organelles and Their Functions

The two major types of cells are prokaryotic cells and eukaryotic cells. (See Figure 5.1.) **Prokaryotic cells** are simpler in structure. Bacteria are prokaryotic cells. **Eukaryotic cells** contain membrane-bound organelles and are much more complex than prokaryotic cells. Animals, plants, fungi, and protists are all composed of eukaryotic cells.

All cells (both prokaryotic and eukaryotic) contain genetic material, **ribosomes**, **cytosol**, and a **plasma membrane**. In prokaryotes, the genetic material (DNA) is a circular chromosome located in the center of the cell in an area called the **nucleoid** region. Bacteria may also contain extra genetic material outside of the chromosome, which is contained in small circular pieces of DNA called **plasmids**. In eukaryotes, DNA is packaged into linear chromosomes that are contained in a membrane-bound **nucleus**.

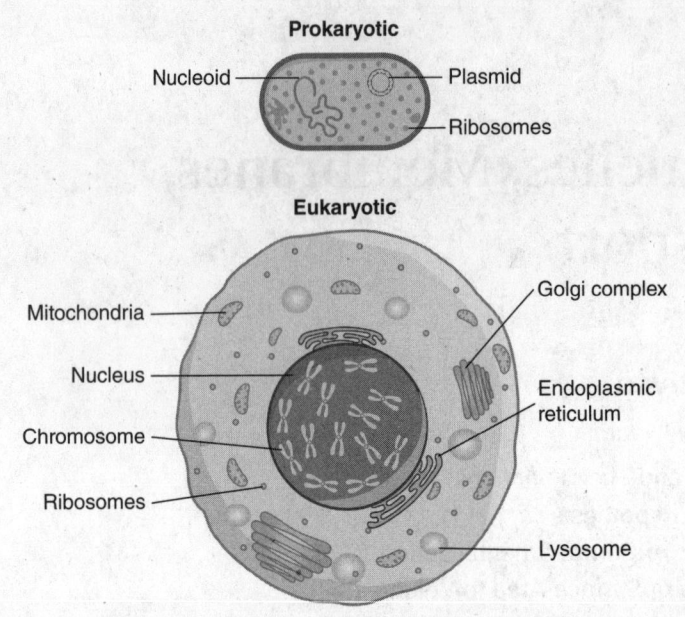

Figure 5.1 Prokaryotic Cell vs. Eukaryotic Cell

Ribosomes are non-membrane subcellular structures that function in protein synthesis and are found in both prokaryotic and eukaryotic cells. Ribosomes are made of proteins and **ribosomal RNA (rRNA)**. Both prokaryotic and eukaryotic ribosomes have a large subunit and a small subunit, but the sizes of these subunits differ slightly, as shown in Figure 5.2.

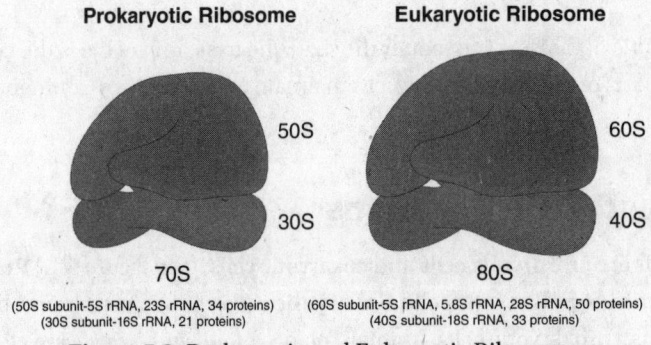

Prokaryotic Ribosome
50S
30S
70S
(50S subunit-5S rRNA, 23S rRNA, 34 proteins)
(30S subunit-16S rRNA, 21 proteins)

Eukaryotic Ribosome
60S
40S
80S
(60S subunit-5S rRNA, 5.8S rRNA, 28S rRNA, 50 proteins)
(40S subunit-18S rRNA, 33 proteins)

Figure 5.2 Prokaryotic and Eukaryotic Ribosomes

During translation, ribosomes assemble amino acids into polypeptide chains according to the mRNA sequence. Free ribosomes are found in the cytosol in both prokaryotes and eukaryotes. In eukaryotes, bound ribosomes are found on the membrane of the rough endoplasmic reticulum.

The **endoplasmic reticulum** is a series of membrane channels in eukaryotic cells, as shown in Figure 5.3. Rough endoplasmic reticulum has ribosomes bound to its membranes and functions in protein synthesis. Smooth endoplasmic reticulum does not contain ribosomes and functions in the synthesis of lipids and the detoxification of harmful substances in the cell.

TIP

While it IS important to know that the subunits of prokaryotic ribosomes differ slightly from the subunits of eukaryotic ribosomes, you do NOT need to memorize the specific differences.

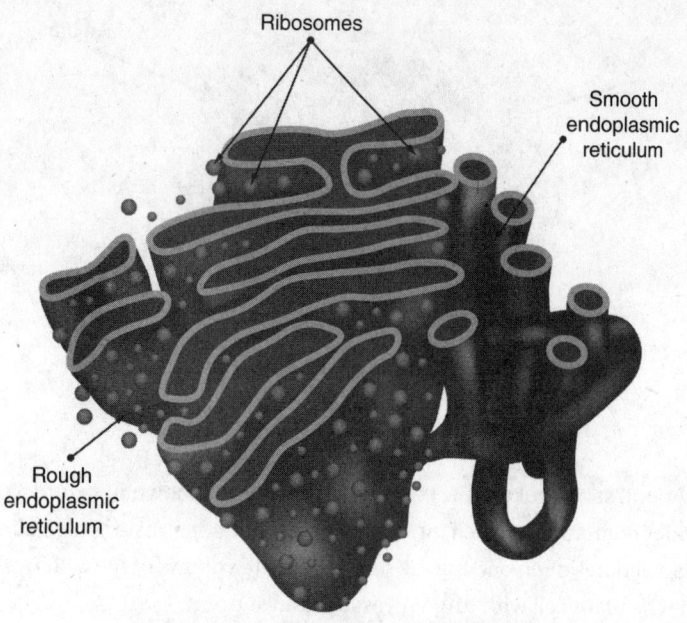

Figure 5.3 Endoplasmic Reticulum

The **Golgi complex** (also known as the Golgi body) is a stack of flattened membrane sacs (called cisternae), as shown in Figure 5.4. The interior of each cisterna is called the lumen and contains the enzymes necessary for the Golgi complex to function. The Golgi complex controls the modification and packaging of proteins for transport. Proteins made on the free ribosomes or the rough endoplasmic reticulum are sent to the Golgi, which modifies the proteins into their final conformation and packages the finished proteins into vesicles for transport throughout the cell. The Golgi complex is often found near the rough endoplasmic reticulum.

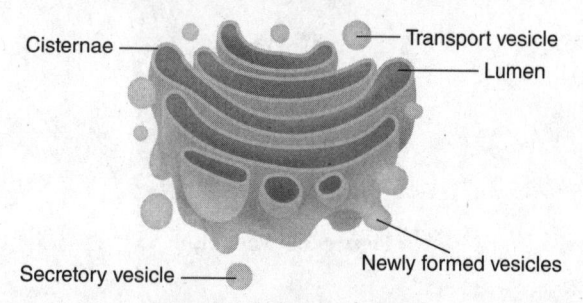

Figure 5.4 Golgi Complex

Lysosomes are membrane-bound sacs, containing hydrolytic enzymes, that function in a variety of cell processes. Lysosomes can help digest macromolecules, break down worn-out cell parts, function in apoptosis, or destroy bacteria and viruses that have entered the cell. (See Figure 5.5.)

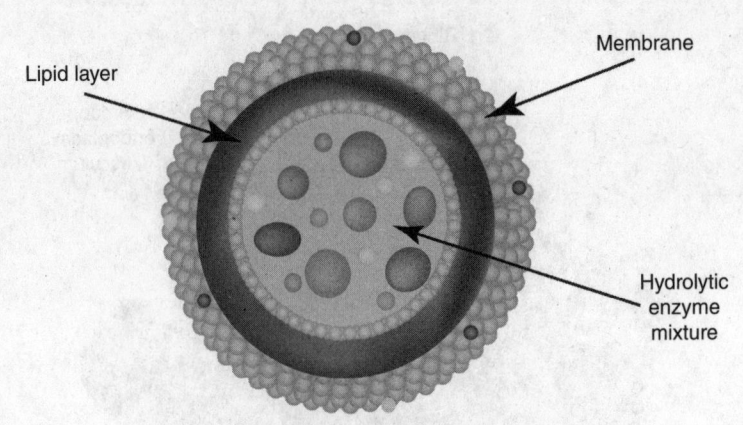

Figure 5.5 Lysosome

Another membrane-bound sac in eukaryotic cells is the **vacuole**, as shown in Figure 5.6. Vacuoles function in food or water storage, water regulation in a cell, or waste storage until the waste can be eliminated from the cell. In well-hydrated plant cells, vacuoles often occupy the majority of the volume of the cell. By filling up space within the cell, vacuoles provide the plant cell with turgor pressure and support.

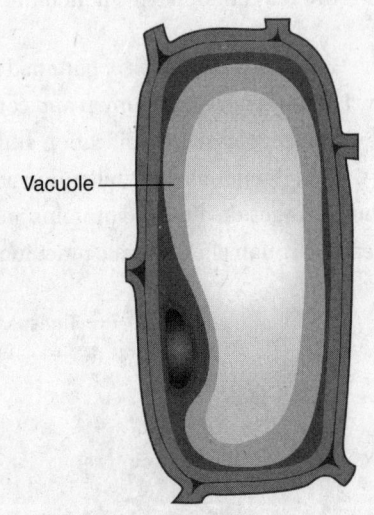

Figure 5.6 Vacuole

Mitochondria produce energy for the cell. The mitochondria have double membranes, with a smooth outer membrane and a folded inner membrane. (See Figure 5.7.) These folds on the inner membrane increase the surface area available for energy production during cellular respiration, which will be discussed in more detail in Chapter 9. The double-membrane structure of the mitochondria also allow them to create the proton gradients that are necessary for ATP production. The center of the mitochondria is an enzyme-containing fluid called the **matrix**. The reactions of the Krebs cycle (the citric acid cycle) occur in the matrix of the mitochondria. Mitochondria also contain their own mitochondrial DNA (mtDNA). Chapter 14 of this book will explain why mtDNA is inherited in a non-Mendelian fashion. Mitochondria also contain their own ribosomes.

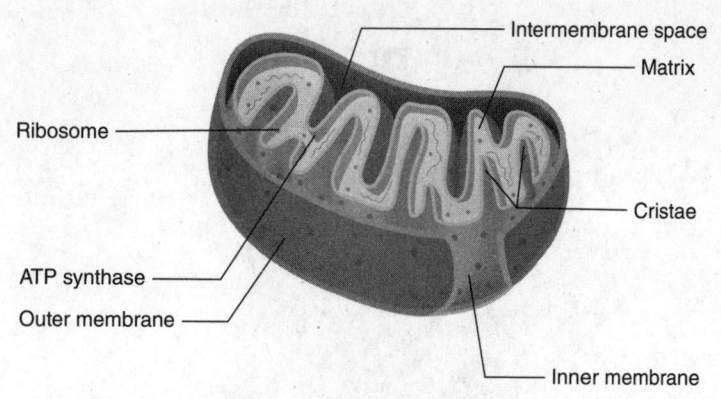

Figure 5.7 Mitochondrion

Chloroplasts are found in plants and algae that carry out photosynthesis. Like the mitochondria, chloroplasts also have a double-membrane structure. Chloroplasts have a smooth outer membrane and pancake-shaped membranous sacs called **thylakoids** that are stacked into structures called **grana** (singular: *granum*). The liquid inside the chloroplast that surrounds the grana is called **stroma**. (See Figure 5.8.) The membranes of the thylakoids function in the light-dependent reactions of photosynthesis, and the enzymes in the stroma function in the light-independent reactions of photosynthesis. These processes will be discussed in detail in Chapter 8. Like mitochondria, chloroplasts also contain their own chloroplast DNA (cpDNA) and their own ribosomes.

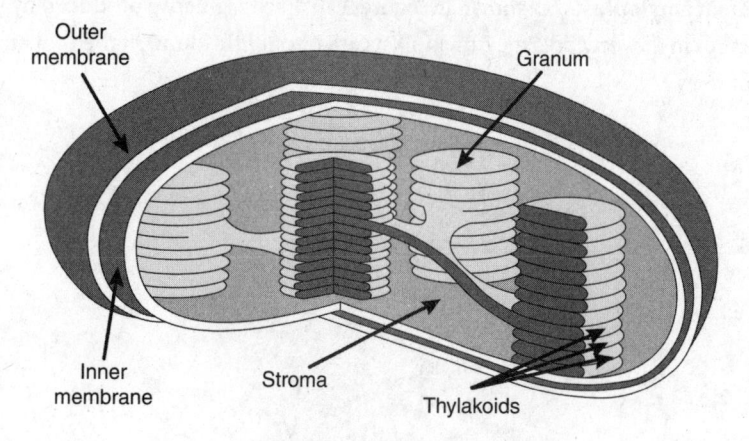

Figure 5.8 Chloroplast

The **centrosome**, as shown in Figure 5.9, is found in animal cells and helps the microtubules assemble into the spindle fibers needed in cell division. Defects in centrosome function have been associated with dysregulation of the cell cycle in some cancers.

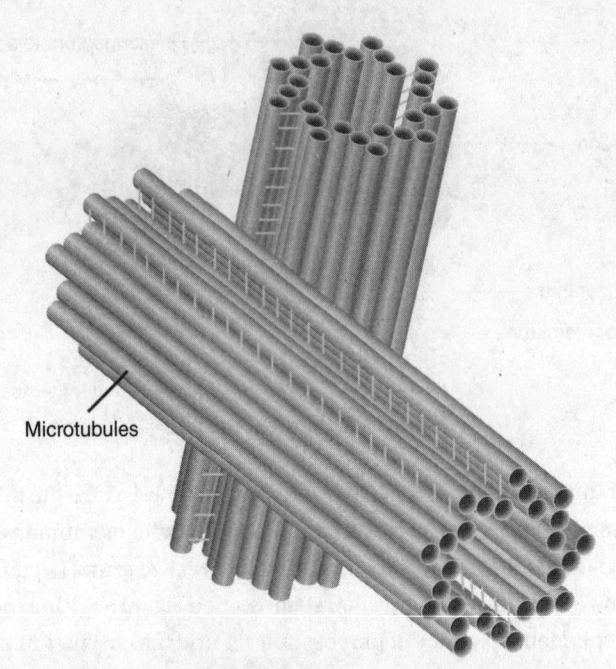

Figure 5.9 Centrosome

Plant cells may contain **amyloplasts**, as shown in Figure 5.10. Excess glucose produced by photosynthesis is stored as starch molecules in the amyloplasts. Amyloplasts are frequently found in the root and tubers of starchy vegetables, such as potatoes.

Figure 5.10 Amyloplast

There are several other structures found in both plant and animal cells. The **peroxisome** helps oxidize molecules and break down toxins in the cell. The **nucleolus** is not a membrane-bound organelle; that term refers to the region in the nucleus where ribosomes are assembled. **Cytoskeleton** fibers help give cells their shape and can be used to move items in the cell. Figure 5.11 shows the cellular organelles that are typically found in plant and animal cells.

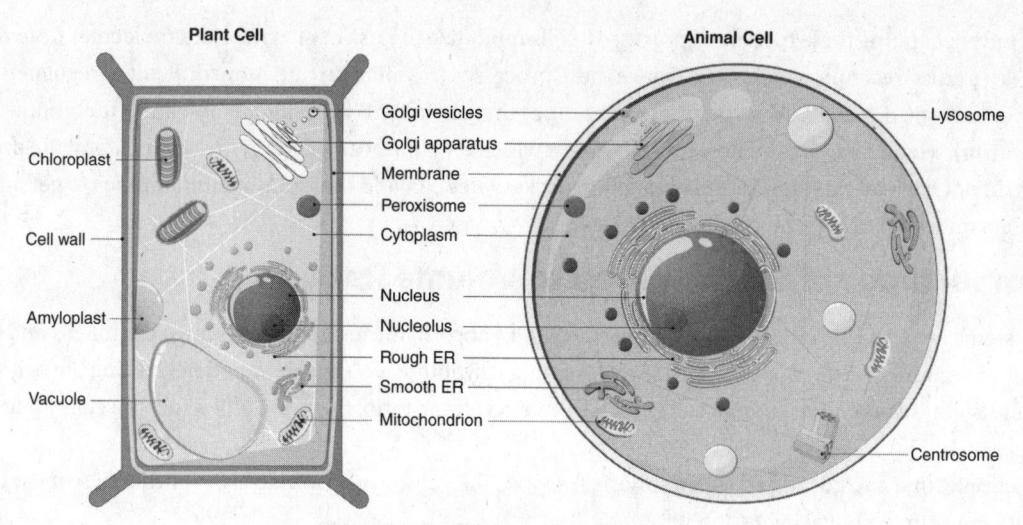

Figure 5.11 Plant and Animal Cells

Together, the nuclear envelope (a.k.a. the nuclear membrane), endoplasmic reticulum, Golgi, lysosomes, vacuoles, transport vesicles, and plasma membrane are considered part of the **endomembrane system**. The functions of the endomembrane system are to modify, package, and transport polysaccharides, lipids, and proteins to their final destinations within the cell, or to prepare them for transport out of the cell. You will learn more about the endomembrane system in Chapter 16.

Endosymbiosis Hypothesis

How did the membrane-bound organelles in eukaryotes originate? The leading theory to explain the origin of membrane-bound organelles is the endosymbiosis hypothesis. The **endosymbiosis hypothesis** states that membrane-bound organelles, such as mitochondria and chloroplasts, were once free-living prokaryotes that were absorbed into other larger prokaryotes. These prokaryotes became interdependent on each other. The smaller prokaryotes that were engulfed by the larger prokaryotes evolved to become membrane-bound organelles. As with any theory, it is important to understand the evidence for the endosymbiosis hypothesis:

- Mitochondria and chloroplasts have their own DNA. Mitochondrial and chloroplast DNA is circular, similar to that of prokaryotic DNA.
- Mitochondria and chloroplasts have their own ribosomes, which are similar in structure to prokaryotic ribosomes.
- Mitochondria and chloroplasts reproduce by binary fission, similar to how bacteria reproduce.

> There are examples of similar relationships in modern organisms. The paramecium *Paramecium bursaria* is a eukaryote that swallows photosynthetic green algae but does not digest their chloroplasts. Instead, these chloroplasts remain active for months. When *P. bursaria* swims into the light, the chloroplasts it swallowed perform photosynthesis, supplying food to the paramecium. This process is called kleptoplasty (*klepto-* means "to steal").

Next is a discussion of how compartmentalization in cells increases their efficiency.

The Advantages of Compartmentalization

Membrane-bound organelles in eukaryotic cells allow different parts of the cells to specialize their functions, and this specialization allows for greater efficiency within the cell. These membrane-bound organelles form specialized compartments. This compartmentalization allows the cell to separate the enzymes involved in different metabolic processes. By separating these metabolic processes, cells minimize the risk of enzymes and molecules from different processes cross-reacting, which would make these processes less efficient and more difficult to regulate.

Some eukaryotic organelles have internal membranes that are folded (for example, the inner membrane of the mitochondria), which provides greater surface area on which reactions can occur in the organelle. While prokaryotic cells do not have membrane-bound organelles, prokaryotes can fold their plasma membranes to get more surface area on which to perform metabolic processes.

The Importance of Surface Area to Volume Ratios

Living systems need to efficiently eliminate waste products, absorb nutrients, and exchange chemicals and energy with their environment. The ratio of a cell's surface area to its volume is a key factor in determining the cell's ability to accomplish these tasks. The larger a cell's surface area to volume ratio, the more efficiently the cell can accomplish these tasks.

For example, first consider a cell with a spherical shape. The volume of a sphere is proportional to the cube of its radius, as shown in the following formula:

$$\text{Volume of a Sphere: } V = \frac{4}{3}\pi r^3$$

The surface area of a sphere is proportional to the square of its radius, as shown in the following formula:

$$\text{Surface Area of a Sphere: } SA = 4\pi r^2$$

Dividing the surface area of a sphere formula by the volume of a sphere formula gives you the surface area to volume ratio:

$$\text{Surface Area to Volume Ratio of a Sphere} = \frac{4\pi r^2}{\frac{4}{3}\pi r^3} = \frac{3}{r}$$

> **TIP**
> You do NOT need to memorize the formulas for surface area or volume. These formulas are on the AP Biology Equations and Formulas sheet that will be provided to you during the AP Biology exam.

Note that since r (the radius) is in the denominator, as the radius *increases*, the surface area to volume ratio will *decrease*. All materials that are exchanged between a cell and its environment must pass through the surface area of the cell. As the radius of a cell increases, the amount of surface area available per unit volume for exchange of materials gets smaller and smaller. Eventually, this will limit the cell's ability to efficiently exchange materials with its environment and ultimately impact how big the cell can grow. Larger cells have a lower surface area to volume ratio and a less efficient exchange of materials with the cell's environment. For this reason, it can be advantageous for an organism to be made of many smaller cells rather than fewer large ones.

One way the surface area to volume ratio can be increased is by folding membranes, as seen in the cristae of the inner mitochondrial membrane. The membrane folds on the inner mitochondrial membrane allow for greater surface area for the electron transport chains that are required for cellular respiration. The lining of the human intestine has many folds called villi that increase the surface area available for nutrient absorption.

Structure of the Plasma Membrane

Plasma membranes are critical in allowing cells to maintain an internal environment that is favorable to life. Plasma (or cell) membranes are selectively permeable, which means that some materials can cross the membrane while other materials cannot. This selective permeability allows the cell to maintain its internal environment.

The plasma membrane is made of a **bilayer** of phospholipids. Recall from Chapter 4 that phospholipids have a hydrophilic (or polar) phosphate "head" and two hydrophobic (or nonpolar) "tails." When the bilayer of phospholipids is formed in an aqueous environment, the hydrophobic lipid tails orient themselves away from the water toward the interior of the membrane, while the hydrophilic phosphate heads are oriented facing the aqueous environment on the exterior of the membrane, as shown in Figure 5.12.

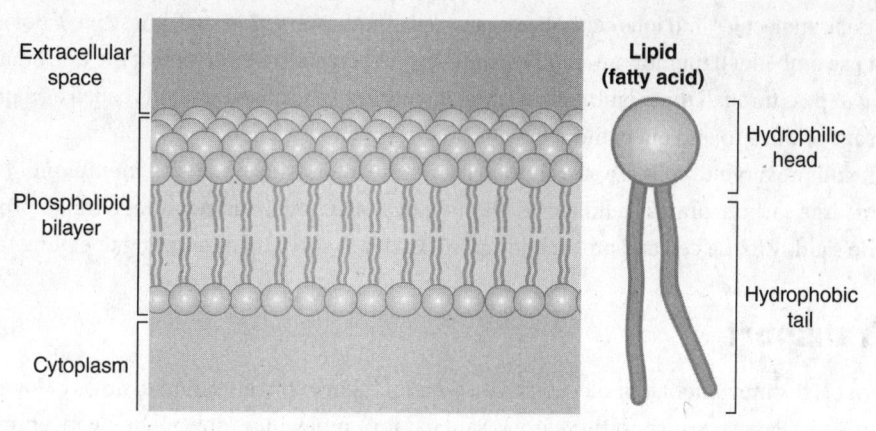

Figure 5.12 Phospholipid

Embedded in this phospholipid bilayer are membrane proteins, modified proteins called **glycoproteins**, modified lipids called **glycolipids**, and steroids. These components of the membrane are mobile and can flow throughout the surface of the plasma membrane, allowing the cell to adapt to changing environmental conditions. Proteins in the plasma membrane have many functions, including transporting materials, participating in cell signaling processes, anchoring the cell to its surroundings, and catalyzing chemical reactions. Hydrophobic parts of membrane proteins are embedded in the lipid bilayer portion of the cell membrane, while hydrophilic portions of proteins are found on the inner or outer surface of the cell membrane. Glycoproteins and glycolipids function in cell recognition. Steroids in the plasma membrane can adjust membrane fluidity in response to changing environmental conditions and the needs of the cell. Individual phospholipids may also move across the surface of the plasma membrane. Because the components of the plasma membrane have mobility, the structure of the membrane is often referred to as a **fluid mosaic model**. (See Figure 5.13.)

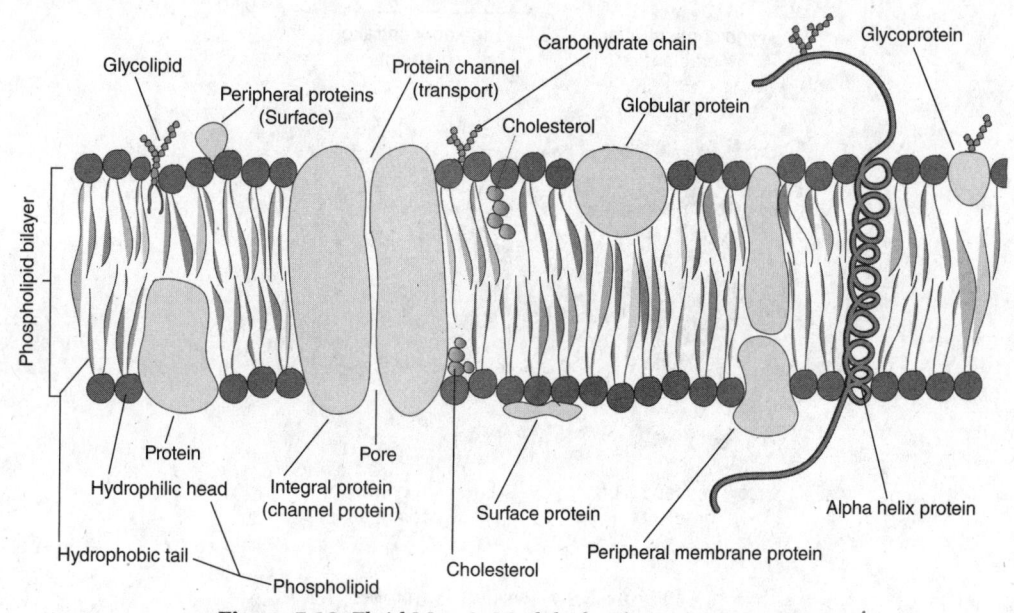

Figure 5.13 Fluid Mosaic Model of a Plasma Membrane

What Can (and Cannot) Cross the Plasma Membrane

The phospholipid bilayer of the cell membrane gives it selective permeability. The hydrophobic lipids of the phospholipids are much larger than the hydrophilic phosphates. Small hydrophobic molecules, such as oxygen (O_2), carbon dioxide (CO_2), and nitrogen (N_2), can easily pass between the phospholipids into and out of the cell.

However, large polar molecules and ions cannot cross the cell membrane unassisted. These large polar and charged molecules must use embedded membrane channels or transport proteins to enter or exit the cell. Small polar molecules, like H_2O, can pass through the membrane in small quantities. Specialized proteins called **aquaporins** allow for most of the passage of water in and out of the cell.

Plants, fungi, and prokaryotic cells are surrounded by a **cell wall** outside of the cell membrane. These cell walls are composed of large carbohydrates (cellulose in plants, glucans in fungi, and peptidoglycan in prokaryotes). The cell walls provide rigidity to the cell and are an additional barrier to substances entering or exiting the cell.

Passive Transport

Passive transport is the movement of molecules from areas of higher concentration to areas of lower concentration. (See Figure 5.14.) Passive transport therefore is said to move molecules "down" their concentration gradient. Passive transport does not require the input of energy. This movement of molecules down their concentration gradient without the input of energy is also called **diffusion**. The diffusion of water down its concentration gradient across a membrane gets a special name, osmosis. This will be discussed in more detail in Chapter 6.

Molecules that are polar or charged may require a membrane protein in order to diffuse across the cell membrane. The process of passive transport that uses a membrane protein is called **facilitated diffusion**. One example is the specialized membrane proteins called aquaporins that allow large quantities of water to move down their concentration gradient. Specialized **channel proteins** can allow the passive transport of ions, such as Ca^{+2} or Cl^{-1}, down their concentration gradients. Because molecules that use facilitated diffusion require a membrane protein in order to cross the membrane, the rate of facilitated diffusion is limited by the number of membrane proteins available and is said to be saturable.

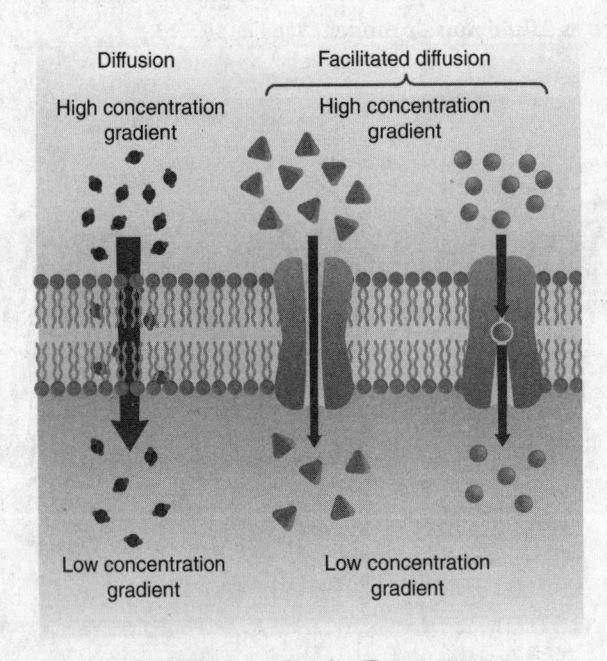

Figure 5.14 Passive Transport

Active Transport

Active transport moves molecules from areas of low concentration to areas of high concentration, as shown in Figure 5.15. This movement of molecules "against" their gradient requires the input of energy.

One example of this is the Na^+/K^+ pump. This membrane protein requires the input of ATP to pump Na^+ ions from their lower concentrations in the cell to an area of higher Na^+ concentration outside the cell. This membrane protein also pumps K⁺ ions from their lower concentrations outside of the cell to an area with higher concentrations of K⁺ ions inside the cell. For every three Na⁺ ions pumped out of the cell, two K⁺ ions are pumped into the cell. This results in a higher concentration of positive ions outside of the cell and helps the cell maintain a membrane potential.

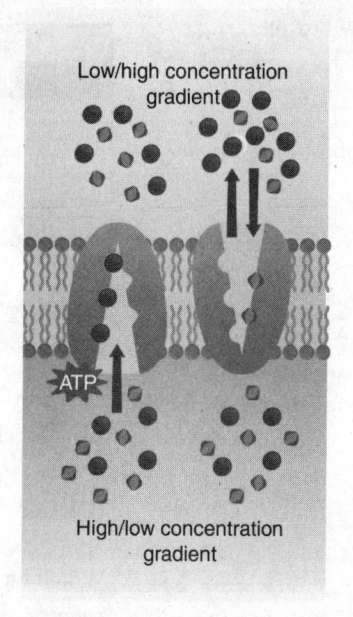

Figure 5.15 Active Transport

Endocytosis and exocytosis also require an input of energy and are forms of active transport. **Endocytosis** is used by the cell to take in water and macromolecules by enfolding them into vesicles formed from the plasma membrane. In **exocytosis**, this process is reversed. Vesicles (that contain molecules to be expelled) are fused with the plasma membrane, which then allows these molecules to be expelled from the cell. This movement of large molecules into or out of the cell requires the input of energy. (See Figure 5.16.)

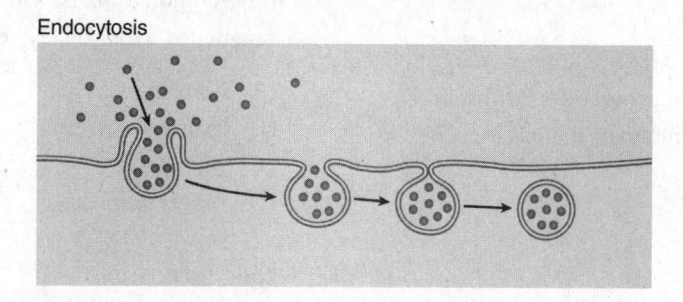

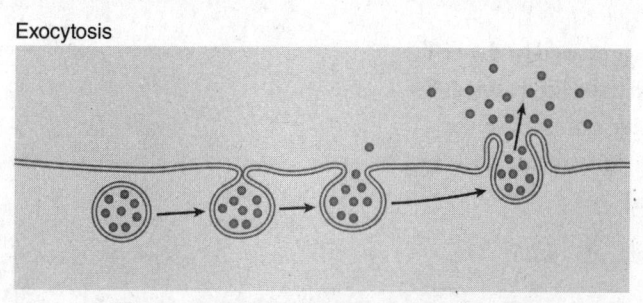

Figure 5.16 Endocytosis and Exocytosis

Practice Questions

Multiple-Choice

1. An animal cell that lacks glycolipids and glycoproteins in its plasma membrane would likely be unable to carry out which of the following functions?

    (A) cell recognition
    (B) maintaining the fluidity of the phospholipid bilayer
    (C) endocytosis
    (D) creating a membrane potential

2. When some bacteria are exposed to antibiotics, the bacteria use ATP to try to pump the antibiotics out of their cells. Which of the following processes is most likely used to do this?

    (A) osmosis
    (B) diffusion
    (C) active transport
    (D) endocytosis

3. Which of the following is NOT a component of the cell membrane?

    (A) phospholipids
    (B) cellulose
    (C) proteins
    (D) glycolipids

4. Muscle cells require large amounts of energy to function. Which cellular organelle is most likely found in high concentrations in muscle cells?

    (A) lysosome
    (B) Golgi complex
    (C) amyloplast
    (D) mitochondria

5. Niemann-Pick disease is caused by the cell's inability to break down large lipid molecules. Which cellular organelle is most likely not functioning properly in an individual with Niemann-Pick disease?

    (A) rough endoplasmic reticulum
    (B) Golgi complex
    (C) lysosomes
    (D) ribosomes

6. A student has access to four dyes that stain different components of cells, as shown in the following table.

Dye	Component of Cells Stained
Nile red	Lipids
Hoechst 33342	Nuclei
DRAQ9	Cytosol
Coomassie blue	Proteins

    Which dye would be the best choice to use to distinguish prokaryotic cells from eukaryotic cells?

    (A) Nile red
    (B) Hoechst 33342
    (C) DRAQ9
    (D) Coomassie blue

7. Which types of molecules pass through the cell membrane unassisted most easily?

    (A) small and hydrophilic
    (B) small and hydrophobic
    (C) large and hydrophilic
    (D) large and hydrophobic

8. A cell is not able to modify and package proteins for secretion from the cell. Which of the following organelles is most likely not functioning correctly?

    (A) ribosomes
    (B) lysosomes
    (C) vacuoles
    (D) Golgi complex

9. Which of the following is NOT evidence that supports the endosymbiosis hypothesis?

   (A) Mitochondria and chloroplasts have their own circular DNA.
   (B) Mitochondria and chloroplasts have their own ribosomes.
   (C) Mitochondria and chloroplasts reproduce by binary fission.
   (D) Mitochondria and chloroplasts are found in all eukaryotic cells.

10. A molecule is moving from an area of higher concentration outside of a cell to an area of lower concentration inside the cell. Which process best describes the movement of this molecule?

    (A) active transport
    (B) diffusion
    (C) endocytosis
    (D) exocytosis

Short Free-Response

11. A student conducts an experiment to investigate which cell shape would allow for the most efficient exchange of materials with its environment. Agar blocks (containing bromothymol blue dye) are cut into three different shapes to model three differently shaped cells. The agar blocks are then placed in a vinegar solution. As the vinegar diffuses into the agar models, the acid in the vinegar will turn the bromothymol blue dye yellow. The time required for each agar model to turn completely yellow is measured and is an indication of the efficiency of movement of materials into the agar models. The shape, volume, and surface area of each agar block is shown in the table.

Shape	Volume	Surface Area
Cylinder	25.1 cm^3	50.2 cm^2
Sphere	33.5 cm^3	50.2 cm^2
Cube	8 cm^3	24 cm^2

Part A

(i) **Describe** the measurement that would best predict the efficiency of each agar block's exchange of materials with its environment.

(ii) **Calculate** that measurement for each agar block.

Part B

(i) **Identify** the independent variable in this experiment.

(ii) **Identify** the dependent variable in this experiment.

Part C

(i) **Predict** which cell would turn completely yellow first.

Part D

(i) **Justify** your prediction from Part C.

12. Apoptosis is programmed cell death. Apoptosis is often triggered by mutations that could cause a cell to form a tumor if the cell continued to grow and multiply.

Part A

(i) **Describe** the organelle in a eukaryotic cell that is most likely to participate in apoptosis.

Part B

(i) **Explain** other functions the organelle from Part A would have in the cell besides participating in apoptosis.

Part C

A mutation causes the enzymes in the organelle from Part A to become nonfunctional.

(i) **Predict** what effects this would have on the cell.

Part D

(i) **Justify** your prediction from Part C.

Long Free-Response

13. Two different molecules, A and B, can enter a cell using passive transport. A cell that did not initially contain molecule A or molecule B was placed in a beaker with a solution containing equal concentrations of both molecules for 30 minutes. The cell was removed from the beaker, and the concentration of each molecule inside the cell was measured. The experiment was repeated with different, but equal, starting concentrations of molecules A and B in the beaker. The data are shown in the table.

Starting Concentration of Molecules A and B in the Beaker (millimolar)	Concentration of Molecule A Inside the Cell After 30 Minutes (millimolar) ± 2 SEM*	Concentration of Molecule B Inside the Cell After 30 Minutes (millimolar) ± 2 SEM
40	20 ± 2.1	20 ± 2.5
80	40 ± 1.9	40 ± 1.8
120	60 ± 3.0	50 ± 2.9
160	80 ± 3.2	50 ± 2.7

*Standard Error of the Mean

Part A

(i) **Describe** two types of passive transport.

Part B

(i) Using the axes provided, **construct** a graph of this data. Include 95% confidence intervals.

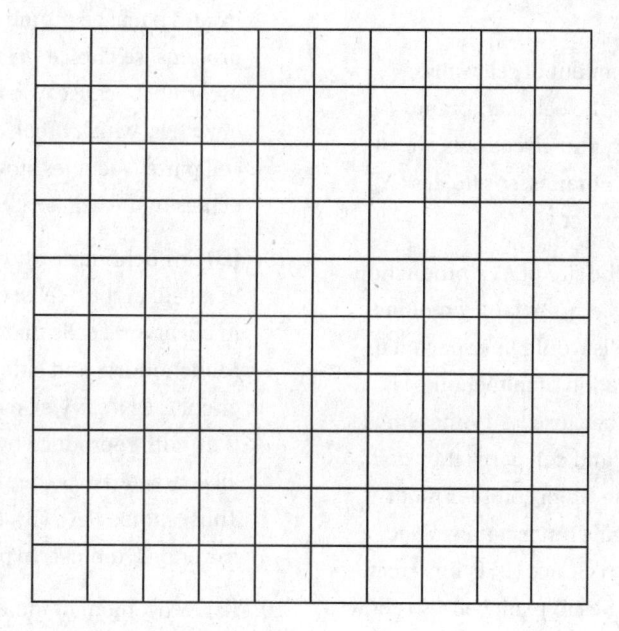

Part C

(i) Based on the data, **make a claim** about which molecule (A or B) uses simple diffusion and which molecule uses facilitated diffusion.

(ii) **Justify** your claim with evidence from the data.

Part D

This experiment is repeated with the addition of a molecule that irreversibly binds to transport proteins in the cell membrane.

(i) **Make a prediction** about what, if any, changes this would lead to in the data.

(ii) **Justify** your prediction.

Answer Explanations

Multiple-Choice

1. **(A)** Glycolipids and glycoproteins are involved in cell recognition, so if a cell lacked these molecules, cell recognition functions would be impaired. Choice (B) is incorrect because cholesterol, not glycoproteins or glycolipids, is involved in maintaining membrane fluidity. Endocytosis is a process that requires energy to bring material into the cell, so choice (C) is incorrect. The Na^+/K^+ pump uses active transport to create the membrane potential, so choice (D) is also incorrect.

2. **(C)** Processes that require the use of ATP to move molecules in or out of a cell are called active transport. Choices (A) and (B) are incorrect because osmosis and diffusion are both forms of passive transport. Endocytosis does use energy and is a form of active transport, but that process moves molecules into the cell, not out of the cell. So choice (D) is incorrect.

3. **(B)** Cellulose is a component of cell walls; it is not a component of the cell membrane. Phospholipids, proteins, and glycolipids are all components of cell membranes, so choices (A), (C), and (D) are all incorrect.

4. **(D)** Mitochondria are the site of ATP production for the cell. Muscle cells require large amounts of energy, so muscle cells would be expected to have a higher concentration of mitochondria. Choice (A) is incorrect because lysosomes break down macromolecules and cell parts; they do not produce energy. The Golgi complex modifies and packages proteins but is not involved in energy production, so choice (B) is incorrect. Amyloplasts store starches in plant cells, so choice (C) is incorrect.

5. **(C)** Lysosomes help break down large molecules, so a defect in the lysosomes could be a likely cause of Niemann-Pick disease. The rough endoplasmic reticulum and the ribosomes serve functions in protein synthesis; they do not break down molecules. So choices (A) and (D) are incorrect. Choice (B) is incorrect because the Golgi complex modifies and packages proteins.

6. **(B)** Eukaryotic cells have nuclei; prokaryotic cells do not. Thus, Hoechst 33342 could distinguish between eukaryotic and prokaryotic cells. Both eukaryotic and prokaryotic cells contain lipids, cytosol, and proteins. Therefore, the dyes that stain these (Nile red, DRAQ9, and Coomassie blue, respectively) would not distinguish between eukaryotic and prokaryotic cells, making choices (A), (C), and (D) incorrect.

7. **(B)** Small, hydrophobic molecules can slide between the phospholipids in the cell membrane and therefore can cross the cell membrane most easily. Choice (A) is incorrect because hydrophilic molecules could not get past the long hydrophobic lipid tails of the phospholipids that make up the cell membrane's bilayer. Large molecules cannot pass through the cell membrane unassisted, so choices (C) and (D) are incorrect.

8. **(D)** The function of the Golgi complex is to modify and package proteins. Ribosomes synthesize proteins, so choice (A) is incorrect. Choice (B) is incorrect because lysosomes contain hydrolytic enzymes, which break down macromolecules and cell parts. Vacuoles store water and other molecules in the cell, so choice (C) is incorrect.

9. **(D)** Mitochondria and chloroplasts are *not* found in all eukaryotic cells; chloroplasts are only found in eukaryotic cells that perform photosynthesis. Mitochondria and chloroplasts do have their own circular DNA as well as their own ribosomes, and they both reproduce by binary fission, which are all characteristics they share with prokaryotes. Thus, choices (A), (B), and (C) are all evidence of the endosymbiosis hypothesis.

10. **(B)** Movement of molecules from an area of higher concentration to an area of lower concentration is diffusion, a passive transport process. Choices (A), (C), and (D) all involve active transport and are therefore incorrect.

Short Free-Response

11. (A-i) The measurement that would best predict the efficiency of each agar block's exchange of materials with its environment is the surface area to volume ratio.

    (A-ii) For the cylinder, the ratio is $\frac{50.2}{25.1} = 2$. For the sphere, the ratio is $\frac{50.2}{33.5} = 1.5$. For the cube, the ratio is $\frac{24}{8} = 3$.

    (B-i) The independent variable is the shape of the agar block.

    (B-ii) The dependent variable is the time it takes for the vinegar to diffuse completely into the cell and turn it yellow.

    (C-i) The cube would turn completely yellow first.

    (D-i) The greater a cell's surface area to volume ratio, the more efficiently it will exchange materials with its environment. Since the cube has the greatest surface area to volume ratio, it will exchange materials the most efficiently and turn completely yellow first.

12. (A-i) The lysosome is most likely to participate in apoptosis because it contains hydrolytic enzymes that can break down parts of the cell.

    (B-i) Other functions of the lysosome are to assist in digestion by breaking down large macromolecules and destroying bacteria or viruses that invade the cell.

    (C-i) The hydrolytic enzymes in the lysosome are crucial for the functioning of the lysosome. If those enzymes were not functional, the lysosome could not perform its activities, thus reducing, or even preventing, the cell's ability to function.

    (D-i) If the hydrolytic enzymes were nonfunctional, the lysosome could not assist in digesting large molecules nor could the lysosome break down waste products in the cell. The cell could simultaneously starve, since it couldn't digest its food, and be poisoned by its own waste products. In addition, the cell's ability to defend itself against invading pathogens would be compromised.

Long Free-Response

13. (A-i) Two types of passive transport are diffusion and facilitated diffusion. In diffusion, a molecule moves from an area of high concentration to an area of low concentration, and no energy is required. In facilitated diffusion, a molecule moves from an area of high concentration to an area of low concentration with the assistance of a transport protein. Again, no energy is required.

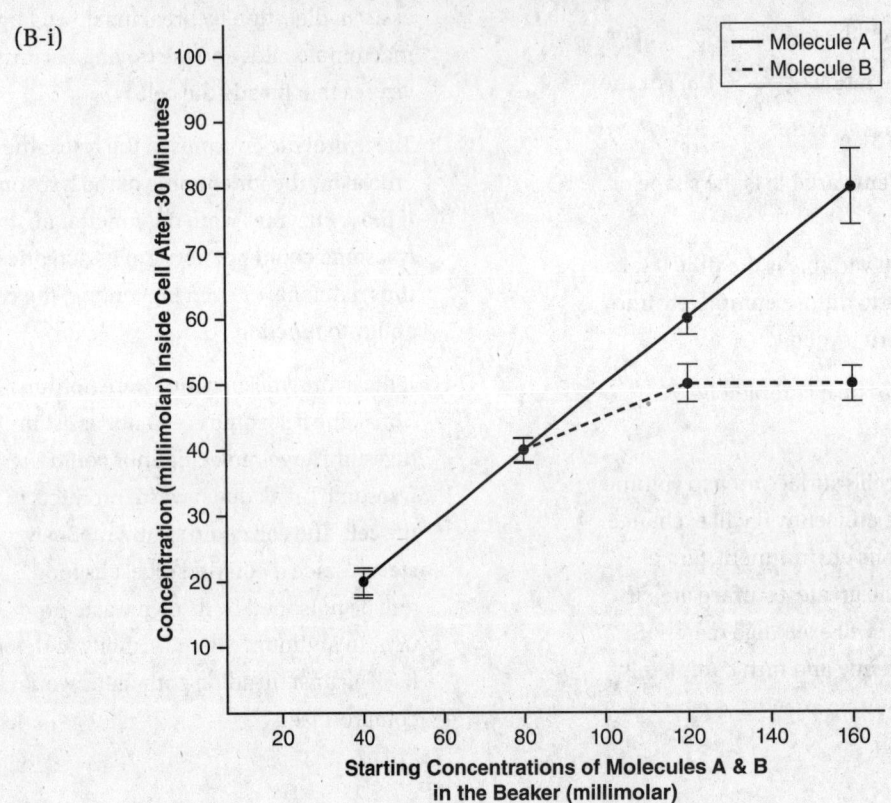

(B-i)

[Graph: Concentration (millimolar) Inside Cell After 30 Minutes vs. Starting Concentrations of Molecules A & B in the Beaker (millimolar). Molecule A (solid line) increases linearly from ~20 at 40 mM to ~80 at 160 mM. Molecule B (dashed line) increases from ~20 at 40 mM to ~50 at 120 mM and plateaus at ~50 at 160 mM. Error bars shown at each data point.]

(C-i and ii) Molecule A is using simple diffusion, and molecule B is using facilitated diffusion. The rate of facilitated diffusion is limited by the number of transport proteins for the molecule in the cell membrane. The rate of simple diffusion is not limited by the number of transport proteins in the cell membrane and will continue to increase as the concentration difference across the cell membrane increases. In the experiment, as the concentration of each molecule in the beaker increases, the amount of molecule A entering the cell continues to increase, but at higher concentrations, the amount of molecule B entering the cell no longer increases. This indicates that molecule B is likely using a transport protein to enter the cell and is therefore using facilitated diffusion.

(D-i and ii) Since molecule A is simply diffusing across the membrane, the addition of a molecule that binds to transport proteins will likely have no effect on the data for molecule A. However, since molecule B requires a transport protein to enter the cell, the addition of this new molecule would likely prevent molecule B from entering the cell.

6

Movement of Water in Cells

Learning Objectives

In this chapter, you will learn:
→ Water Potential
→ Osmolarity and Its Regulation

Overview

Maintaining water balance in cells is a critical component of homeostasis, and life depends upon it. This chapter provides an overview of water potential, discusses how to calculate water potential, and explains how differences in water potential affect the movement of water into and out of cells.

Water Potential

In an introductory biology course, you may have learned the terms hypotonic, hypertonic, and isotonic. These three terms focus on the relative concentrations of *solute* in a solution. A **hypotonic** solution has a lower concentration of solute. A **hypertonic** solution has a higher concentration of solute. An **isotonic** solution has the same concentration of solute as that of another solution.

Since hypotonic, hypertonic, and isotonic are *relative* terms, how a solution is characterized with these terms is dependent on what you are comparing the solution to. For example, a cup of coffee may be considered hypertonic if it is being compared to a cup of water. However, the same cup of coffee would be considered hypotonic if compared to a cup of coffee to which two sugar cubes have been added.

An advantage of using water potential to describe solutions is that water potential is not a relative term; rather, water potential can be calculated using a few simple equations. The calculated water potential values of solutions make comparing these solutions more exact. This helps in predicting how water will move between solutions.

Water potential can be defined as the potential energy of water in a solution, or the ability of water to do work. The more water there is in a solution, the higher its water potential will be. The less water there is in a solution, the lower its water potential will be. Water flows from areas of higher water potential to areas of lower water potential. This is similar to how a ball at the top of a hill, which has higher potential energy, will roll down a hill to an area of lower potential energy. (See Figure 6.1.)

> **TIP**
>
> Learning prefixes can help you remember terms and decode the definitions of unfamiliar terms. The prefix *hyper-* means "more," the prefix *hypo-* means "less," and the prefix *iso-* means "same."

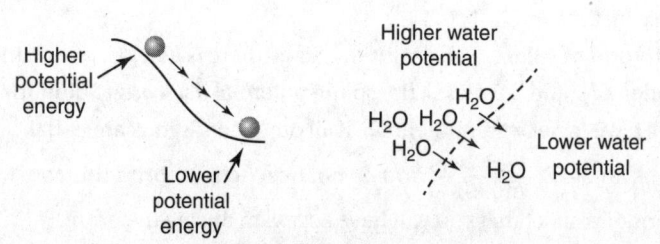

Figure 6.1 Water Potential

Water potential focuses on the concentration of *water* in a solution. If there is more solute in a solution, the water is less concentrated in that solution. A hypertonic solution that has more solute will have less concentrated water and a *lower* water potential. A hypotonic solution has less solute and therefore more concentrated water and a *higher* water potential. Since water flows from areas of higher water potential to areas of lower water potential, water will flow from hypotonic solutions (which have a *higher* water potential) to hypertonic solutions (which have a *lower* water potential).

Calculating Water Potential

The water potential of a solution has two components: a component that depends on the amount of solute and a component that depends on pressure, as shown in the following equation:

$$\Psi = \Psi_s + \Psi_p$$

Ψ_s is the water potential due to the solute concentration (**solute potential**). Ψ_p is the water potential due to the pressure on the system (**pressure potential**).

When you drink a liquid through a straw, you are using negative pressure potential. As you suck air from the straw, you create a negative pressure in the straw. The liquid in your drink then moves from an area of higher pressure potential (in the cup) into an area of lower pressure potential (in the straw). An example of positive pressure potential would be Super Soaker water toys. In those toys, you create positive pressure inside the water toy by pumping air into the toy. When you pull the trigger on the water toy, water flows out of the higher pressure potential environment in the toy and into the lower pressure potential environment outside of the toy.

Most biological systems are open to the atmosphere and are in pressure equilibrium with their environment. So in most cases, the Ψ_p is 0 and the water potential will depend solely on the solute potential. In this case, the water potential becomes simply:

$$\Psi = \Psi_s$$

The solute potential depends on the concentration of the solute, how many particles the solute forms when in solution, and the temperature of the solution. The solute potential is calculated with this formula:

$$\Psi_s = -iCRT$$

The **ionization constant** is i, which is a function of how many particles or ions a solute will form in solution. For covalent compounds, which do not separate into ions in solution (for example, glucose and sucrose), the ionization constant is 1 since one molecule of glucose will form one particle in solution. For ionic compounds, which do separate into ions in solution (for example, sodium chloride NaCl or calcium chloride $CaCl_2$), the ionization constant depends upon how many ions the compound will form in solution. NaCl forms two ions when in solution, Na^+ and Cl^-, so the ionization constant for NaCl is 2. $CaCl_2$ forms three ions when in solution, Ca^{2+} and two Cl^- ions, so the ionization constant for $CaCl_2$ is 3.

C refers to the **concentration of solute** in the solution. Since there is a negative sign in the solute potential formula, as the concentration of solute *increases*, the solute potential *decreases*. Solutions with more solute (higher solute concentrations) will have lower water potentials if all other conditions are equal.

R is the **pressure constant**, $0.0831 \frac{\text{liters-bars}}{\text{mole-K}}$. You do not need to memorize this constant; it will be on the AP Biology Equations and Formulas sheet that you will have access to during the exam.

> **TIP**
> Remember, hypertonic solutions have more solute and less water, so their water potential is lower. Hypotonic solutions have less solute and more water, so their water potential is higher.

> **TIP**
> You do not need to memorize the formula for water potential; it will be on the AP Biology Equations and Formulas sheet.

T is the **temperature** of the solution. The temperature is in Kelvin, NOT degrees Celsius. There are two reasons why temperature must be stated in Kelvin:

- The pressure constant, R, uses temperature units of Kelvin, so consistent units for temperature are needed throughout the formula.
- The Kelvin temperature scale is an absolute scale, meaning there are no negative numbers in the Kelvin scale. Using negative temperatures in this formula would result in incorrect water potential calculations.

Converting temperatures from Celsius to Kelvin is simple: Kelvin = °C + 273.

The reference solution for solute potential is distilled water. Because distilled water has no solutes, $C = 0$, and the solute potential for distilled water is 0 bars at any temperature. As soon as any amount of solute is added to distilled water, the solution's solute potential will become negative. The more solute that is added, the more negative the solute potential becomes.

Try the following sample calculation for water potential.

A solution of 0.50 molar glucose is at a temperature of 21°C and open to the atmosphere. What is its water potential?

Because the solution is open to the atmosphere, its pressure potential is zero. So its total water potential is its solute potential.

$$\Psi = \Psi_s + \Psi_p$$
$$\Psi = \Psi_s + 0$$
$$\Psi = \Psi_s$$

Glucose is a covalent compound, so $i = 1$. Convert 21°C to Kelvin:

$$21°C + 273 = 294K$$

Using the formula for solute potential:

$$\Psi_s = -iCRT$$
$$\Psi_s = -(1)\left(0.50 \frac{\text{moles}}{\text{liter}}\right)\left(0.0831 \frac{\text{liters-bars}}{\text{mole-K}}\right)(294K)$$
$$\Psi_s = -12.2 \text{ bars}$$
$$\Psi = -12.2 \text{ bars}$$

Here is an example in which the pressure potential is not 0 bars:

A plant cell has a pressure potential of 1.0 bars and a solute potential of −6.50 bars. What is its water potential?

$$\Psi = \Psi_s + \Psi_p$$
$$\Psi = -6.50 \text{ bars} + 1.0 \text{ bars}$$
$$\Psi = -5.50 \text{ bars}$$

Osmolarity and Its Regulation

Osmolarity is the total concentration of solutes in a solution. Living organisms need to closely regulate their internal solute concentration and their water potential. If an organism became too dehydrated, it might die. If too much water moved into a cell, the pressure potential from the water moving into the cell could cause the cell to burst.

A paramecium living in freshwater has a higher internal solute concentration and a lower water potential than its environment. Because water flows from areas of higher water potential to areas of lower water potential, water will be constantly flowing from the higher water potential in the freshwater environment into the paramecium. If this were unregulated, excess water would build up inside the paramecium and eventually the cell would burst.

To counter this, paramecia have a specialized organelle, a **contractile vacuole**, in which this excess water entering the cell is stored and then pumped out of the cell. This allows the paramecium to maintain its internal solute concentration.

The cells in a saltwater fish have a lower solute concentration and a higher water potential than their marine environment. In this scenario, water would be constantly flowing out of the fish cells (which have a higher water potential) into the saltwater surrounding the fish (which has a lower water potential). To counteract this, the saltwater fish is constantly drinking large quantities of saltwater. The fish retains the water, and specialized salt-secreting organs in the fish excrete the excess solute. This allows the saltwater fish to regulate its internal solute concentration and water potential.

Practice Questions

Multiple-Choice

1. The solute potential of distilled water is

   (A) negative.
   (B) zero.
   (C) positive.
   (D) dependent on the temperature.

2. A solution has a solute concentration of 0.25 moles per liter and is at a temperature of 37°C. The ionization constant of the solute is 1. What is the solute potential of this solution?

   (A) −0.64 bars
   (B) −0.77 bars
   (C) −6.44 bars
   (D) −7.70 bars

3. A cell has a solute potential of −5.42 bars and a pressure potential of 0.48 bars. What is its total water potential?

   (A) −5.42 bars
   (B) −4.94 bars
   (C) 0.48 bars
   (D) 4.94 bars

4. A blood cell with a water potential of −7.7 bars is placed in distilled water. Which of the following correctly describes what will occur?

   (A) Water will flow out of the blood cell because the blood cell has a higher water potential than distilled water.
   (B) Water will flow into the blood cell because the blood cell has a higher water potential than distilled water.
   (C) Water will flow out of the blood cell because the blood cell has a lower water potential than distilled water.
   (D) Water will flow into the blood cell because the blood cell has a lower water potential than distilled water.

5. A plant cell with a solute potential of −4.0 bars and a pressure potential of 0.5 bars is placed into a solution with a water potential of −5.0 bars. What will happen to the plant cell in this solution?

   (A) Water will flow into the plant cell because the plant cell has a total water potential that is lower than that of the surrounding solution.
   (B) Water will flow into the plant cell because the plant cell has a total water potential that is higher than that of the surrounding solution.
   (C) Water will flow out of the plant cell because the plant cell has a total water potential that is lower than that of the surrounding solution.
   (D) Water will flow out of the plant cell because the plant cell has a total water potential that is higher than that of the surrounding solution.

6. Four solutions of covalent compounds are in beakers that are open to the atmosphere. Which of the following solutions has the highest water potential?

   (A) 0.5 molar glucose at a temperature of 21°C
   (B) 0.75 molar fructose at a temperature of 21°C
   (C) 1.0 molar sucrose at a temperature of 21°C
   (D) 1.25 molar lactose at a temperature of 21°C

7. A freshwater fish is placed in a saltwater aquarium. Predict the most likely effect this will have on the fish, and justify your prediction.

   (A) The fish will gain water because it likely has a lower water potential than its new surroundings.
   (B) The fish will lose water because it likely has a lower water potential than its new surroundings.
   (C) The fish will gain water because it likely has a higher water potential than its new surroundings.
   (D) The fish will lose water because it likely has a higher water potential than its new surroundings.

8. Potato slices that are immersed in distilled water for 24 hours become stiff and hard. Potato slices that are immersed in 0.5 molar sucrose solution for 24 hours become limp and soft. Which of the following is the most logical conclusion based on this information?

(A) The potato slices are hypotonic to both distilled water and the 0.5 molar sucrose solution.
(B) The potato slices are hypertonic to both distilled water and the 0.5 molar sucrose solution.
(C) The potato slices are hypotonic to distilled water and hypertonic to the 0.5 molar sucrose solution.
(D) The potato slices are hypertonic to distilled water and hypotonic to the 0.5 molar sucrose solution.

9. Human cells have an approximate NaCl concentration of 0.15 moles per liter. Seawater has an approximate NaCl concentration of 0.45 moles per liter. Which of the following is the most likely effect if a person drank seawater?

(A) The lower water potential in the human cells would cause water to flow out of the human cells into the seawater that was consumed.
(B) The lower water potential in the human cells would cause NaCl to flow out of the human cells into the seawater that was consumed.
(C) The higher water potential in the human cells would cause water to flow out of the human cells into the seawater that was consumed.
(D) The higher water potential in the human cells would cause NaCl to flow out of the human cells into the seawater that was consumed.

10. A dandelion plant is growing in an environment with a temperature of 21°C. The water potential in the root cells of a dandelion plant is −1.2 bars. If a 0.1 molar sodium chloride solution at 21°C was poured on the dandelion roots, what would be the most likely result?

(A) Sodium chloride would move into the dandelion cells, raising the water potential of the dandelion cells.
(B) Water would flow out of the dandelion cells because the water potential in the dandelion cells would be higher than the water potential of the 0.1 molar sodium chloride solution.
(C) There would be no effect on the dandelion cells because they would be open to the atmosphere and have a pressure potential of zero.
(D) Water would flow into the dandelion cells because the water potential in the dandelion cells would be lower than the water potential of the 0.1 molar sodium chloride solution.

Short Free-Response

11. The paramecium *Paramecium aurelia* has a contractile vacuole, which it uses to pump excess water out of its cell. *P. aurelia* was placed in four different salt solutions with concentrations of 0.02 molar, 0.04 molar, 0.08 molar, and 0.10 molar salt. The number of contractions of the contractile vacuole per minute was measured over a 10-minute period.

    Part A
    (i) **Describe** how the water potential of the surrounding solutions would affect the rate of contraction of the contractile vacuole.

    Part B
    (i) **Identify** the independent variable in this experiment.
    (ii) **Identify** the dependent variable in this experiment.

    Part C
    As a follow-up experiment, *P. aurelia* is placed in a beaker that contains distilled water.
    (i) **Predict** what effect, if any, this would have on the rate of contraction of the contractile vacuole.

    Part D
    (i) **Justify** your prediction from Part C using your knowledge of water potential.

12. If a person is in the hospital with severe dehydration, often an intravenous infusion of physiological saline (0.9% saline) will be administered.

    Part A
    (i) **Describe** how the water potential of a person's cells would be affected by severe dehydration.

    Part B
    (i) **Explain** why 0.9% saline is used to rehydrate a person with severe dehydration and why distilled water is not used.

    Part C
    During strenuous exercise under extreme heat, some athletes are advised to consume salt tablets.
    (i) **Predict** whether consuming a salt tablet would lead to water loss or water conservation in the athlete's body cells.

    Part D
    (i) **Justify** your prediction from Part C using your knowledge of water potential.

Long Free-Response

13. A student conducts an experiment with four different root vegetables (carrots, beets, parsnips, and potatoes). Five cubes of equal sizes and surface areas are cut from each of the vegetables, and their masses are recorded. Each cube is placed in a beaker that contains a 0.35 molar sucrose solution for 24 hours. After 24 hours, the cubes were removed from the solutions and weighed, and the percent change in mass for each cube was calculated. The mean percent change in mass for each vegetable and the standard errors of the mean are shown in the table.

Vegetable	Mean Percent Change in Mass	Standard Error of the Mean
Carrots	+7.5%	1.0%
Beets	+21.5%	2.5%
Parsnips	−15.5%	1.5%
Potatoes	−3.5%	0.5%

Part A

(i) **Explain** why some of the vegetables would have a positive percent change in mass while others would have a negative percent change in mass during the course of the experiment.

Part B

(i) **Construct** an appropriately labeled graph that shows the mean percent change in mass for each vegetable. Include 95% confidence intervals.

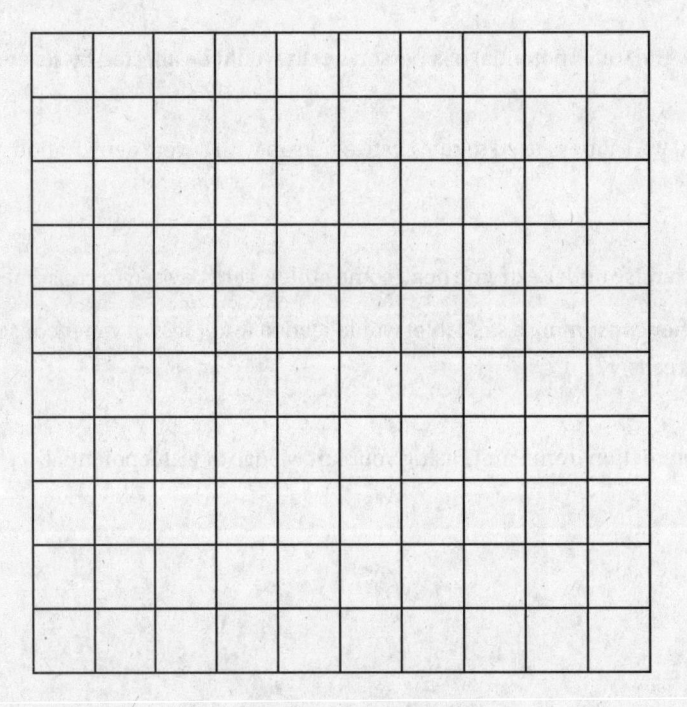

Part C

(i) **Analyze** the data to determine which vegetable has a water potential closest to the water potential of the 0.35 molar sucrose solution used in the experiment.

(ii) **Justify** your answer with evidence from the experiment.

Part D

It is determined that the sugar content in each vegetable is the major determinant of the vegetable's water potential. Turnips have a higher sugar content than carrots but a lower sugar content than beets.

(i) If the experiment were repeated using turnip cubes, **predict** the percent change in mass for the turnip cubes.

(ii) **Justify** your prediction with evidence from the experiment.

Answer Explanations

Multiple-Choice

1. **(B)** The solute potential of distilled water is defined as zero bars; it is the reference point to which other solutions are compared. Since the solute potential is zero, it is neither negative nor positive, so choices (A) and (C) are incorrect. The solute potential of distilled water is 0 bars, regardless of its temperature, so choice (D) is incorrect.

2. **(C)** $\Psi_s = -iCRT$. In this example, $i = 1$, $C = 0.25$ moles per liter, $R = 0.0831 \frac{\text{liters-bars}}{\text{mole-K}}$, and $T = 37 + 273 = 310\text{K}$.

    $\Psi_s = -(1)\left(0.25 \frac{\text{moles}}{\text{liter}}\right)\left(0.0831 \frac{\text{liters-bars}}{\text{mole-K}}\right)(310\text{K})$

    $= -6.44$ bars. Choices (B) and (D) are incorrect because the Celsius temperature was used in the calculation instead of the Kelvin temperature. Avoid this common mistake by remembering to always convert the temperature to Kelvin in water potential calculations. Choice (D) also incorrectly used a value of 2.5 moles per liter instead of the correct value of 0.25 moles per liter in the calculation. Avoid this type of mistake by carefully inputting numbers into your calculator when working on calculations. In choice (A), an incorrect value of 0.025 moles per liter was used instead of 0.25 moles per liter. Again, be careful when inputting numbers into your calculator. If time allows, you may want to repeat calculations to confirm you inputted numbers correctly.

3. **(B)** Total water potential is $\Psi = \Psi_s + \Psi_p$. Using the value -5.42 bars for Ψ_s and 0.48 bars for Ψ_p, then $\Psi = -5.42$ bars $+ 0.48$ bars $= -4.94$ bars. Choice (A) is incorrect because it is the value of Ψ_s alone and doesn't take into account the effect of Ψ_p on the total water potential. Choice (C) is incorrect because it is the value of Ψ_p alone and doesn't take into account the effect of Ψ_s on the total water potential. Choice (D) is incorrect because the total water potential is negative in this case, not positive.

4. **(D)** Water flows from areas of higher water potential to areas of lower water potential. Distilled water has a water potential of 0 bars. If the blood cell has a water potential of -7.7 bars, water will flow from the distilled water, which has a higher water potential, into the blood cell, which has a lower water potential. Choices (A) and (C) are incorrect because water will not flow out of the blood cell. Choice (B) is incorrect because even though that choice correctly states that water will flow into the blood cell, it incorrectly states that the blood cell has a higher water potential than distilled water.

5. **(D)** The total water potential of the plant cell is the sum of the solute potential and the pressure potential, in this case $\Psi = -4.0$ bars $+ 0.5$ bars $= -3.5$ bars, which is higher than the water potential of the surrounding solution (-5.0 bars). Water flows from higher water potential to lower water potential, so water will flow out of the plant cell into the surrounding solution. Choices (A) and (B) are incorrect because water will not flow into the plant cell. Choice (C) correctly states that water will flow out of the plant cell but incorrectly states that the plant cell has a lower water potential than that of the surrounding solution.

6. **(A)** The solutions are in beakers that are open to the atmosphere, so Ψ_p is 0 and the total water potential will be equal to the solute potential. The formula for solute potential is

    $$\Psi_s = -iCRT$$

    All four solutions are at the same temperature, and all four solutes are covalent compounds and have ionization constants that equal 1. R is constant for all four solutions. So the water potential will be inversely proportional to the concentration of the solute (if there is less concentrated solute, the water potential will be higher). The least concentrated solute in this question is 0.5 molar, so choice (A) would have the highest water potential. Note that if you read this question carefully and understand how each variable affects water potential, you don't have to actually calculate the water potential for each solution.

7. **(D)** A fish that is adapted to a freshwater environment likely has a water potential that is more similar to freshwater than to saltwater. The freshwater fish would likely have a higher water potential than the saltwater. Water flows from areas of higher water potential to areas of lower water potential, so water would flow out of the fish into its new surroundings. Choices (A) and (C) are incorrect because the fish would likely lose water, not gain water. Choice (B) is incorrect because it is not likely that the freshwater fish would have a lower water potential than its new saltwater environment.

8. **(D)** Potato slices that are stiff and hard gained water, and potato slices that are limp and soft lost water. Since potato slices became stiff and hard in distilled water, distilled water has a higher water potential than the potato, and the potato is hypertonic to distilled water. Since the potato slices became limp and soft in 0.5 molar sucrose solution, the potatoes have a higher water potential than the 0.5 molar sucrose solution and are hypotonic to the sucrose solution. Choice (A) is incorrect because the potato slices are hypertonic to distilled water. The potato slices are hypotonic to the 0.5 molar sucrose solution, so choice (B) is incorrect. Choice (C) is incorrect because the potato slices are *hyper*tonic to distilled water and *hypo*tonic to the 0.5 molar sucrose solution.

9. **(C)** Human cells with an NaCl concentration of 0.15 moles per liter would have a higher water potential than seawater that has an NaCl concentration of 0.45 moles per liter. So water would flow out of the human cells into the seawater with the lower water potential. Choices (A) and (B) are incorrect because the human cells would have a *higher* water potential than the seawater. Choice (D) incorrectly states that NaCl would flow out of the human cells; it is in fact water that would flow out of the human cells.

10. **(B)** The water potential of a 0.1 molar NaCl solution at 21°C is −4.88 bars, which is calculated as follows:

$$\Psi = -(2)\left(0.1 \frac{\text{moles}}{\text{liter}}\right)\left(0.0831 \frac{\text{liters-bars}}{\text{mole-K}}\right)(294K)$$
$$= -4.88 \text{ bars}$$

Since the dandelion root cells have a water potential of −1.2 bars, water would flow from the higher water potential in the dandelion root cells into the lower water potential of the NaCl solution. Choice (A) is incorrect because if NaCl moved into the dandelion cells, it would lower, not raise, the water potential of those cells. The dandelion cells are open to the atmosphere and have a pressure potential of 0, but choice (C) fails to take into account the effect of the lower solute potential of the surrounding sodium chloride solution on the dandelion cells. Therefore, choice (C) is incorrect. Choice (D) incorrectly states that the water potential in the dandelion cells would be lower than that of the NaCl solution.

Short Free-Response

11. (A-i) If the water potential of the surrounding solution was higher than that of the paramecium, more water would enter the paramecium's cell and the contractile vacuole would have to pump more often to remove the excess water from the cell.

    (B-i) The independent variable is the concentration of the salt solution.

    (B-ii) The dependent variable is the number of contractions of the contractile vacuole per minute.

    (C-i) The contractile vacuole would have to pump more times per minute to remove the excess water.

    (D-i) Distilled water would have a higher water potential than *P. aurelia*. So more water would move into the paramecium and the contractile vacuole would have to work harder to remove the excess water.

12. (A-i) When people are severely dehydrated, their cells have less water and the water potential of their cells is lower.

    (B-i) If distilled water were used to rehydrate a person, too much water would enter the person's cells and there would be a risk of the blood cells bursting from the pressure caused by the excess water. A saline solution of 0.9%, which is the same concentration of saline that is normally seen in body cells, can rehydrate a severely dehydrated person without running the risk of bursting the blood cells.

    (C-i) Consuming a salt tablet would lead to water conservation.

    (D-i) Consuming salt lowers the water potential of the person's body cells, so less water would leave the cells as sweat or urine.

Long Free-Response

13. (A-i) Vegetables with a lower water potential than the surrounding 0.35 molar sucrose solution would gain water and mass. Vegetables with a higher water potential than the surrounding 0.35 molar sucrose solution would lose water and mass.

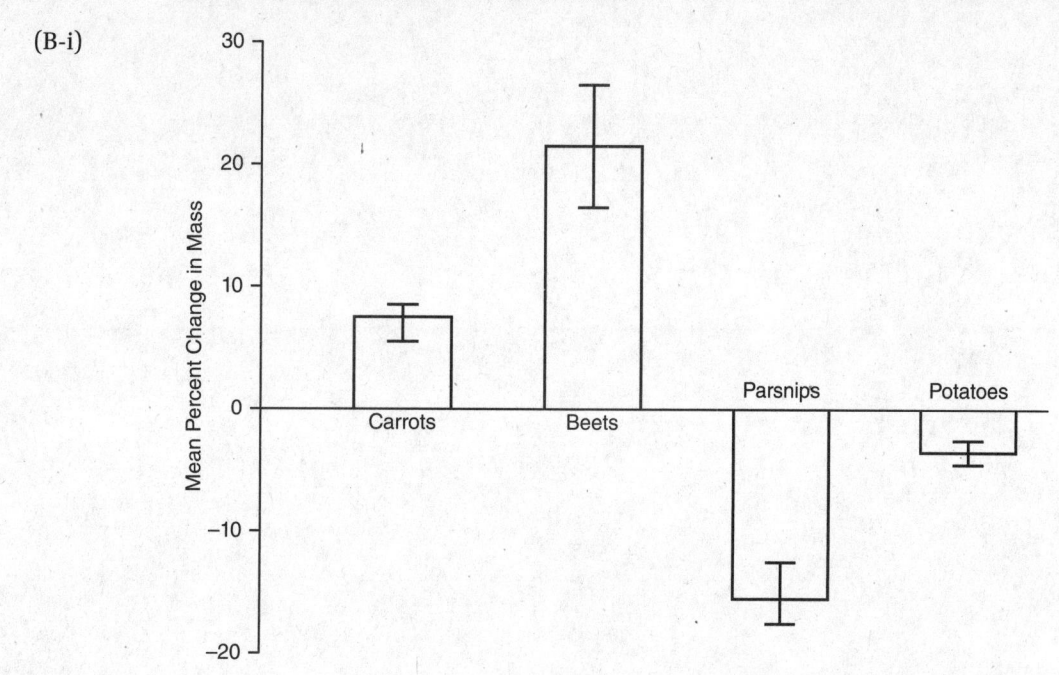

(B-i)

[Bar graph showing Mean Percent Change in Mass for four vegetables: Carrots (~+7.5%), Beets (~+21.5%), Parsnips (~−15.5%), Potatoes (~−3.5%), with error bars.]

(C-i and ii) The vegetable with a water potential that is closest to that of the 0.35 molar sucrose solution would have the smallest percent change in mass. If a vegetable had the same water potential as the 0.35 molar sucrose solution, its cells would be isotonic to the sucrose solution, and you would expect to find a 0% change in mass. The smaller the percent change in mass, the closer the vegetable's water potential is to that of the 0.35 molar sucrose solution. The potato has a water potential that is closest to that of the 0.35 molar sucrose solution because its percent change in mass (−3.5%) is the smallest and closest to zero.

(D-i and ii) Turnips would have a percent change in mass that is greater than +7.5% but less than +21.5%. Since turnips have a higher sugar content than carrots, the water potential of the turnip cells would be less than that of the carrot cells. So the turnips would gain more water from the surrounding 0.35 molar sucrose solution than the carrots did (more than 7.5%). Since the turnips have a lower sugar content than the beets, the water potential of the turnip cells would be higher than the water potential of the beet cells. So the turnips would gain less water from the surrounding 0.35 molar sucrose solution than the beets did (less than 21.5%).

UNIT 3
Cellular Energetics

7

Enzymes

Learning Objectives

In this chapter, you will learn:
- → Enzyme Structure and Function
- → Environmental Factors that Affect Enzyme Function
- → Activation Energy in Chemical Reactions
- → Energy and Metabolism/Coupled Reactions

Overview

Living cells are complex chemical factories that carry out the chemical reactions necessary to support life. These chemical reactions proceed at the rates required to support life because of the catalytic action of enzymes. This chapter will review the structure and function of enzymes and how environmental factors influence the rate of enzyme-catalyzed chemical reactions.

Enzyme Structure and Function

Many of the chemical reactions needed to support living systems happen too slowly to meet the changing needs of organisms. Catalysts speed up chemical reactions. **Enzymes** are biological catalysts. Most enzymes are made of proteins, which have a three-dimensional tertiary structure that is specific to their function. (**Ribozymes** are biological catalysts that are made of RNA.)

The **active site** of an enzyme interacts with the **substrate** (or reactant). The shape of the active site on the enzyme is specific to the shape of the substrate, as shown in Figure 7.1. The substrate must be able to fit into the active site to interact with the enzyme. If there are any charged R-groups on amino acids within the active site of the enzyme, there must be compatible charges on the substrate. For example, an active site that contained positively charged amino acids would repel any positively charged molecules, even if the molecule's shape could fit in the enzyme's active site.

> **NOTE**
> Increasing the concentration of substrate will increase the rate of an enzyme-catalyzed reaction, up until all of the active sites become saturated with substrate.

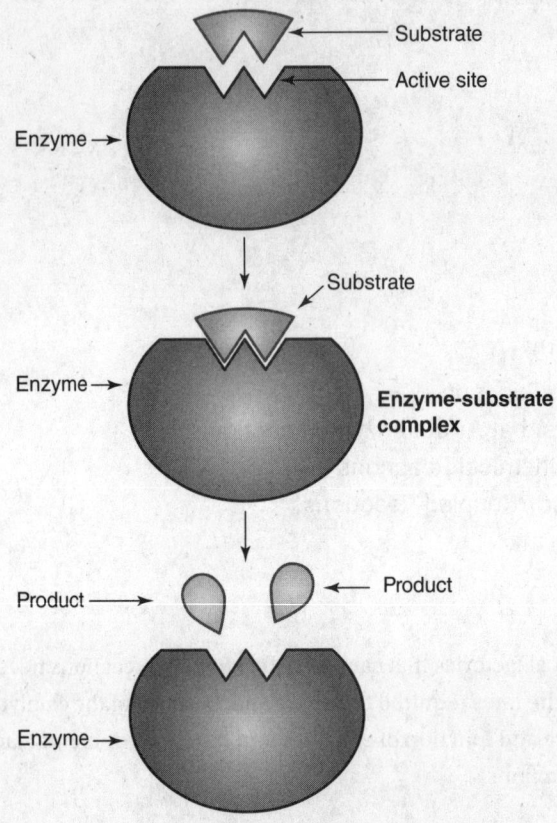

Figure 7.1 Enzyme Action

Environmental Factors that Affect Enzyme Function

Enzymes catalyze reactions most efficiently at optimum temperatures and pHs that are specific to the enzyme. If the temperature in the environment is too low, the rate of collisions between the enzyme and its substrate will be reduced, and the reaction will slow down. If the temperature is too high, bonds that hold the enzyme together may be disrupted, and the shape of the enzyme can be altered. Similarly, a pH that is too far from optimum can disrupt bonds in the enzyme and result in a change in its tertiary structure. Changes to the ionic environment of an enzyme can also disrupt bonds in the enzyme. A change to an enzyme's structure is called **denaturation**, and this can limit the enzyme's ability to catalyze chemical reactions. Sometimes, but not always, denaturation can be reversed when the environment returns to more optimum conditions.

Competitive inhibitors are similar in shape to substrates and compete with substrates for the active site of an enzyme, see Figure 7.2. This competition lowers the rate of enzyme-catalyzed reactions. The effect of competitive inhibitors can be diluted by adding higher concentrations of substrate, creating an environment where the substrate can outcompete the competitive inhibitor.

Noncompetitive (or **allosteric**) **inhibitors** do not bind to the active site but rather bind to a different site on the enzyme (called the **allosteric site**), see Figure 7.2. The binding of the noncompetitive inhibitor to the allosteric site changes the shape of the enzyme, affecting its function. Because the noncompetitive inhibitor does not bind to the active site of the enzyme, adding higher concentrations of substrate does not affect the action of a noncompetitive inhibitor. Noncompetitive inhibitors can function in feedback mechanisms, adjusting the rate of chemical reactions in the cell to suit changing environmental conditions.

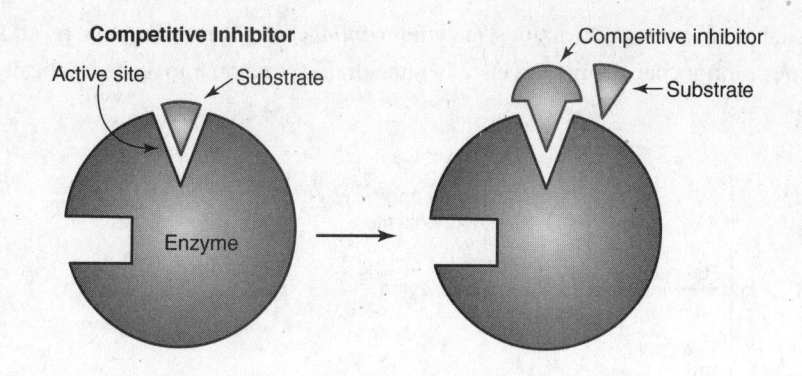

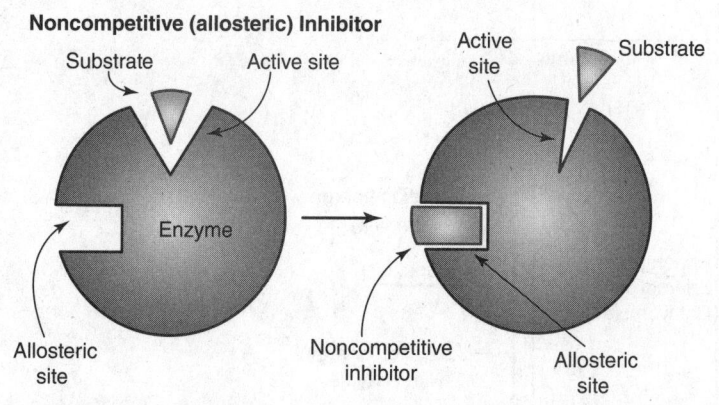

Figure 7.2 Competitive Inhibitors vs. Noncompetitive Inhibitors

Cofactors (inorganic molecules) and **coenzymes** (organic molecules) increase the efficiency of enzyme-catalyzed reactions, usually by binding to the active site or the substrate, which enhances the binding of the substrate to the active site.

> Many of the vitamins and minerals in the foods you eat function as coenzymes and cofactors. For example, B vitamins help your body make the electron carrier NAD, which is important in cellular respiration, which will be discussed in Chapter 9.

Activation Energy in Chemical Reactions

All molecules have a given amount of free energy (G). The chemical reactions necessary for life involve changes in molecules. Chemical reactions can be endergonic or exergonic. **Endergonic** reactions have products with a higher free energy level than its reactants and are considered energetically unfavorable. **Exergonic** reactions have products with a lower free energy level than its reactants and are considered energetically favorable.

All chemical reactions require an input of energy to reach a transition state to get the reaction started. The **activation energy (E_A)** is the difference between the energy level of the reactants and the transition state of the reaction. (See Figure 7.3.) Higher activation energies result in slower chemical reactions; lower activation energies allow chemical reactions to proceed at a faster rate. The enzymes speed up chemical reactions by lowering the activation energy of the reaction.

Enzymes can lower the activation energy of a reaction in a number of ways:

1. Bringing substrates together in the proper orientation for a reaction to occur
2. Destabilizing chemical bonds in the substrate by bending the substrate
3. Forming temporary ionic or covalent bonds with the substrate

> **TIP**
> Think of activation energy as being analogous to a speed bump in a parking lot. The higher the speed bump is, the slower the car needs to proceed over it. The higher the activation energy, the slower the chemical reaction.

While enzymes can lower the activation energy of reactions, enzymes *cannot* change an endergonic reaction into an exergonic reaction. Enzymes cannot change an energetically unfavorable reaction into an energetically favorable reaction.

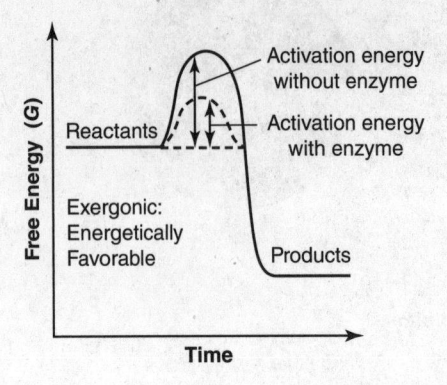

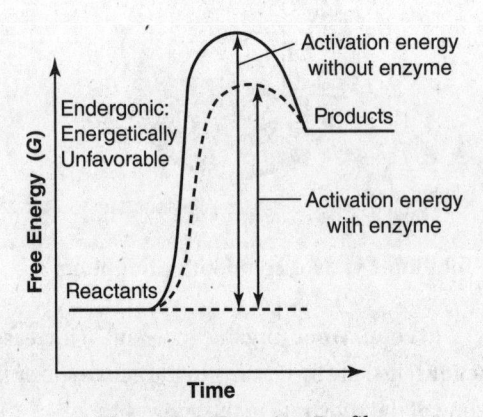

Figure 7.3 Reaction Profiles

Energy and Metabolism/Coupled Reactions

The **First Law of Thermodynamics** states that energy cannot be created or destroyed, only transformed from one form to another. The **Second Law of Thermodynamics** states that with each energy transformation, the disorder (entropy) of a system increases. Therefore, living organisms need a constant input of energy to power cellular processes and maintain order in living systems. The energy input into the cell must be greater than the energy requirements of the cell in order to maintain life. Processes that release energy can be paired (or coupled) with processes that require energy. These **coupled reactions**, as shown in Figure 7.4, occur in multiple steps to allow for the controlled transfer of energy between molecules, leading to more efficiency.

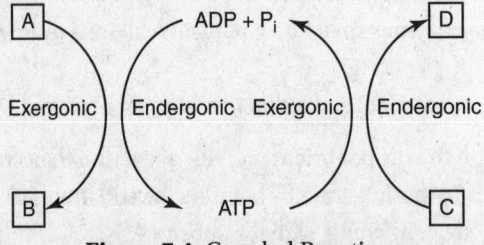

Figure 7.4 Coupled Reactions

Coupling an exergonic reaction with an endergonic reaction allows the energy released by the exergonic reaction to "drive" the endergonic reaction. For example, the breakdown of ATP into ADP and a phosphate group (P_i) is exergonic and releases approximately 30 kilojoules of energy per mole of ATP:

$$ATP \rightarrow ADP + P_i + 30 \text{ kJ}$$

The reaction that combines glucose and fructose to form sucrose requires approximately 27 kilojoules of energy per mole of sucrose formed:

$$\text{glucose} + \text{fructose} + 27 \text{ kJ} \rightarrow \text{sucrose}$$

Coupling these two reactions together shows that the exergonic breakdown of ATP into ADP releases more than enough energy to power the formation of sucrose from glucose and fructose:

$$\text{glucose} + \text{fructose} + ATP \rightarrow \text{sucrose} + ADP + P_i + 3 \text{ kJ}$$

Many of the endergonic chemical reactions that are required by living systems are powered by coupling them with exergonic reactions, such as the breakdown of ATP.

Practice Questions

Multiple-Choice

Questions 1–3

Catalase, an enzyme found in aerobic organisms, catalyzes the following reaction:

$$2H_2O_2 \rightarrow 2H_2O + O_2$$

A filter paper disk is saturated with the enzyme catalase and then placed at the bottom of a beaker of H_2O_2. As the reaction proceeds, oxygen bubbles will cling to the paper and eventually the paper will float to the top of the liquid. By measuring the time it takes for the catalase-saturated disk to float, one can compare the relative rates of decomposition of H_2O_2 under different experimental conditions.

An experiment was performed using catalase extracted from potatoes, with varying concentrations of H_2O_2. Multiple trials were conducted, and the means and the standard errors of the mean are shown in the following table:

Concentration of H_2O_2	Mean Time for Disks to Float (seconds)	Standard Error of the Mean (seconds)
1%	92.6	6.2
3%	32.5	5.9
6%	15.1	4.5

1. Which of the following would be a suitable control for this experiment?

    (A) changing the temperature of the H_2O_2 in one of the beakers
    (B) changing the solution the paper disk is soaked in from catalase to water in one of the beakers
    (C) changing the source of the catalase from potato to liver
    (D) changing the pH of the H_2O_2 in one of the beakers

2. Predict what would happen if the catalase solution was boiled before performing the experiment, and justify your prediction.

    (A) The average time for the disks to float would increase because boiling denatured the enzymes.
    (B) The average time for the disks to float would increase because enzyme-catalyzed reactions are always slower at higher temperatures.
    (C) The average time for the disks to float would decrease because boiling increased the number of molecular collisions between the enzymes and substrates.
    (D) The average time for the disks to float would increase because enzyme-catalyzed reactions are always faster at higher temperatures.

3. A student graphs this data, showing 95% confidence intervals. Based on the 95% confidence intervals, which concentrations of substrate are *least* likely to have statistically significant differences?

    (A) 1% and 3%
    (B) 1% and 6%
    (C) 3% and 6%
    (D) None of the substrate concentrations used are likely to have statistically significant differences.

---

4. BCR-ABL is an enzyme found in cancer cells in chronic myelogenous leukemia. ATP binds to the active site of BCR-ABL, which then stimulates cell division in cancer cells. Which of the following would most likely slow the rate of cell division in cancer cells with the BCR-ABL enzyme?

    (A) adding a cofactor of BCR-ABL
    (B) adding a coenzyme of BCR-ABL
    (C) adding a competitive inhibitor of BCR-ABL
    (D) adding a transcription factor of BCR-ABL

Questions 5–7

Refer to the following figure.

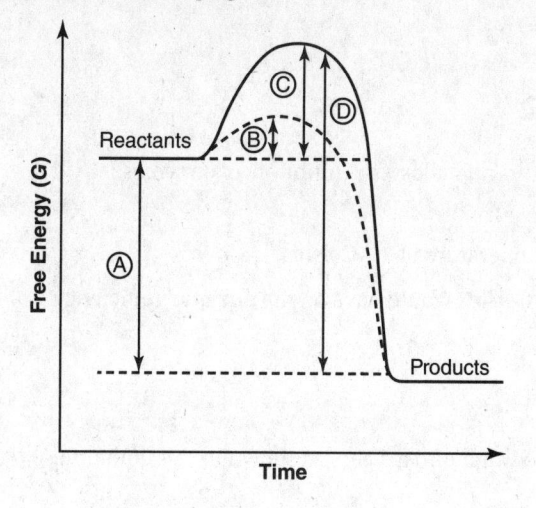

5. Which of the following represents the activation energy of an enzyme-catalyzed reaction?

    (A) A
    (B) B
    (C) C
    (D) D

6. Which of the following represents the activation energy of an uncatalyzed reaction?

    (A) A
    (B) B
    (C) C
    (D) D

7. Which of the following represents the overall free energy change in the reaction?

    (A) A
    (B) B
    (C) C
    (D) D

8. Which of the following correctly describes how enzymes increase the rate of a chemical reaction?

    (A) Enzymes decrease the overall free energy change of the reaction.
    (B) Enzymes decrease the activation energy of the reaction.
    (C) Enzymes increase the free energy of the reactants of the reaction.
    (D) Enzymes increase the free energy of the products of the reaction.

9. Which of the following correctly describes the differences between competitive inhibitors and noncompetitive inhibitors?

    (A) Competitive inhibitors bind to the active site of the enzyme; noncompetitive inhibitors bind to the allosteric site of the enzyme.
    (B) Competitive inhibitors bind to the substrate; noncompetitive inhibitors bind to the products.
    (C) The effects of a noncompetitive inhibitor can be mitigated by adding large amounts of substrate.
    (D) Competitive inhibitors increase the rate of enzyme-catalyzed reactions; noncompetitive inhibitors slow the rate of enzyme-catalyzed reactions.

10. Which of the following statements about enzymes is true?

    (A) Enzymes can change endergonic reactions into exergonic reactions.
    (B) Both proteins and RNA can have catalytic functions.
    (C) Enzymes always function optimally at a pH of 7.
    (D) Enzymes are consumed during chemical reactions.

Short Free-Response

11. Enzymes are important biological molecules.

    Part A
    (i) **Describe** how an enzyme interacts with a substrate.

    Part B
    (i) **Explain** the difference between competitive inhibitors and allosteric inhibitors of enzymes.

    Part C
    An enzyme has its maximum efficiency at an optimum temperature of 25° Celsius.
    (i) **Predict** what effect a decrease in temperature to 5° Celsius would have on the enzyme's efficiency.

    Part D
    (i) **Justify** your prediction from Part C.

12. The enzyme catalase breaks down hydrogen peroxide into water and oxygen, as shown in this equation:

    $$2H_2O_{2\,(l)} \rightarrow 2H_2O_{(l)} + O_{2\,(g)}$$

    An experiment was performed to measure the amount of oxygen bubbles produced at two different temperatures. The data are shown in the table.

Time (minutes)	Milliliters of Oxygen Produced at 37° Celsius	Milliliters of Oxygen Produced at 45° Celsius
0	0	0
5	2.5	0.5
10	5.2	0.8
15	7.7	1.0
20	10.3	1.1

    Part A
    (i) **Calculate** the rate of the enzyme-catalyzed reaction at both temperatures for the final 10 minutes of the experiment.

    Part B
    (i) **Predict** which temperature (37°C or 45°C) is closer to the optimum temperature for this enzyme.

    Part C
    (i) **Justify** your prediction from Part B using the data provided.

    Part D
    In a follow-up experiment, the reaction was performed at a temperature of 50°C and no oxygen was produced.
    (i) **Explain** why the enzyme would not function at 50°C.

Long Free-Response

13. Amylase is an enzyme that catalyzes the breakdown of amylose into its glucose subunits. The activity of amylase was measured at 37° Celsius and a pH of 7, both of which are optimum conditions for the activity of this enzyme. The production of glucose was measured both with and without the presence of compound X. Data are shown in the following table:

Time (minutes)	Cumulative Amount of Glucose Produced (millimoles) in the Absence of Compound X	Cumulative Amount of Glucose Produced (millimoles) in the Presence of Compound X
0	0	0
5	5.6	7.5
10	12.0	16.7
15	17.0	24.0
20	22.1	31.4

Part A

(i) On the axes provided, **construct** an appropriately labeled graph of this data.

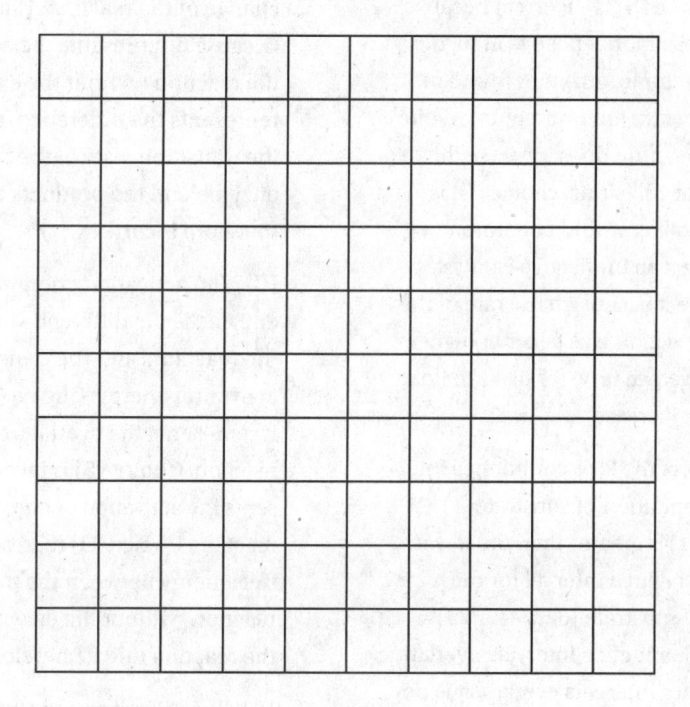

Part B

(i) Based on the data and your graph from Part A, **identify** compound X as either: a cofactor, a competitive inhibitor, or a noncompetitive inhibitor of amylase.

(ii) **Justify** your answer with evidence from the data.

Part C

(i) **Construct** an additional line on your graph from Part A that represents your prediction as to the expected experimental results at a temperature of 10° Celsius in the absence of compound X.

Part D

(i) **Explain** your prediction from Part C, stating why you placed the additional line where you did.

Answer Explanations

Multiple-Choice

1. **(B)** Using a paper disk soaked in water would demonstrate the rate of reaction without the enzyme and would provide a basis for comparison. Choice (A) is incorrect because changing the temperature in just one of the beakers would introduce another variable into the experiment. Choice (C) is incorrect because this change would only lead to comparing the effectiveness of liver catalase versus that of potato catalase. Changing the pH in just one of the beakers would also introduce another variable into the experiment, so choice (D) is also incorrect.

2. **(A)** Boiling denatures most enzymes, making them ineffective, and thus the reaction time would increase. Choice (B) is incorrect because some enzymes may be more effective at higher temperatures (for example, enzymes found in bacteria that live in warm environments). While increasing the temperature does increase the number of molecular collisions, choice (C) is incorrect because boiling would denature the enzyme, making it an ineffective catalyst. Choice (D) is incorrect because if the rate of the enzyme-catalyzed reaction was faster at higher temperatures, the average time for disks to float would decrease, not increase.

3. **(C)** The upper limit of the 95% confidence interval for the 6% concentration of substrate $(15.1 + 2(4.5) = 24.1)$ is greater than the lower limit of the 95% confidence interval for the 3% concentration of substrate $(32.5 - 2(5.9) = 20.7)$, so their 95% confidence intervals overlap. When 95% confidence intervals overlap, it is not possible to say there is a statistically significant difference between the two groups. Thus, choice (C) is the correct answer. The 95% confidence intervals for the 1% and 3% substrate concentrations do not overlap, so choice (A) is incorrect. Similarly, choice (B) is incorrect because the 95% confidence intervals for the 1% and 6% substrate concentrations do not overlap. Choice (D) is incorrect because the 95% confidence intervals for the 1% and 3% and the 1% and 6% substrate concentrations do not overlap and likely have statistically significant differences between them.

4. **(C)** Competitive inhibitors bind to the active site of an enzyme. Therefore, adding a competitive inhibitor of BCR-ABL would block ATP from binding to its active site. Choices (A) and (B) are both incorrect because cofactors and coenzymes enhance enzyme function. Choice (D) is incorrect because a transcription factor would not affect ATP's ability to bind to the enzyme.

5. **(B)** The activation energy in the presence of an enzyme is the difference in free energy between the reactants and the transition state in the presence of the enzyme. Choice (A) is incorrect because it represents the overall free energy change of the reaction. Choice (C) is incorrect because it represents the activation energy of the reaction without the enzyme. Choice (D) represents the difference in free energy between the transition state of the reaction without the enzyme and the products of the reaction and is therefore incorrect.

6. **(C)** The activation energy in the absence of an enzyme is the difference in free energy between the reactants and the transition state of the higher activation energy. Choice (A) is incorrect because it represents the overall free energy change of the reaction. Choice (B) is incorrect because it represents the activation energy of an enzyme-catalyzed reaction. Choice (D) represents the difference in free energy between the transition state of the reaction without the enzyme and the products of the reaction and is therefore incorrect.

7. **(A)** The overall free energy change of the reaction is the difference between the free energy of the products less the free energy of the reactants. Choice (B) is incorrect because it represents the activation energy of an enzyme-catalyzed reaction. Choice (C) is incorrect because it represents the activation energy of the reaction without the enzyme. Choice (D) represents the difference in free energy between the transition state of the reaction without the enzyme and the products of the reaction and is therefore not the right answer.

8. **(B)** Enzymes increase the rate of reactions by reducing the activation energy of the reaction. Choice (A) is incorrect because enzymes never affect the overall free energy change of the reaction. Choices (C) and (D) are incorrect because enzymes cannot change the free energy of the reactants, nor can they change the free energy of the products of a reaction.

9. **(A)** This statement accurately describes where competitive and noncompetitive inhibitors bind on enzymes. Choice (B) is incorrect because neither competitive inhibitors nor noncompetitive inhibitors bind to substrates or products. Choice (C) is incorrect because adding substrate can mitigate the effects of a competitive inhibitor, not a noncompetitive inhibitor. Both types of inhibitors reduce, not increase, the rate of enzyme-catalyzed reactions, so choice (D) is also incorrect.

10. **(B)** Enzymes are made of protein, ribozymes are made of RNA, and both have a catalytic effect on chemical reactions. Choice (A) is incorrect because enzymes never affect whether a reaction is endergonic or exergonic. Not all enzymes have an optimum pH of 7, so choice (C) is incorrect. Choice (D) is incorrect because enzymes are never consumed in the reactions they catalyze.

Short Free-Response

11. (A-i) Enzymes interact with a specific substrate that has properties (such as shape and charge) that are compatible with those at the enzyme's active site.

    (B-i) Competitive inhibitors bind to enzymes at the active site, whereas allosteric inhibitors bind to enzymes at the allosteric site.

    (C-i) A decrease in temperature to 5° Celsius would decrease the enzyme's efficiency.

    (D-i) Decreases in temperature reduce the number of molecular collisions between the substrate and the enzyme, reducing the number of chemical reactions and the enzyme's efficiency. Decreases in temperature may also alter the tertiary structure of the enzyme, altering its active site and reducing its catalytic ability.

12. (A-i) At 37°C, the rate of the enzyme-catalyzed reaction for the final 10 minutes of the experiment was

    $$\frac{(10.3 - 5.2) \text{ milliliters}}{10 \text{ minutes}} = 0.51 \frac{\text{milliliters}}{\text{minutes}}.$$

    At 45°C, the rate of the enzyme-catalyzed reaction for the final 10 minutes of the experiment was

    $$\frac{(1.1 - 0.8) \text{ milliliters}}{10 \text{ minutes}} = 0.3 \frac{\text{milliliters}}{\text{minutes}}.$$

    (B-i) The optimum temperature for this enzyme is closer to 37°C.

    (C-i) The rate of the reaction is greater at 37°C than at 45°C, so 37°C is closer to the optimum temperature for this enzyme.

    (D-i) At 50°C, the enzyme was probably denatured. High temperatures can denature enzymes, changing their shape so that they no longer function.

Long Free-Response

13. (A-i)

[Graph: Cumulative Amount of Glucose Produced (millimoles) vs Time (minutes), showing two curves — dashed line "Presence of compound X" reaching ~32 mmol at 20 min, solid line "Absence of compound X" reaching ~22 mmol at 20 min.]

(B-i) Compound X is likely a cofactor.

(B-ii) Cofactors increase enzyme efficiency. Because the amount of product produced in the presence of compound X is greater than the amount of product produced in the absence of compound X, compound X is most likely a cofactor of amylase.

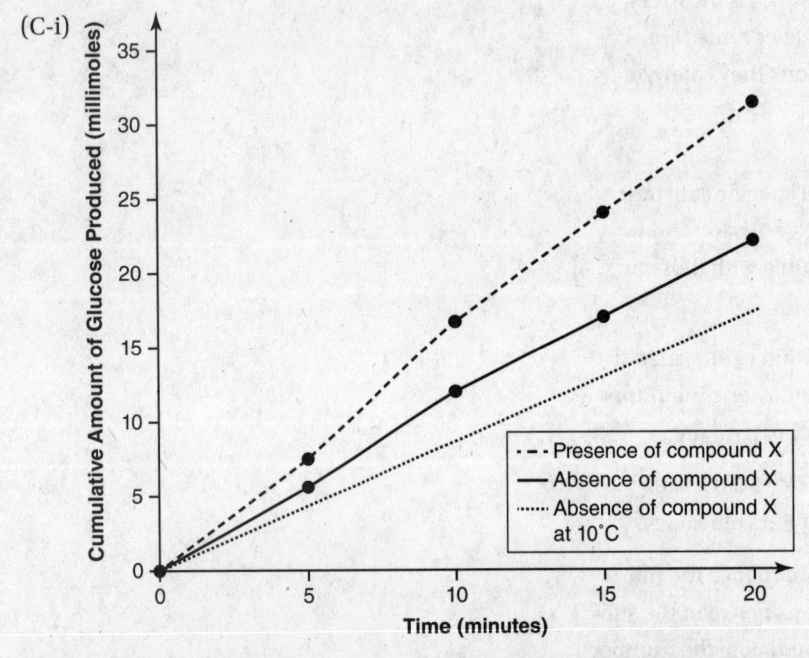

(C-i)

[Graph: Same as above with an additional dotted line "Absence of compound X at 10°C" reaching ~17 mmol at 20 min, lower than the other two lines.]

(D-i) At reduced temperatures, there are fewer molecular collisions between the enzyme and the substrate as a result of the reduced kinetic energy at lower temperatures. Therefore, it is reasonable to predict that the reaction rate at 10°C would be slower than that of either 37°C measurement and thus that line is lower on the graph than the other two lines.

8
Photosynthesis

Learning Objectives

In this chapter, you will learn:

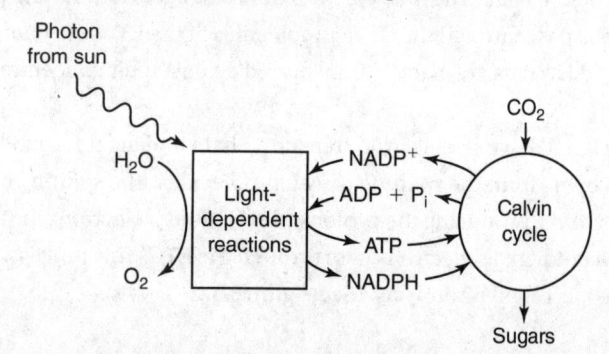

→ Light-Dependent Reactions
→ Light-Independent Reactions (The Calvin Cycle)

Overview

Some organisms consume other organisms to obtain organic molecules—these organisms are called **heterotrophs**. Organisms that can produce their own organic molecules from inorganic molecules are called **autotrophs**. Autotrophs that use light energy to power this process are called **photoautotrophs**. This chapter will review the process that photoautotrophs use to produce organic molecules: photosynthesis.

Before studying all the reactions associated with photosynthesis, it is important to understand the basics of this process. Here is the overall equation for photosynthesis:

$$6CO_2 + 6H_2O \xrightarrow{\text{light energy}} C_6H_{12}O_6 + 6O_2$$

Note that during photosynthesis, the carbon atoms from carbon dioxide gain hydrogen atoms (carbon is reduced) and the oxygen atoms in water lose hydrogen atoms (oxygen is oxidized).

Photosynthesis can be divided into two main parts—the **light-dependent reactions** and the **light-independent reactions** (also known as the **Calvin cycle**), as shown in Figure 8.1. The light-dependent reactions use energy from sunlight to split water, producing oxygen gas, protons, and high-energy electrons. Oxygen gas is released into the atmosphere. The protons and high-energy electrons are used to power the production of ATP and NADPH (which are sent to the light-independent reactions). The light-independent reactions use this ATP and NADPH, along with carbon dioxide, to produce sugars. The light-independent reactions then send ADP, P_i, and NADP$^+$ back to the light-dependent reactions so that photosynthesis can continue. In this way, the two parts of photosynthesis are interdependent.

> **TIP**
>
> An easy way to remember the difference between oxidation and reduction is the mnemonic device **OILRIG**: **O**xidation **I**s **L**osing hydrogen atoms; **R**eduction **I**s **G**aining hydrogen atoms.

Figure 8.1 Light-Dependent Reactions and Light-Independent Reactions (The Calvin Cycle)

In plants, photosynthesis occurs in the chloroplasts. Chloroplasts have an outer membrane, which is filled with a liquid called stroma. Floating in the stroma are stacks of membranous sacs called grana, and each individual sac is called a thylakoid. The stroma is the location of the light-independent reactions, and the thylakoid is the site of the light-dependent reactions.

Some prokaryotes (for example, cyanobacteria) also perform photosynthesis. However, prokaryotes do not possess membrane-bound organelles, such as the chloroplasts. In photosynthetic prokaryotes, the light-dependent reactions of photosynthesis occur on infoldings of the plasma membrane and the light-independent reactions occur in the cytosol.

Light-Dependent Reactions

In the light-dependent reactions, light energy is used to drive the production of ATP. This is called **photophosphorylation**. Light energy excites the electrons in the chloroplast to higher energy levels. As these excited electrons move through the chloroplast, energy is released. At the end of the light-dependent reactions, $NADP^+$ accepts these electrons, forming NADPH, which is a source of reducing power for the light-independent reactions.

Chlorophyll is a light-absorbing pigment that captures the energy of photons from the sun. Chlorophylls are the primary light-absorbing pigments in photosynthesis. Chlorophylls are found in photosystems I and II (PSI and PSII). A **photosystem** is composed of proteins, chlorophyll, and other light-absorbing pigments called accessory pigments. PSI and PSII contain different types of chlorophyll that absorb the most light energy at slightly different wavelengths (700 nm and 680 nm, respectively). Photosystems are located in the thylakoid membrane of the chloroplast and are connected by an electron transport chain (ETC). See Figure 8.2.

> **TIP**
>
> As you review the light-dependent reactions, focus on the role of light energy and the path of the excited electrons during the process.

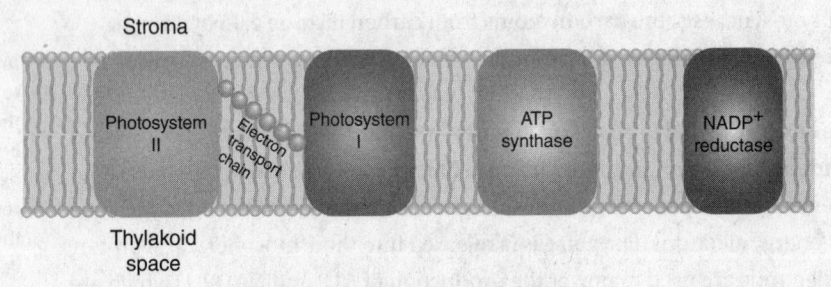

Figure 8.2 PSI, PSII, and the ETC

The energy in the photons is used to boost electrons in chlorophyll to a higher energy level in PSII. These electrons from PSII are passed from one protein carrier to another in a series of redox (reduction-oxidation) reactions that are analogous to "falling down a hill." The final electron donor in the electron transport chain passes the electron to PSI. As the electrons pass through the carrier molecules of the ETC, the energy that is released is used to create a proton gradient, and H^+ ions are actively transported against their concentration gradient across the thylakoid membrane.

The electrons from PSII that fell down the electron transport chain and are now on PSI need to be replaced on PSII. The electrons in PSII come from the splitting of water molecules. The splitting of water molecules strips electrons from the hydrogen atoms, producing the protons (the H^+ ions), electrons for PSII, and oxygen gas. The protons will be used to form a gradient as electrons pass through the ETC. This process, which is driven by the energy from the sun (photons), is called **photolysis**. (See Figure 8.3.)

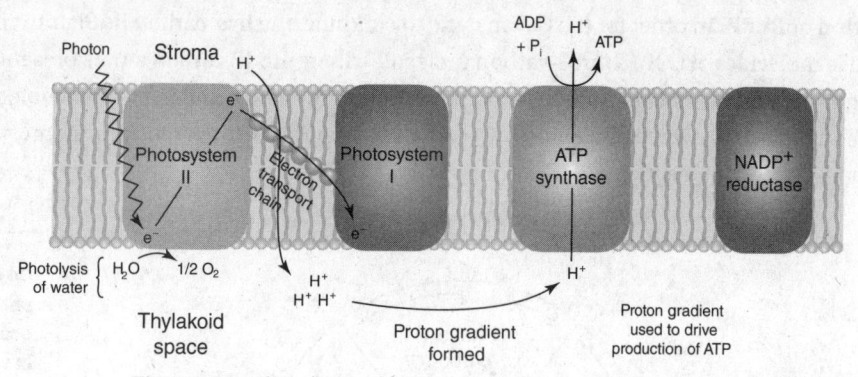

Figure 8.3 Photolysis of Water and the Production of ATP

The proton gradient generated by the photolysis of water and the ETC powers the production of ATP by the enzyme **ATP synthase**. (See Figure 8.3.) The process of using a proton gradient and ATP synthase to produce ATP is called **chemiosmosis**. Chemiosmosis is also used in mitochondria to generate ATP during cellular respiration, which will be reviewed in Chapter 9.

The electron from the ETC (that is now on PSI) is boosted by another photon of light energy from the sun. The electron again passes through a series of carriers, although much shorter than that of the ETC, where it is finally transferred, along with a proton, to $NADP^+$ by the enzyme $NADP^+$ reductase. This produces a molecule of NADPH as shown in Figure 8.4, which will provide reducing power for the light-independent reactions.

> Chemiosmosis requires a proton gradient, so a membrane is needed to separate the protons into a gradient. In eukaryotes, this gradient is generated across the thylakoid membrane. In prokaryotes, this gradient is formed in the infoldings of the plasma membrane.

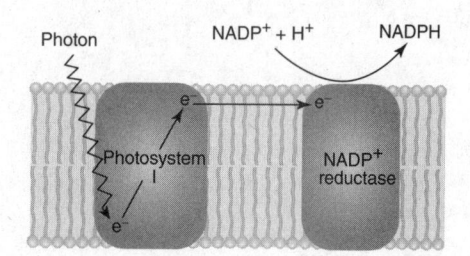

Figure 8.4 PSI and the Production of NADPH

Light-Independent Reactions (The Calvin Cycle)

The light-independent reactions (also known as the Calvin cycle) occur in the stroma of the chloroplast, which is the liquid surrounding the stacks of thylakoids. The Calvin cycle, as shown in Figure 8.5, is a complicated, multi-step process that can be broken down into three main parts:

1. **Fixation of carbon:** *Fixation* means turning a biologically unusable form into a usable form. In the fixation of carbon, the enzyme ribulose-bisphosphate-carboxylase (Rubisco, for short) adds one molecule of carbon dioxide to the five-carbon molecule ribulose-bisphosphate (RuBP, for short). This produces a six-carbon intermediate that is unstable, which then breaks down further into two three-carbon molecules.
2. **Reduction:** Now the ATP and NADPH from the light-dependent reactions are used to reduce the three-carbon molecules. The energy to do this comes from the ATP, and the NADPH provides the hydrogen atoms (reducing power). A three-carbon molecule called glyceraldehyde-3-phosphate (simply known as G3P) is produced at the end of this process. G3P can be used to make sugars, but some of the G3P is used in the final part of the Calvin cycle: regeneration.

3. **Regeneration of RuBP:** In order for the Calvin cycle to continue, the five-carbon RuBP must be regenerated. For every five molecules of G3P (a three-carbon molecule), there are 15 carbon atoms present. Using ATP from the light-dependent reactions, these five G3P molecules rearrange and form three molecules of RuBP (a five-carbon molecule), which also contain 15 carbon atoms. This process requires energy, which comes from the light-dependent reactions.

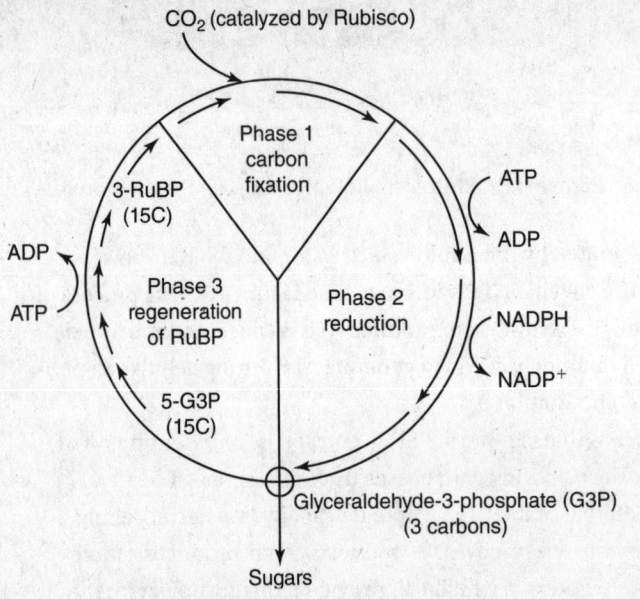

Figure 8.5 The Calvin Cycle

> **TIP**
>
> C4 and CAM photosynthesis occurs in plants in warmer climates, but that is beyond the scope of the AP Biology exam. Focus on the light-dependent reactions and the Calvin cycle when preparing for test day.

Practice Questions

Multiple-Choice

1. In prokaryotes, where in the cell do the light-dependent reactions of photosynthesis occur?

    (A) on the thylakoid membrane
    (B) on the plasma membrane
    (C) in the cytoplasm
    (D) in the nucleoid region

Questions 2 and 3

A scientist, who is studying photosynthesis, places a plant in an environment where the oxygen atoms in carbon dioxide are labeled with radioactive ^{18}O but the oxygen atoms in water have nonradioactive ^{16}O.

2. Based on this scenario, predict which product of photosynthesis will contain radioactive ^{18}O.

    (A) NADPH
    (B) glucose
    (C) oxygen gas
    (D) ATP

3. Based on this scenario, predict which product of photosynthesis will contain nonradioactive ^{16}O.

    (A) NADPH
    (B) glucose
    (C) oxygen gas
    (D) ATP

4. Which of the following lists the three major parts of the light-independent reactions (the Calvin cycle)?

    (A) carbon fixation, electron transport chain, reduction
    (B) carbon fixation, reduction, production of ATP
    (C) carbon fixation, reduction, regeneration of RuBP
    (D) reduction, regeneration of RuBP, electron transport chain

5. Which of the following correctly lists the products of the light-dependent reactions?

    (A) ADP, $NADP^+$, oxygen
    (B) ATP, NADPH, oxygen
    (C) G3P, ADP, $NADP^+$
    (D) G3P, ATP, NADPH

6. Which of the following correctly lists the products of the Calvin cycle?

    (A) ADP, $NADP^+$, oxygen
    (B) ATP, NADPH, oxygen
    (C) G3P, ADP, $NADP^+$
    (D) G3P, ATP, NADPH

7. Which of the following statements correctly identifies the locations of the major parts of photosynthesis in plant cells?

    (A) Light-dependent reactions occur in the stroma; light-independent reactions occur in the thylakoid.
    (B) Light-dependent reactions occur in the matrix; light-independent reactions occur in the thylakoid.
    (C) Light-dependent reactions occur in the thylakoid; light-independent reactions occur in the matrix.
    (D) Light-dependent reactions occur in the thylakoid; light-independent reactions occur in the stroma.

8. If a thylakoid membrane is punctured so that molecules can freely flow between the thylakoid and the stroma, which of the following processes of photosynthesis will be most directly affected?

    (A) the formation of G3P
    (B) the generation of a proton gradient
    (C) the absorption of light
    (D) the fixation of carbon

9. Which of the following best describes a relationship between the light-dependent reactions and the Calvin cycle?

   (A) The light-dependent reactions supply the Calvin cycle with oxygen, and the Calvin cycle returns carbon dioxide to the light-dependent reactions.
   (B) The light-dependent reactions supply the Calvin cycle with ATP, and the Calvin cycle returns ADP to the light-dependent reactions.
   (C) The light-dependent reactions supply the Calvin cycle with carbon dioxide, and the Calvin cycle returns oxygen to the light-dependent reactions.
   (D) The light-dependent reactions supply the Calvin cycle with $NADP^+$, and the Calvin cycle returns NADPH to the light-dependent reactions.

10. In the light-dependent reactions, the final electron acceptor in the electron transport chain is

    (A) oxygen.
    (B) water.
    (C) NAD^+.
    (D) $NADP^+$.

Short Free-Response

11. The sea slug *Elysia crispata* eats photosynthetic algae. However, after consuming the algae, the chloroplasts from the algae are incorporated into the sea slug's own cells, give the sea slug a green color, and remain functional for up to four months. This phenomenon is called kleptoplasty.

    Part A

    (i) **Describe** the role of photosynthetic algae in ecosystems.

    Part B

    (i) **Explain** why kleptoplasty would give *Elysia crispata* a survival advantage.

    Part C

    An oil spill on the surface of the water reduces the intensity of light in *Elysia crispata's* habitat.

    (i) **Predict** the effect this would have on *Elysia crispata's* survival.

    Part D

    (i) **Justify** your prediction from Part C.

12. The light-dependent and light-independent reactions of photosynthesis exchange materials and are interdependent, as shown in the following figure.

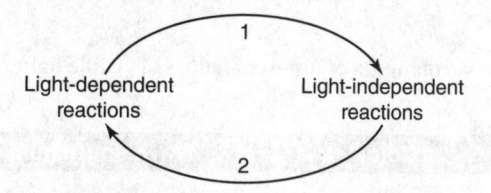

    Part A

    (i) **Identify** the molecules that the light-dependent reactions provide to the light-independent reactions.

    (ii) **Label** arrow 1 with those molecules.

    Part B

    (i) **Identify** the molecules that the light-independent reactions return to the light-dependent reactions.

    (ii) **Label** arrow 2 with those molecules.

    Part C

    An inhibitor of the enzyme Rubisco is added to a plant cell.

    (i) **Explain** which part of photosynthesis would be most directly affected by this inhibitor.

    Part D

    A student claims that the light-independent reactions of photosynthesis would stop if a plant were to be kept in the dark for a long period of time.

    (i) Use your knowledge about photosynthesis to **support the student's claim**.

Long Free-Response

13. The following graph shows the absorption spectrum for the pigments chlorophyll *a*, chlorophyll *b*, and carotenoids.

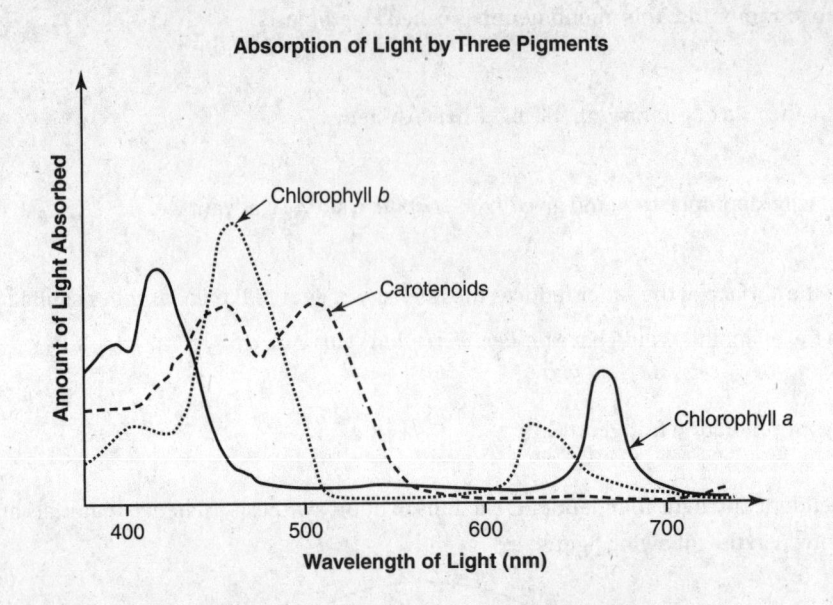

Absorption of Light by Three Pigments

The following table shows the wavelengths of different colors of visible light.

Color	Wavelength (nm)
Violet	380–450
Blue	450–495
Green	495–570
Yellow	570–590
Orange	590–620
Red	620–750

Part A

(i) For each of the three pigments, **identify** the color of light that will be most absorbed by that pigment.

(ii) Use the graph to **justify** your answer.

Part B

A mutation causes a plant to lose its ability to produce the pigments chlorophyll *a* and *b*. A student wants to design an experiment to study the rate of photosynthesis in the plant with this mutation.

(i) **Identify** an appropriate control for the experiment.

(ii) **Identify** the independent variable in the experiment.

(iii) **Identify** the dependent variable in the experiment.

Part C

The following figure graphs the rate of photosynthesis (as measured by oxygen production) in control plants and in plants with the mutation described in Part B.

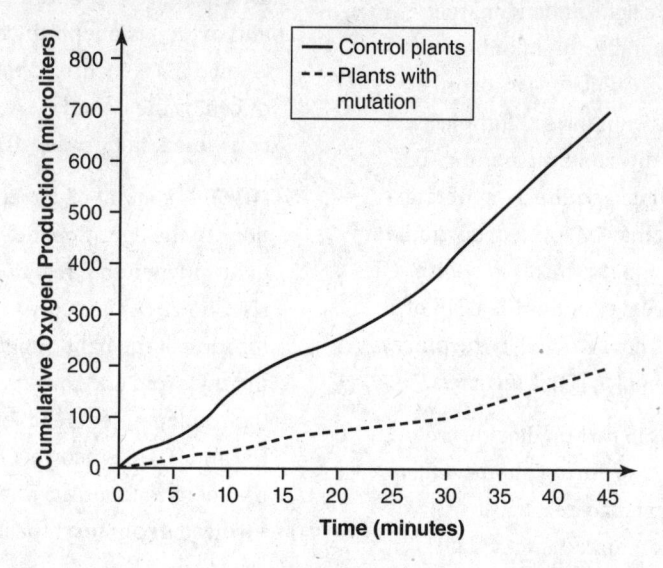

(i) **Calculate** the rate of photosynthesis in both sets of plants during the first 30 minutes of the experiment.

Part D

Plants that use carotenoids as their primary photosynthetic pigment are grown under three different wavelengths of light: 450 nm, 500 nm, and 550 nm.

(i) **Predict** which group of plants will perform the least amount of photosynthesis.

(ii) **Justify** your prediction.

Answer Explanations

Multiple-Choice

1. **(B)** In prokaryotes, the light-dependent reactions of photosynthesis occur on infoldings of the plasma membrane. Choice (A) is incorrect because prokaryotes do not have chloroplasts, so they do not have a thylakoid membrane. The light-dependent reactions require a membrane so that a proton gradient may be formed; no gradient can be formed in the liquid cytosol, so choice (C) is incorrect. The nucleoid contains the DNA of a prokaryotic cell and is not involved in the process of photosynthesis, so choice (D) is incorrect.

2. **(B)** The oxygen atoms in carbon dioxide are incorporated into glucose during photosynthesis. Choices (A) and (D) are incorrect because no oxygen atoms are incorporated into NADPH or ATP during photosynthesis. The oxygen gas released by photosynthesis is a result of the photolysis of water, so choice (C) is also incorrect.

3. **(C)** The oxygen gas released by photosynthesis is a product of the photolysis of water. Choices (A) and (D) are incorrect because no oxygen atoms are incorporated into NADPH or ATP during photosynthesis. Choice (B) is incorrect because the oxygen atoms in carbon dioxide are incorporated into glucose during photosynthesis.

4. **(C)** The three main parts of the light-independent reactions (the Calvin cycle) are carbon fixation, reduction, and regeneration of RuBP. Choices (A) and (D) are incorrect because there is no electron transport chain involved in the Calvin cycle. ATP is consumed, not produced, in the Calvin cycle, so choice (B) is also incorrect.

5. **(B)** The products of the light-dependent reactions are ATP, NADPH, and oxygen gas. Choices (A) and (C) are incorrect because ADP and NADP$^+$ are consumed, not produced, in the light-dependent reactions. G3P is a product of the light-independent reactions, which further rules out choice (C) and eliminates choice (D).

6. **(C)** The products of the Calvin cycle are ADP, NADP$^+$, and G3P. Choice (A) is incorrect because oxygen is not produced by the Calvin cycle. Choice (B) is incorrect because ATP and NADPH are consumed, not produced, by the Calvin cycle, and oxygen is not produced by the Calvin cycle. Choice (D) is incorrect because while G3P *is* produced by the Calvin cycle, ATP and NADPH are consumed, not produced, by the Calvin cycle.

7. **(D)** This statement correctly identifies the locations of the light-dependent reactions and the light-independent reactions of photosynthesis. Choice (A) is incorrect because the correct location of the light-dependent reactions is the thylakoid and the correct location of the light-independent reactions is the stroma. Choices (B) and (C) are incorrect because the matrix is part of mitochondria, not chloroplasts, and is not involved in photosynthesis.

8. **(B)** If the thylakoid membrane is punctured, it would not be possible to generate a proton gradient since the protons would be able to diffuse freely across the thylakoid membrane. Choices (A) and (D) occur during the Calvin cycle, so (A) and (D) would not be directly affected by the puncturing of the thylakoid membrane. The absorption of light by pigments would not be affected by the puncturing of the thylakoid membrane, so choice (C) is also incorrect.

9. **(B)** The light-dependent reactions supply ATP to the Calvin cycle, which then returns ADP back to the light-dependent reactions. Choice (A) is incorrect because the oxygen produced by the light-dependent reactions is released to the atmosphere and the Calvin cycle consumes, not produces, carbon dioxide. Choice (C) is incorrect because the carbon dioxide needed for the Calvin cycle is absorbed from the atmosphere and the Calvin cycle does not produce oxygen. Choice (D) is incorrect because the light-dependent reactions send NADPH to the Calvin cycle and the Calvin cycle returns NADP$^+$ to the light-dependent reactions.

10. **(D)** NADP$^+$ is the final electron acceptor in the light-dependent reactions. NADP$^+$ becomes NADPH after accepting electrons. Choice (A) is incorrect because oxygen is the final electron

acceptor in cellular respiration, not photosynthesis. Water does not function as an electron acceptor, so choice (B) is incorrect. NAD⁺ is not used in photosynthesis; it is used in cellular respiration. Thus, choice (C) is incorrect.

Short Free-Response

11. (A-i) Photosynthetic algae are autotrophs—they make their own food and are consumed by other organisms.

    (B-i) Kleptoplasty would give *Elysia crispata* a survival advantage because the chloroplasts that remained functional in the *Elysia crispata*'s cells would provide the sea slug with food.

    (C-i) Reducing the intensity of light in *Elysia crispata*'s habitat would have a negative impact on *Elysia crispata*'s survival.

    (D-i) Photosynthesis requires light energy, so reducing the light intensity would reduce the amount of photosynthesis the chloroplasts would perform and the amount of food from those chloroplasts that would be available to *Elysia crispata*. The sea slug might become more dependent on consuming other food sources.

12. (A-i) The light-dependent reactions provide ATP and NADPH to the light-independent reactions.

    (A-ii)

    ATP and NADPH
    Light-dependent reactions → (1) → Light-independent reactions
    Light-independent reactions → (2) → Light-dependent reactions

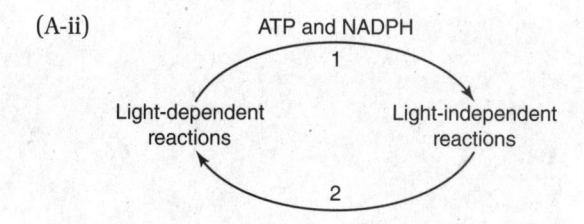

    (B-i) The light-independent reactions return ADP and NADP⁺ to the light-dependent reactions.

    (B-ii)

    Light-dependent reactions → (1) → Light-independent reactions
    Light-independent reactions → (2) → Light-dependent reactions
    ADP and NADP⁺

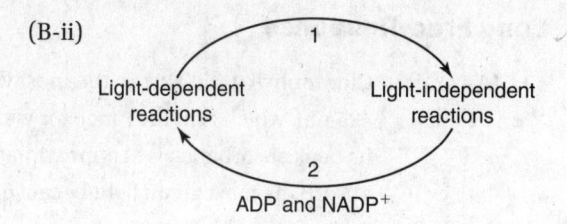

    (C-i) Rubisco is an enzyme that is used in the light-independent reactions (the Calvin cycle) of photosynthesis, so the light-independent reactions would be most directly affected by an inhibitor of Rubisco.

    (D-i) The light-independent reactions of photosynthesis require ATP and NADPH that are produced by the light-dependent reactions of photosynthesis. If a plant was kept in the dark for a long period of time, the light-dependent reactions would stop and the ATP and NADPH that are required for the light-independent reactions would not be produced. Thus, the light-independent reactions would stop when their supply of ATP and NADPH was depleted.

Long Free-Response

13. **(A-i and ii)** Chlorophyll *a* will absorb the most violet light because its peak absorbance is at approximately 425 nm, which is in the range for violet. Chlorophyll *b* will absorb the most blue light because its peak absorbance is at approximately 480 nm, which is in the range for blue. The carotenoids absorb the most green light because their peak absorbance is at approximately 510 nm, which is in the range for green.

   (B-i) An appropriate control would be a plant without the mutation that could produce chlorophyll *a*, chlorophyll *b*, and carotenoids.

   (B-ii) The independent variable would be the presence or absence of chlorophyll *a* and chlorophyll *b*.

   (B-iii) The dependent variable would be the rate of photosynthesis.

   (C-i) Rate of photosynthesis in the control plants = 400 microliters oxygen/30 minutes = 13.3 microliters oxygen/minute

   Rate of photosynthesis in the plants with the mutation = 100 microliters oxygen/30 minutes = 3.3 microliters oxygen/minute

   (D-i and ii) The plants grown under light with a wavelength of 550 nm will perform the least amount of photosynthesis because carotenoids absorb the least amount of light energy at 550 nm. According to the first graph, the absorbance is greater at 450 nm and 500 nm than it is at 550 nm, so less photosynthesis will occur at 550 nm.

9

Cellular Respiration

Learning Objectives

In this chapter, you will learn:
- → Glycolysis
- → Oxidation of Pyruvate
- → Krebs Cycle (Citric Acid Cycle)
- → Oxidative Phosphorylation
- → Fermentation

Overview

Chapter 8 reviewed how photosynthetic organisms harness light energy and store it in the chemical bonds of organic molecules. This chapter will review how living organisms release the energy stored in the chemical bonds of organic molecules to drive the processes necessary for life.

The overall equation for cellular respiration is:

$$C_6H_{12}O_6 + 6O_2 \rightarrow 6CO_2 + 6H_2O + ATP$$

Cellular respiration includes the following cellular processes: **glycolysis**, the **oxidation of pyruvate**, the **Krebs cycle** (also known as the **citric acid cycle**), and **oxidative phosphorylation**, all of which will be discussed in more detail later in this chapter.

The presence or absence of oxygen determines which cellular processes a living organism can use to obtain energy from food. **Anaerobic** organisms, which do not have access to or do not require oxygen, can perform glycolysis and fermentation. **Aerobic** organisms that do not have access to oxygen can also perform glycolysis and fermentation. In addition, in the presence of oxygen, aerobic organisms can also perform the oxidation of pyruvate, the Krebs cycle, and oxidative phosphorylation. Table 9.1 summarizes the differences between anaerobic organisms and aerobic organisms in the presence or absence of oxygen.

Table 9.1 Cellular Processes Performed by Anaerobic and Aerobic Organisms

Cellular Process	Anaerobic Organisms (in the Presence or Absence of Oxygen)	Aerobic Organisms (in the Presence of Oxygen)	Aerobic Organisms (in the Absence of Oxygen)
Glycolysis	✓	✓	✓
Oxidation of Pyruvate		✓	
Krebs Cycle		✓	
Oxidative Phosphorylation		✓	
Fermentation	✓	✓	✓

Aerobic organisms can perform more metabolic processes than anaerobic organisms can perform. This allows aerobic organisms to extract much more energy from organic compounds than anaerobic organisms can.

First is a review of the major cellular processes of cellular respiration, which is then followed by a review of fermentation.

Glycolysis

Glycolysis occurs in the cytosol of the cell. Since all living organisms have cytosol, all living organisms can perform glycolysis. Glycolysis was probably one of the first metabolic processes to evolve. This theory is supported by the fact that enzymes, which catalyze the steps of glycolysis, are highly conserved and are found in all living organisms.

The six-carbon molecule glucose enters glycolysis, along with two molecules of the electron carrier NAD^+. During glycolysis, the glucose molecule is oxidized (loses hydrogen atoms and their electrons), and each NAD^+ is reduced (gains a hydrogen atom and its electrons) to NADH. Remember that whenever one molecule is oxidized, another molecule must be reduced. During cellular respiration, the molecules that contain carbon are oxidized and the electron carriers NAD^+ and FAD^+ are reduced.

Two molecules of ATP are required in the early steps of glycolysis. However, four molecules of ATP are produced by glycolysis, resulting in a net gain of two ATP molecules.

At the end of glycolysis, the six-carbon glucose molecule is cleaved into two three-carbon pyruvate molecules.

Glycolysis is a multistep process that involves many enzyme-catalyzed steps and intermediates. When studying for the AP Biology exam, focus on where each process in cellular respiration occurs and what the inputs and outputs are for each process. Figure 9.1 and Table 9.2 summarize the location, inputs, and outputs of glycolysis.

> As you review each cellular process, it is not necessary to memorize every step and enzyme involved! Focus on the flow of electrons and energy through the process. For each process (glycolysis, oxidation of pyruvate, the Krebs cycle, oxidative phosphorylation, and fermentation), know where it happens and what the inputs and outputs are for each.

TIP

Remember to follow the carbons: While you do not need to know the structures of glucose and pyruvate, know how many carbons are in each molecule.

Figure 9.1 Glycolysis

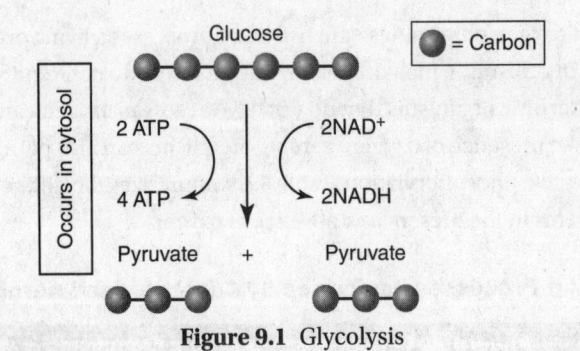

Table 9.2 Glycolysis

Location	Inputs	Outputs
Cytosol	Glucose (6C)	2 Pyruvate (3C)
	2 NAD^+	2 NADH
	2 ATP	4 ATP

Oxidation of Pyruvate

The next step in cellular respiration occurs in the mitochondria. The three-carbon pyruvate molecule must be modified in order to enter the mitochondria. Pyruvate is oxidized (loses a hydrogen atom and its electrons), and the electron carrier NAD^+ is reduced (gains a hydrogen atom and its electrons) and becomes NADH. As this happens, one of the carbons in the pyruvate molecule is released as carbon dioxide, leaving behind a two-carbon acetyl group. **Coenzyme A** attaches to this two-carbon acetyl group. Coenzyme A will deliver the acetyl group to the Krebs cycle.

Each molecule of glucose that enters glycolysis will generate two molecules of pyruvate, so the oxidation of pyruvate will occur twice for each molecule of glucose that entered glycolysis. Figure 9.2 and Table 9.3 summarize the location, inputs, and outputs of the oxidation of pyruvate.

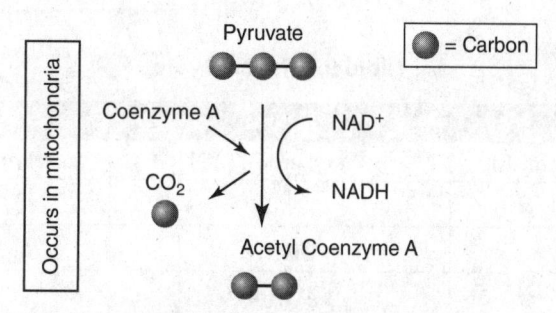

Figure 9.2 Oxidation of Pyruvate

Table 9.3 Oxidation of Pyruvate

Location	Inputs	Outputs
Mitochondria	Pyruvate (3C)	Acetyl group (2C)
	NAD^+	Carbon dioxide (1C)
		NADH

Krebs Cycle (Citric Acid Cycle)

The Krebs cycle (also known as the citric acid cycle) occurs in the matrix (the liquid center) of the mitochondria. Coenzyme A brings the two-carbon acetyl group to the Krebs cycle, where it is initially attached to a four-carbon intermediate, forming a six-carbon molecule. This six-carbon molecule goes through a series of enzyme-catalyzed reactions, during which two more carbon dioxide molecules are released and the four-carbon intermediate is regenerated. At the completion of the cycle, all of the carbon that was originally in the glucose molecule at the start of glycolysis has been released as carbon dioxide. (See Figure 9.3 and Table 9.4.)

During one turn of the Krebs cycle, four electron carriers are reduced. Three molecules of NAD^+ are reduced to NADH, and one molecule of FAD^+ is reduced to $FADH_2$. In addition, one molecule of ATP is produced through **substrate-level phosphorylation** (the direct addition of a phosphate group to ADP without the use of an electron transport chain or chemiosmosis).

> There are two methods of ATP production you need to understand for the AP Biology exam: substrate-level phosphorylation and oxidative phosphorylation. Substrate-level phosphorylation is a simpler process, but oxidative phosphorylation allows for the production of greater amounts of ATP.

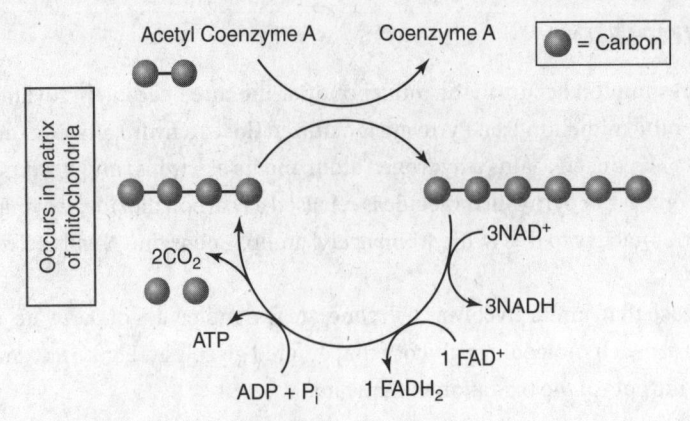

Figure 9.3 Krebs Cycle

Table 9.4 Krebs Cycle

Location	Inputs	Outputs
Matrix of the mitochondria	Acetyl group (2C)	2 carbon dioxides (1C each)
	3 NAD$^+$	3 NADH
	1 FAD$^+$	1 FADH$_2$
	1 ADP + P$_i$	1 ATP

Before moving on to oxidative phosphorylation, review Table 9.5 for a recap of what has been generated for each glucose molecule that entered cellular respiration.

Table 9.5 Total Products of Glycolysis, Oxidation of Pyruvate, and the Krebs Cycle for Each Glucose Molecule that Entered Cellular Respiration

Molecule	Glycolysis	Oxidation of Pyruvate	Krebs Cycle	Totals
ATP	2 (net)	0	2	4
NADH	2	2	6	10
FADH$_2$	0	0	2	2
CO$_2$	0	2	4	6

> **NOTE**
>
> Glycolysis takes in a six-carbon glucose and produces two three-carbon pyruvates. For this reason, while glycolysis occurs once per glucose molecule, both the oxidation of pyruvate and the Krebs cycle occur twice per glucose molecule.

When reviewing Table 9.5, notice that all six carbons in the glucose molecule that entered cellular respiration have been released as carbon dioxide. Four molecules of ATP have been produced by substrate-level phosphorylation (two in glycolysis and two in the Krebs cycle). A total of 12 high-energy electron carriers (10 NADH and two FADH$_2$) have been produced and will enter the next stage of cellular respiration: oxidative phosphorylation.

Oxidative Phosphorylation

Oxidative phosphorylation involves the **electron transport chain (ETC)** and **chemiosmosis**, both of which occur on the inner membrane of the mitochondria. Oxidative phosphorylation yields the vast majority of ATP in cellular respiration.

The electron carriers (NADH and FADH$_2$) that were generated during glycolysis, the oxidation of pyruvate, and the Krebs cycle bring their electrons to the **electron transport chain** on the inner mitochondrial membrane. (See Figure 9.4.) As these electron carriers deliver their hydrogen atoms and electrons to the ETC, NADH and FADH$_2$ are oxidized to NAD$^+$ and FAD$^+$, respectively. NAD$^+$ and FAD$^+$ can then be reused in the earlier processes of cellular respiration.

As the electrons travel through the electron transport chain, their potential energy decreases, and energy is released. This energy is used to pump protons (H$^+$) out of the matrix and into the intermembrane space of the mitochondria, creating a proton gradient. The concentration of protons in the intermembrane space can be 1,000 times that of the matrix!

At the end of the electron transport chain, molecular oxygen (O$_2$) combines with four protons (H$^+$) and four electrons (e$^-$) to form two water molecules. This makes oxygen the final, or terminal, electron acceptor during cellular respiration.

> **TIP**
> Remember that *NADH* is one of the electron carriers used in cellular respiration, whereas *NADPH* is the electron carrier used in photosynthesis.

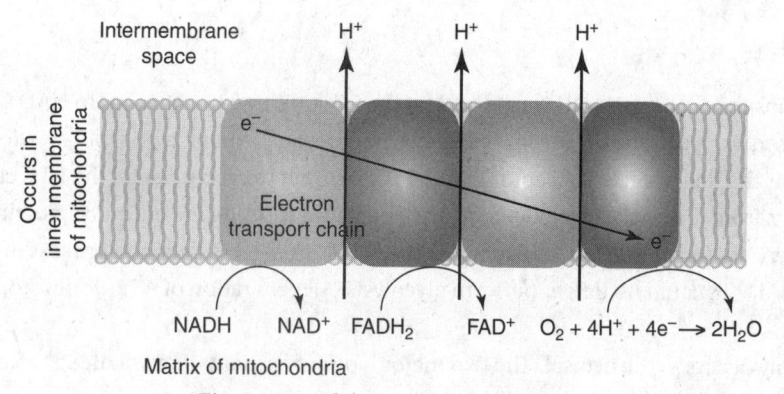

Figure 9.4 Electron Transport Chain

The proton gradient created by the electron transport chain is used to drive ATP synthesis. Using a proton gradient to drive the production of ATP is called **chemiosmosis**, as shown in Figure 9.5. The enzyme **ATP synthase** catalyzes this process. ATP synthase is located on the inner membrane of the mitochondria. Protons flow from an area of higher concentration in the intermembrane space to an area of lower concentration in the matrix through a channel in the ATP synthase enzyme. This flow of protons down their concentration gradient through ATP synthase causes a shape change in the enzyme. This change in shape allows ATP synthase to catalyze the production of ATP.

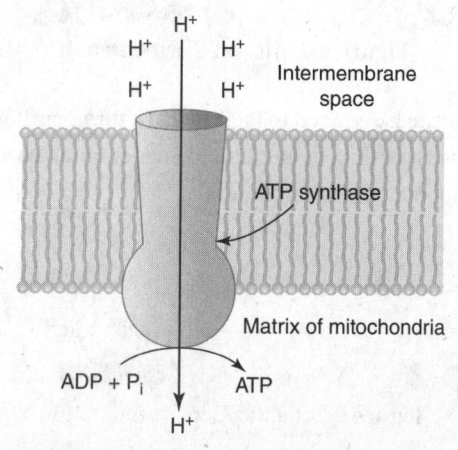

Figure 9.5 Chemiosmosis Produces ATP

Ideally, each NADH that enters the electron transport chain can generate as many as three ATPs. Each FADH$_2$ that enters the electron transport chain can generate as many as two ATPs; this is because FADH$_2$ has less potential energy than NADH and enters the electron transport chain at a later point than NADH. Using the totals from previous processes of cellular respiration (as presented in Table 9.5), the number of ATPs that can be generated by oxidative phosphorylation can be calculated as follows:

$$10 \text{ NADH} \times 3 \text{ ATP} = 30 \text{ ATP}$$

$$2 \text{ FADH}_2 \times 2 \text{ ATP} = 4 \text{ ATP}$$

Therefore, oxidative phosphorylation can generate 34 ATPs. Compare that to the net gain of two ATPs from glycolysis and the yield of two ATPs from the Krebs cycle to see how much more productive oxidative phosphorylation is with regard to ATP production! However, in living organisms, metabolic processes rarely work with 100% efficiency. For example, some membranes may be "leaky," and some protons may cross the inner membrane of the mitochondria without going through ATP synthase. So actual yields from this process may vary.

Fermentation

During oxidative phosphorylation, NADH is oxidized (at the electron transport chain) to NAD$^+$, which can then be returned to be used in glycolysis. However, if oxygen is not present, oxidative phosphorylation cannot occur. (Recall that oxygen is the final electron acceptor in the ETC. Without oxygen present, the ETC cannot release its low-energy electrons from the final carrier, blocking the chain and shutting the entire system down.) In **anaerobic** conditions, cells carry out **fermentation** to regenerate the NAD$^+$ needed to keep the process of glycolysis going. If a cell ran out of NAD$^+$, it could no longer perform glycolysis. All generation of ATP would stop, resulting in the death of the cell.

Fermentation only occurs in the cytosol. The two major types of fermentation you need to know are shown in Figures 9.6 and 9.7: alcohol fermentation and lactic acid fermentation.

In **alcohol fermentation**, pyruvate is reduced to an alcohol (typically the two-carbon alcohol ethanol) and carbon dioxide, and NADH is oxidized to NAD$^+$. An example of this is the fermentation that yeast undergoes as bread dough rises. As yeast ferments the sugars in the uncooked bread dough, alcohol and carbon dioxide are produced. The carbon dioxide gas causes the bread dough to "rise," and the alcohol is evaporated by high temperatures during the baking of the bread dough.

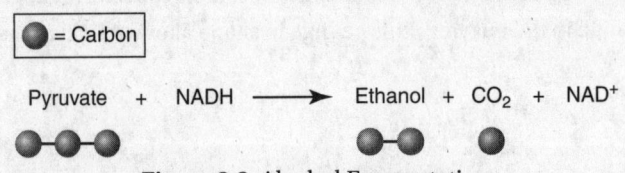

● = Carbon

Pyruvate + NADH ⟶ Ethanol + CO$_2$ + NAD$^+$

Figure 9.6 Alcohol Fermentation

In **lactic acid fermentation**, pyruvate is reduced to lactic acid (a three-carbon molecule), and NADH is oxidized to NAD$^+$. No carbon dioxide is produced. Lactic acid fermentation can occur in muscle cells if they do not have enough oxygen to carry out oxidative phosphorylation.

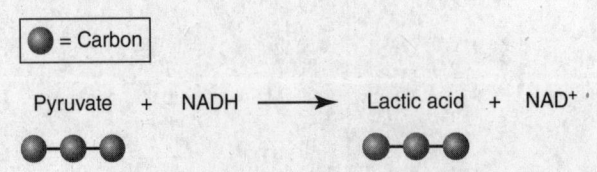

● = Carbon

Pyruvate + NADH ⟶ Lactic acid + NAD$^+$

Figure 9.7 Lactic Acid Fermentation

Practice Questions

Multiple-Choice

1. What is the primary purpose of fermentation?

   (A) to generate ATP
   (B) to generate pyruvate
   (C) to generate NAD$^+$
   (D) to generate carbon dioxide

2. Which of the following processes does NOT release carbon dioxide?

   (A) glycolysis
   (B) oxidation of pyruvate
   (C) Krebs cycle
   (D) alcohol fermentation

3. Which of the following statements about glycolysis is NOT correct?

   (A) Glycolysis occurs in the cytosol.
   (B) Glycolysis can be carried out by all living organisms.
   (C) The enzymes of glycolysis are highly conserved.
   (D) Glycolysis produces a proton gradient.

4. A mutation in mitochondrial DNA causes the creation of a pore in the mitochondrial membrane through which protons can freely pass. Which of the following processes would most likely be disrupted by this mutation?

   (A) glycolysis
   (B) Krebs cycle
   (C) chemiosmosis
   (D) fermentation

5. During oxidative phosphorylation, _____ is _____, and oxygen is _____.

   (A) NADH; oxidized; produced
   (B) NADH; oxidized; reduced
   (C) NAD$^+$; reduced; oxidized
   (D) NAD$^+$; reduced; produced

6. Which of the following processes produce ATP?

   (A) glycolysis, oxidation of pyruvate, Krebs cycle
   (B) glycolysis, Krebs cycle, chemiosmosis
   (C) glycolysis, Krebs cycle, fermentation
   (D) oxidation of pyruvate, Krebs cycle, fermentation

7. What is the role of oxygen in cellular respiration?

   (A) to combine with carbon and electrons to form carbon dioxide
   (B) to combine with protons and electrons to form water
   (C) to remove carbon from glucose to form pyruvate
   (D) to remove carbon from pyruvate to form alcohol

8. Where are the electron transport chain and ATP synthase located?

   (A) in the cytosol
   (B) on the outer membrane of the mitochondria
   (C) in the inner membrane of the mitochondria
   (D) in the mitochondrial matrix

9. Which of the following choices correctly describes the flow of electrons in cellular respiration?

   (A) glucose → Krebs cycle → oxygen → NAD$^+$
   (B) glucose → NAD$^+$ → electron transport chain → oxygen
   (C) glucose → electron transport chain → pyruvate → oxygen
   (D) glucose → NAD$^+$ → electron transport chain → carbon dioxide

10. Cyanide is a poison that blocks the movement of electrons down the electron transport chain. Which of the following would be the most immediate result of cyanide poisoning?

    (A) No pyruvate would be produced.
    (B) No NADH would form.
    (C) No ATP would be produced.
    (D) Chemiosmosis would not occur.

Short Free-Response

11. NAD⁺ has an important role as an electron carrier in glycolysis.

    Part A

    (i) **Identify** the process of *aerobic* cellular respiration during which a cell replenishes its supply of NAD⁺.

    Part B

    (i) **Identify** the process of *anaerobic* respiration during which a cell replenishes its supply of NAD⁺.

    Part C

    (i) If a cell's supply of NAD⁺ ran out, **predict** the effect that would have on glycolysis within that cell.

    Part D

    (i) **Justify** your prediction from Part C.

12. In eukaryotes, ATP synthase is found in the inner membrane of the mitochondria. In prokaryotes, which do not have mitochondria, it is found on infoldings of the cell membrane.

    Part A

    (i) **Make a claim** as to why ATP synthase is found on the cell membrane in prokaryotes.

    Part B

    (i) **Justify** your claim from Part A using your knowledge of how ATP synthase works.

    Part C

    (i) **Explain** why ATP synthase would not be effective if it was found in the cytosol of a cell.

    Part D

    (i) **Describe** how the structure of the mitochondria allows for the formation of a proton gradient.

Long Free-Response

13. *Saccharomyces cerevisiae* (also known as baker's yeast) is a eukaryotic organism. A solution containing 5% yeast and 10% glucose is placed in a transfer pipette. A weight is placed on the transfer pipette, and the apparatus is placed in water in a large test tube, as shown in the following figure.

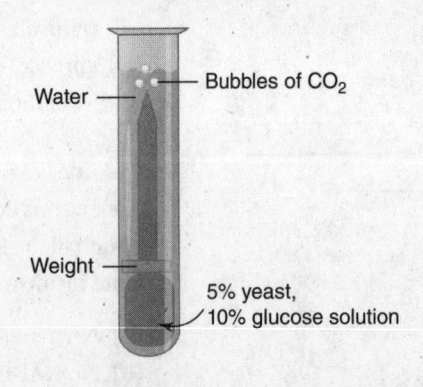

This process is repeated two more times, and each test tube is placed in a beaker of water at different temperatures (5° Celsius, 20° Celsius, and 35° Celsius). To measure the amount of cellular respiration, the cumulative number of bubbles of carbon dioxide released over a 10-minute period is recorded. The data are shown in the following table.

Temperature	Time (minutes)									
	1	2	3	4	5	6	7	8	9	10
5° Celsius	0	0	0	0	0	1	1	1	1	2
20° Celsius	0	0	2	2	3	4	4	5	6	6
35° Celsius	0	2	3	4	6	6	8	9	10	10

Part A

(i) **Identify** the independent variable in this experiment.

(ii) **Identify** the dependent variable in this experiment.

Part B

(i) On the axes provided, **construct** an appropriately labeled graph of the experimental data.

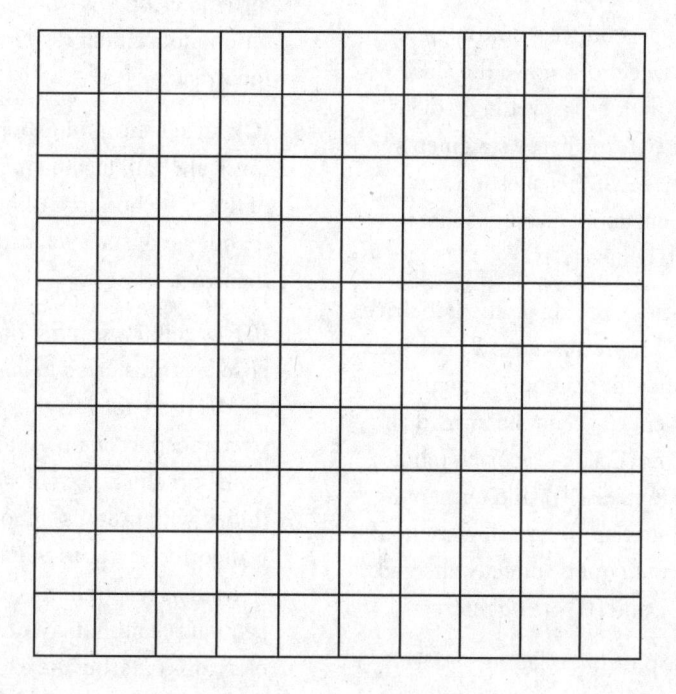

Part C

(i) **Calculate** the average rate of carbon dioxide production per minute at each temperature.

Part D

A fourth experimental apparatus is placed in a beaker of water at a temperature of 10° Celsius.

(i) Add a line to your graph from Part B that **predicts** the rate of cellular respiration at 10° Celsius.

(ii) **Justify** your prediction.

Answer Explanations
Multiple-Choice

1. **(C)** Fermentation oxidizes NADH into NAD^+ so that NAD^+ can be used in glycolysis. Choice (A) is incorrect because fermentation does not produce ATP. Pyruvate is a reactant used in fermentation, not a product of fermentation, so choice (B) is incorrect. While carbon dioxide is a product of alcohol fermentation, carbon dioxide is not always produced by fermentation (for example, lactic acid fermentation does not produce carbon dioxide). So choice (D) is incorrect.

2. **(A)** Glycolysis does not produce carbon dioxide. Choices (B), (C), and (D) all produce carbon dioxide, so all three of those choices are incorrect.

3. **(D)** Glycolysis does not produce a proton gradient; the movement of electrons down the electron transport chain produces a proton gradient. Choices (A), (B), and (C) are all true statements about glycolysis. Notice that this question was asking for the statement about glycolysis that is NOT correct, which is choice (D).

4. **(C)** Chemiosmosis uses a proton gradient to drive the production of ATP. If protons were able to pass freely through the inner mitochondrial membrane, a proton gradient could not be formed and chemiosmosis could not occur. Glycolysis (choice (A)), the Krebs cycle (choice (B)), and fermentation (choice (D)) do not require a proton gradient, so those processes would not be directly affected. Thus, choices (A), (B), and (D) are not correct.

5. **(B)** During oxidative phosphorylation, NADH is *oxidized* to NAD^+ and oxygen is *reduced* to water. Choice (A) is incorrect because oxygen is not produced in oxidative phosphorylation. Choice (C) is incorrect because, during oxidative phosphorylation, NAD^+ is not reduced and oxygen is not oxidized (it is reduced). Choice (D) is incorrect because NAD^+ is not reduced and oxygen is not produced during oxidative phosphorylation.

6. **(B)** The vast majority of ATP produced in cellular respiration is produced by chemiosmosis (approximately 34 ATPs per glucose molecule!), and glycolysis and the Krebs cycle each produce a net gain of 2 ATPs per glucose molecule. Choice (A) is incorrect because the oxidation of pyruvate does not produce any ATP. Fermentation does not produce ATP, so choices (C) and (D) are incorrect.

7. **(B)** Oxygen is the final (or terminal) electron acceptor in cellular respiration; oxygen combines with electrons and protons to form water. While carbon dioxide is a product of cellular respiration, oxygen does not combine with carbon during cellular respiration to produce carbon dioxide. Thus, choice (A) is incorrect. Choice (C) is incorrect because oxygen neither removes carbon from glucose nor does it form pyruvate during cellular respiration. Removing a carbon from pyruvate to form an alcohol is what happens during fermentation, not cellular respiration, so choice (D) is incorrect.

8. **(C)** The electron transport chain and ATP synthase are both located in the inner membrane of the mitochondria. Choices (A), (B), and (D) do not list the correct location and are therefore incorrect.

9. **(B)** In cellular respiration, the electrons in glucose are transferred to NAD^+ to form NADH. NADH then delivers these electrons to the electron transport chain. At the end of the electron transport chain, oxygen accepts these electrons (along with protons) to form water. Choice (A) is incorrect because the final electron acceptor in cellular respiration is oxygen, not NAD^+. In cellular respiration, pyruvate is formed prior to electrons entering the electron transport chain, so choice (C) is incorrect. Choice (D) is incorrect because carbon dioxide does not receive electrons during cellular respiration.

10. **(D)** If electrons could not travel down the electron transport chain, a proton gradient would not be produced and chemiosmosis could not occur. Choice (A) is incorrect because the production of pyruvate does not require the electron transport chain. Choice (B) is incorrect because the ETC regenerates NAD^+, not NADH. Limited amounts of ATP are produced in glycolysis and the Krebs cycle, so choice (C) is also incorrect.

Short Free-Response

11. (A-i) In aerobic cellular respiration, a cell's supply of NAD^+ is replenished during oxidative phosphorylation (when NADH delivers its electrons to the electron transport chain).

    (B-i) In anaerobic respiration, a cell's supply of NAD^+ is replenished during fermentation (when NADH donates its electrons to a pyruvate molecule).

    (C-i) If a cell's supply of NAD^+ ran out, the cell would not be able to undergo glycolysis.

    (D-i) Glycolysis requires the input of NAD^+; if there were no NAD^+ available, glycolysis would halt. Living organisms require the input of energy, so the cell would eventually die.

12. (A-i) The enzyme ATP synthase is found on the infoldings of the cell membrane in prokaryotes because the enzyme needs a proton gradient to function.

    (B-i) ATP synthase uses a proton gradient to drive the synthesis of ATP through chemiosmosis. To create a proton gradient, there must be a way to separate the protons. In eukaryotes, this separation occurs on either side of the inner mitochondrial membrane. Since prokaryotes do not have mitochondria, they separate protons across the only membrane they have, the cell membrane. So that is where ATP synthase is located in prokaryotes.

    (C-i) If ATP synthase was found in the cytosol of the cell, there would be no way to separate the protons and create a proton gradient in the cytosol. So ATP synthase could not function.

    (D-i) The mitochondria contain outer membranes and inner membranes. The electron transport chain is located on the inner membrane, and it pumps protons into the intermembrane space between the two membranes. ATP synthase, which is also located on the inner membrane, can then use this proton gradient to power the production of ATP.

Long Free-Response

13. (A-i) The independent variable is temperature.

    (A-ii) The dependent variable is the cumulative number of bubbles of carbon dioxide produced.

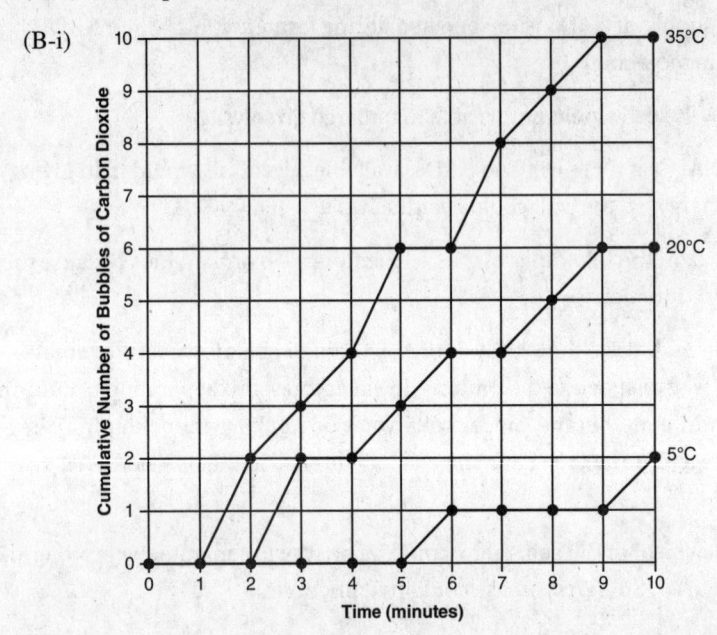

    (B-i)

    (C-i) Rate for 5°C = $\frac{2 \text{ bubbles}}{10 \text{ minutes}} = \frac{0.2 \text{ bubbles}}{\text{minute}}$

    Rate for 20°C = $\frac{6 \text{ bubbles}}{10 \text{ minutes}} = \frac{0.6 \text{ bubbles}}{\text{minute}}$

    Rate for 35°C = $\frac{10 \text{ bubbles}}{10 \text{ minutes}} = \frac{1.0 \text{ bubble}}{\text{minute}}$

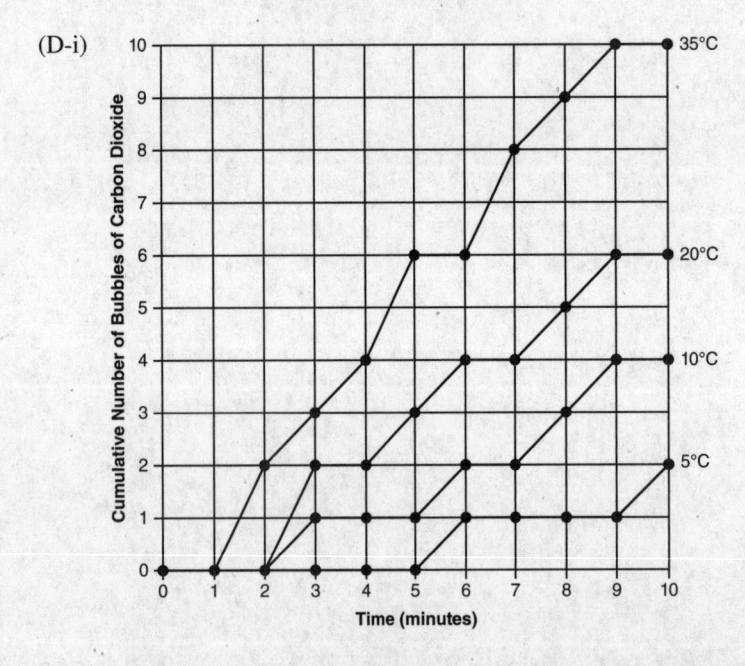

    (D-i)

    (D-ii) The line for 10°C should be between the lines for 5°C and 20°C (higher than the line for 5°C and lower than the line for 20°C). This is because the rate of cellular respiration at 10°C will be higher than the rate at the colder temperature of 5°C and lower than the rate at the warmer temperature of 20°C.

UNIT 4
Cell Communication and Cell Cycle

10
Cell Communication and Signaling

Learning Objectives

In this chapter, you will learn:
- → Types of Cell Signaling
- → Signal Transduction
- → Disruptions in Signal Transduction Pathways
- → Feedback Mechanisms

Overview

Biological systems interact, exchange information about their environments, and respond to this information. Life often depends on responding quickly to changing environmental conditions. This chapter reviews the basics of cell communication and the process of signal transduction.

Types of Cell Signaling

The survival of a living organism depends on the ability of its cell, or cells, to communicate by sending, receiving, and responding to chemical signals. These chemical signals are called **ligands**. Figures 10.1–10.4 show four general types of cell signaling:

1. **Autocrine signaling**—In autocrine signaling, the cell secretes a ligand. This ligand then binds to a receptor on the cell that secreted the ligand, triggering a response within that same cell. The root word *auto* means "self," so this process can be thought of as a cell signaling itself to generate a response. An example of this is a cancer cell, which releases its own growth hormones (the ligands) that stimulate the cancer cell to grow and divide.

> **NOTE**
> Many different types of molecules can function as ligands, including proteins, cholesterol, and ions, to name just a few.

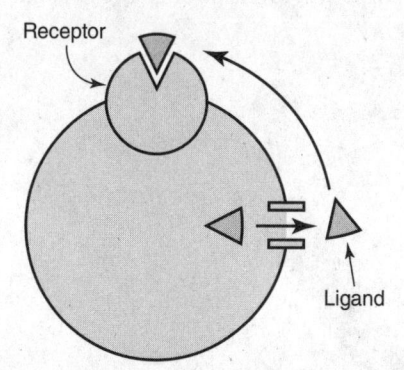

Figure 10.1 Autocrine Signaling

2. **Juxtacrine signaling**—This is signaling that depends on direct contact between the cell that is sending the ligand and the cell that is receiving and responding to it via a surface receptor. Examples of juxtacrine signaling include plasmodesmata in plants (which involve the ligand traveling between channels that connect

adjacent cells) and antigen-presenting cells in the human immune system (which signal helper T cells through direct cell-to-cell contact).

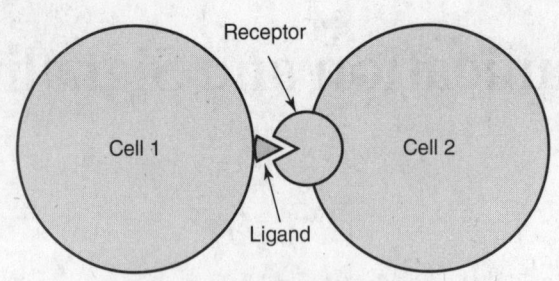

Figure 10.2 Juxtacrine Signaling

3. **Paracrine signaling**—In paracrine signaling, the cell secretes a ligand that travels a short distance, eliciting an effect on cells in the nearby area. These ligands are sometimes referred to as local regulators since they only affect cells in the immediate vicinity of the cell that is sending the signals. Neurotransmitters are local regulators that travel the short distance across a synapse to communicate with nearby cells.

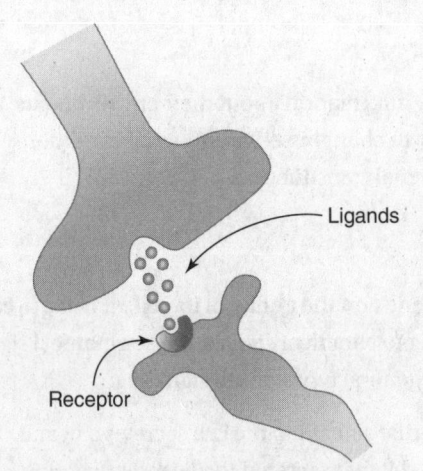

Figure 10.3 Paracrine Signaling

4. **Endocrine signaling**—Some ligands travel a long distance between the sending and receiving cells; this is called endocrine signaling. Ligands that travel a long distance are called **hormones**. Insulin, a hormone that is produced and released by the pancreas, travels through the circulatory system to trigger responses in cells all over the body.

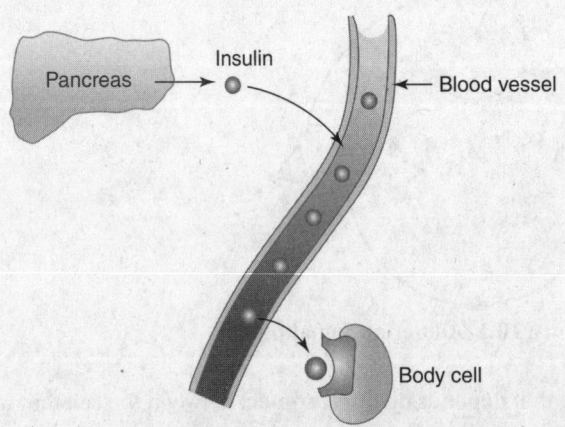

Figure 10.4 Endocrine Signaling

> There are many examples of endocrine signaling in everyday life. For example, the hormone ethylene can travel through the air to trigger the ripening of fruits. Another example is the release of hormones called pheromones into the air by some animals and moths, which help them locate and attract mates over long distances.

Signal Transduction

Signal transduction determines how a cell responds internally to a signal in its environment. Important processes, such as gene expression, cell growth and division, and the release of hormones, depend on signal transduction.

Signal transduction begins with a chemical message or **ligand**. Ligands interact with specific **target cells**, which respond to the presence of the ligand. Ligands may be hydrophilic or hydrophobic. Hydrophilic ligands cannot cross the phospholipid bilayer of the cell membrane and enter the cell. Consequently, hydrophilic ligands interact with receptors located on the cell membrane (**cell membrane receptors**), as shown in Figure 10.5. The binding of the ligand to the cell membrane receptor then triggers a series of chemical reactions inside the cell (the series of chemical reactions are shown as A → B, C → D, and E → F in Figure 10.5). Hydrophobic ligands may enter the cell by sliding between the phospholipids of the cell membrane. These hydrophobic ligands then bind to **intracellular receptors** in the cytosol of the cell, as shown in Figure 10.5. Once bound to the intracellular receptor, the ligand can then cross the nuclear membrane and bind to DNA in the nucleus, changing the expression of genes.

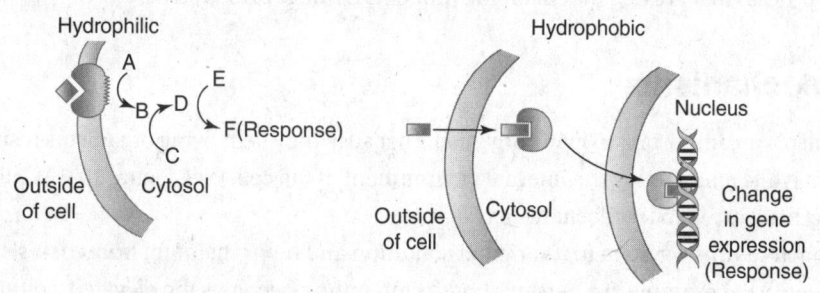

Figure 10.5 Hydrophilic Ligands vs. Hydrophobic Ligands

Signal transduction has three major steps:

1. **Reception:** The ligand binds to a specific receptor on or in the target cell. The receptor may be located on the cell membrane (as is the case for hydrophilic ligands) or in the cytosol of the target cell (as is the case for hydrophobic ligands). Receptors contain ligand-specific binding domains. If a cell does not have the receptor for a specific ligand, the cell will not respond to that ligand. Upon the binding of the ligand to the receptor, the receptor undergoes a conformational (shape) change, which triggers the next step in the process on the inside of the cell. Examples of receptors include G-protein-coupled receptors and receptor tyrosine kinases.

2. **Transduction:** This is the series of chemical reactions (triggered by the binding of the ligand to its receptor) that helps the cell choose the appropriate response. This is often the most complicated part of signal transduction. Possible components of transduction include:
   - Signaling cascades, a series of chemical reactions in which one molecule activates multiple molecules, amplifying the cell's response to a signal (a process called **signal amplification**)
   - **Kinases**, which can transfer phosphate groups to other molecules (which activates those molecules)
   - **Phosphatases**, which can remove phosphate groups from other molecules (which inactivates those molecules)
   - Enzymes, which produce **secondary messengers**; an example of this is the enzyme **adenylyl cyclase**, which produces the secondary messenger cyclic AMP (cAMP) from ATP

3. **Response:** This is the final step of signal transduction and the ultimate result generated by the ligand. Examples of cellular responses include the activation of genes by steroid hormones, the opening of ligand-gated ion channels, and the initiation of cell processes, such as apoptosis (programmed cell death).

Disruptions in Signal Transduction Pathways

Signal transduction pathways refer to the series of chemical reactions that mediate the sensing and processing of stimuli. Disruptions in signal transduction pathways can have profound effects on cells. Since receptors are specific to certain ligands, a mutation in a gene that is coding for a receptor protein could result in a change in shape of the receptor such that it would no longer bind to its specific ligand. Without a functional receptor for the ligand, the cell with the mutated receptor protein would no longer be able to respond to the ligand. Examples of disorders caused by mutations in receptor proteins include androgen insensitivity syndrome (AIS), in which the receptor for testosterone is nonfunctional in gonadal tissue (causing it not to form into gonads during embryonic development), and nephrogenic diabetes insipidus (NDI), in which portions of the structure of the kidneys are insensitive to antidiuretic hormone (ADH) and urine production is affected.

Signal transduction pathways may also be disrupted when molecules in the environment interfere with a ligand's ability to bind to its receptor. For example, the cholera toxin binds to G-protein-coupled receptors in the cell membrane, leading to disruptions in a cell signaling pathway that can cause life-threatening dehydration.

Mutations in the gene for adenylyl cyclase can interfere with a cell's ability to produce the secondary messenger cAMP, disrupting all steps in the signal transduction process that are dependent on that secondary messenger. A disruption to any step in the signal transduction process will affect not only that step but will also affect any subsequent steps in the process that are dependent on the products of the previous steps.

Feedback Mechanisms

Feedback mechanisms are important to living organisms because they help living organisms respond to changes in the environment while maintaining the internal environment of the cell. (See Figure 10.6.) Cell communication and signaling are crucial in feedback mechanisms.

Negative feedback returns a system to its original condition and helps maintain **homeostasis** (the maintenance of a stable state). For example, if a person's body temperature becomes too elevated, cell signaling processes will trigger skin cells to release sweat, which will cool the body and help return it to its normal body temperature. Another example of negative feedback is the control of blood glucose levels by insulin and glucagon. After a sugary snack, if blood sugar levels get too high, the pancreas releases the ligand (hormone) insulin. Insulin binding to its receptor triggers a series of chemical reactions. This causes body cells to absorb glucose from the blood, returning blood glucose levels to the normal range. Conversely, if blood sugar levels get too low, the pancreas releases the ligand (hormone) glucagon, which stimulates liver cells to break down glycogen into glucose, releasing glucose into the blood and again returning blood glucose levels to the normal range. Since insulin and glucagon travel long distances in the bloodstream, this is an example of endocrine signaling.

Positive feedback magnifies cell processes. For example, the hormone oxytocin stimulates contractions of the uterine muscles in labor contractions during childbirth. The contraction of the uterine muscles triggers the production of even more oxytocin, which in turn increases the contractions of the uterine muscles. This positive feedback process causes labor contractions to amplify, getting stronger and stronger during childbirth.

> **NOTE**
>
> Positive feedback and negative feedback are NOT the same! Positive feedback increases the deviation from homeostasis; negative feedback returns a system to homeostasis.

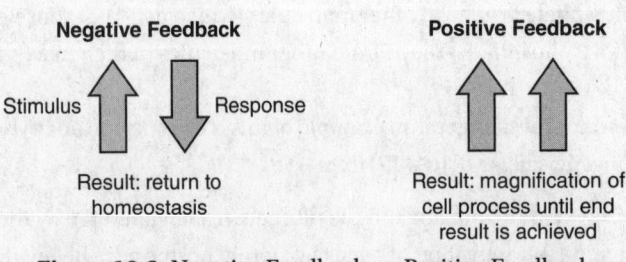

Figure 10.6 Negative Feedback vs. Positive Feedback

Practice Questions

Multiple-Choice

1. Auxin is a plant hormone that triggers cell division. A mutation occurs that deletes the gene for the auxin receptor. Which of the following is the most likely result of this mutation?

   (A) The cells will still divide but at a faster rate.
   (B) The cells will not be able to divide.
   (C) The cells will develop a new receptor for the signaling molecule.
   (D) The cells will not be affected by the lack of the auxin receptor.

2. Cortisol is a hormone produced in response to stress. Hunger is a stressor that can increase cortisol levels. Which of the following is most likely an effect of increased cortisol levels in response to hunger?

   (A) increased activation of the immune response
   (B) increased storage of calcium in bones
   (C) increased reabsorption of water by the kidneys
   (D) increased hydrolysis of glycogen to glucose

3. How do small ligands move between plant cells?

   (A) Receptor proteins on plant cell membranes transport the ligands.
   (B) Ligands pass through plasmodesmata that connect plant cells.
   (C) Ligands can pass through the plant cell membrane unassisted.
   (D) Active transport transfers ligands across the plant cell wall.

4. How do growth factors stimulate cell division?

   (A) Growth factors bind to multiple cells, grouping them in multicellular structures.
   (B) Growth factors bind to cyclic AMP, removing it from the cytosol.
   (C) Growth factors bind to cell membrane receptors, triggering a signal transduction pathway.
   (D) Growth factors bind to cell membrane phospholipids, which results in increased cell division.

5. In some autoimmune disorders, the body produces antibodies that bind to cell surface receptors on target cells, blocking them from interacting with other molecules. Which of the following is the most likely effect of the binding of the antibodies to the receptors?

   (A) There will be increased stimulation of the target cell.
   (B) The target cell will not be able to respond to ligands.
   (C) There will be no effect on the target cell.
   (D) There will be stimulation of gene expression in the target cell.

6. Which of the following best describes the roles of calcium ions and cyclic AMP in the signal transduction process?

   (A) They act as ligands.
   (B) They act as receptor proteins.
   (C) They act as secondary messengers.
   (D) They act as protein kinases.

7. Which of the following removes phosphate groups from other molecules?

   (A) cyclic AMP
   (B) protein kinase
   (C) protein phosphatase
   (D) adenylyl cyclase

8. The hormone insulin travels through the circulatory system to reach target cells. Insulin is involved in which type of cell signaling?

   (A) autocrine signaling
   (B) juxtacrine signaling
   (C) paracrine signaling
   (D) endocrine signaling

9. Neurotransmitters travel short distances across synapses. This is an example of which type of signaling?

   (A) autocrine signaling
   (B) juxtacrine signaling
   (C) paracrine signaling
   (D) endocrine signaling

10. Which of the following is the most likely reason why some cells do not respond to certain ligands?

    (A) Nonresponsive cells lack cyclic AMP.
    (B) Nonresponsive cells lack receptors for the ligand.
    (C) Nonresponsive cells lack the gene for the ligand.
    (D) Nonresponsive cells cannot metabolize the ligand.

Short Free-Response

11. The ligands in signal transduction pathways may be hydrophobic or hydrophilic.

    Part A

    (i) **Describe** where in a cell the receptors for hydrophilic ligands and the receptors for hydrophobic ligands are located.

    Part B

    (i) **Explain** why small hydrophobic ligands can cross the cell membrane unassisted.

    Part C

    A mutation in the gene for adenylyl cyclase renders the enzyme ineffective.

    (i) **Predict** the effect this would have on the cell.

    Part D

    (i) **Justify** your prediction from Part C.

12. An experiment is performed to measure how different cell types respond to a hormone that causes cells to absorb glucose from their environment. Four different cell types (A, B, C, and D) are used. At the start of this experiment, there are eight Petri dishes, each of which contains 100 millimolar glucose solution. Two dishes each contain cell type A, two dishes contain cell type B, two dishes contain cell type C, and the final two dishes contain cell type D, as illustrated in the following figure.

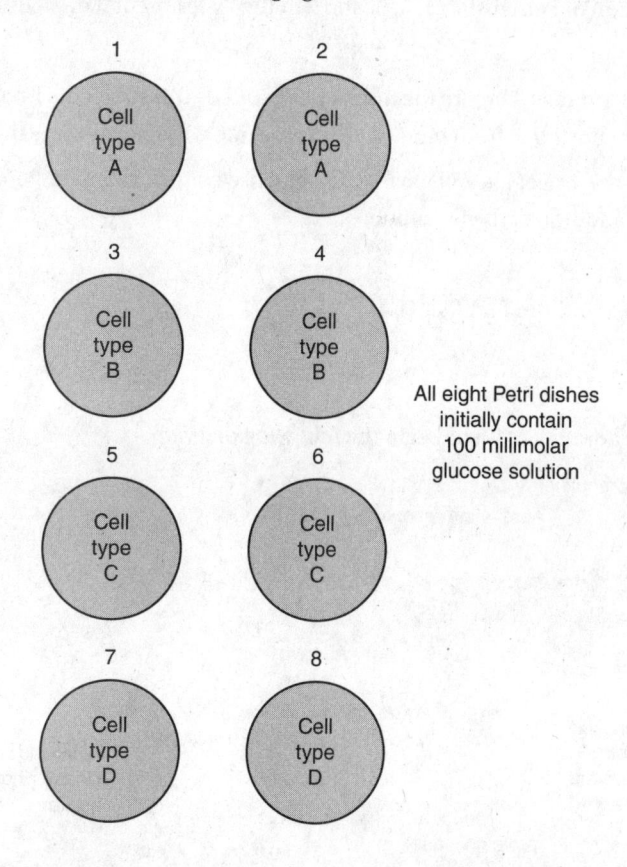

All eight Petri dishes initially contain 100 millimolar glucose solution

The hormone is added to four of the Petri dishes, one of each cell type. Glucose levels are measured in all eight Petri dishes 30 minutes after the addition of the hormone, and those glucose levels are listed in the following table.

Glucose Levels in Four Cell Types								
Petri Dish	1	2	3	4	5	6	7	8
Cell Type	A	A	B	B	C	C	D	D
Hormone Present (+) or Absent (0)	0	+	0	+	0	+	0	+
Glucose Levels 30 Minutes After the Addition of the Hormone (in millimolars)	100	50	100	50	100	100	100	100

Part A

(i) **Describe** the most likely reason why cell types C and D did not respond to the presence of the hormone.

Part B

(i) **Identify** the function of Petri dishes 1, 3, 5, and 7 in the experimental procedure.

Part C

A molecule is added to Petri dish 4 before the hormone is added. This molecule irreversibly binds to this hormone, preventing the hormone from binding to any receptor. The hormone is then added to Petri dish 4.

(i) **Predict** the effect this molecule will have on the glucose concentration in Petri dish 4 at 30 minutes after the addition of the hormone.

Part D

(i) **Justify** your prediction from Part C.

Long Free-Response

13. A signaling pathway for adrenaline is shown in the following diagram.

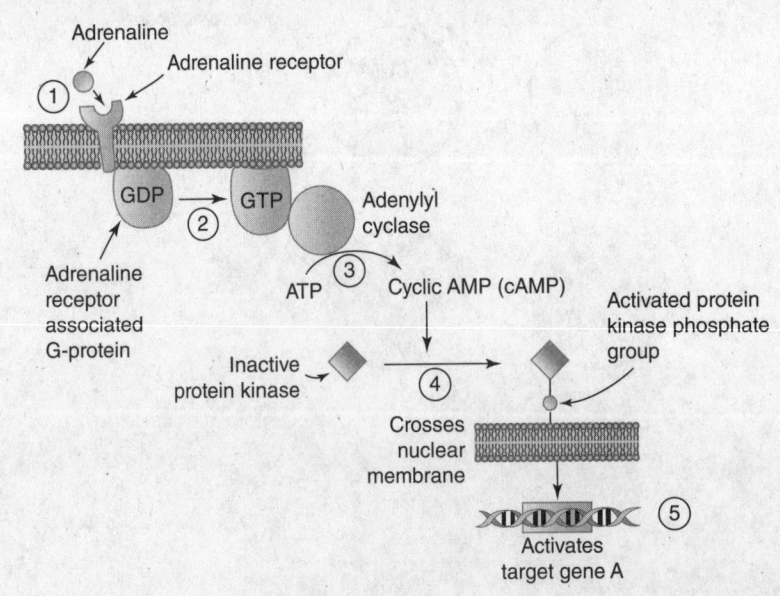

Researchers wanted to measure the effects of three different molecules on the adrenaline signaling pathway. Data are shown in the following table.

	Adrenaline (Alone)	Adrenaline (plus Molecule X)	Adrenaline (plus Molecule Y)	Adrenaline (plus Molecule Z)
G-Protein Activation	Yes	Yes	No	Yes
cAMP Production	Yes	No	No	Yes
Activation of Transcription of Target Gene A	Yes	No	No	No

Part A

(i) **Describe** the parts of the cell signaling process that are represented in each of the five numbered steps in the diagram.

Part B

(i) **Identify** the control in this experimental procedure.

(ii) **Identify** the independent variable in this experimental procedure.

(iii) **Identify** the dependent variables in this experimental procedure.

Part C

(i) **Make a claim** about which molecule(s) most likely interfere(s) with the enzyme adenylyl cyclase.

(ii) **Support your claim** with evidence from the experimental data.

Part D

Cyclic AMP-dependent phosphodiesterase is an enzyme that breaks down cyclic AMP.

(i) **Predict** the numbered step in the diagram that would be most directly affected by adding cAMP-dependent phosphodiesterase to the cell.

(ii) **Justify** your prediction.

Answer Explanations

Multiple-Choice

1. **(B)** If the target cell lacks receptors for the signaling molecule, it will not be able to initiate the signal transduction pathway that triggers cell division. Choice (A) is incorrect because the cells will not divide if they lack the appropriate receptor for the ligand; the signaling pathway for cell division will not be triggered. Choice (C) is incorrect because individual cells do not evolve new receptors. If the cell lacks the receptors for the signaling molecule, the cell will not be able to respond to the signal, so choice (D) is incorrect.

2. **(D)** Cortisol would stimulate the breakdown of glycogen into glucose. This would provide energy, alleviating one of the symptoms of hunger. Choices (A), (B), and (C) are incorrect because the activation of the immune response, increased storage of calcium in bones, and increased reabsorption of water by the kidneys would not increase the energy available to cells.

3. **(B)** Small molecules pass between plant cells through channels called plasmodesmata, an example of juxtacrine signaling. Choices (A) and (C) are incorrect because plant cells have a cell wall surrounding the cell membrane. So molecules cannot pass through plant cell membranes unassisted nor can they bind to cell membrane receptors unless they can pass through the cell wall first. Choice (D) is incorrect because not all ligands require active transport.

4. **(C)** Growth factors bind to receptors, which triggers a signal transduction process that stimulates cell division. Grouping cells together would not trigger cell division, so choice (A) is incorrect. Removing cyclic AMP from the cytosol would most likely disrupt the signaling process that is required to trigger cell division, so choice (B) is incorrect. Choice (D) is incorrect because growth factors bind to receptor proteins, not to membrane phospholipids.

5. **(B)** If an antibody binds to the receptor, the ligand will be prevented from binding to the receptor, and the signaling process will not occur. Choice (A) is incorrect because the result will be decreased, not increased, stimulation of the target cell. The effect of the antibody will be decreased stimulation of the target cell, so choice (C) is also incorrect. Since the antibody is preventing the ligand from binding to the receptor, no stimulation of gene expression could be triggered, making choice (D) incorrect as well.

6. **(C)** Both calcium ions and cyclic AMP function as secondary messengers during cell signaling. Ligands bind to the receptor to start the cell signaling process; neither calcium ions nor cAMP do this, so choice (A) is incorrect. Receptor proteins and protein kinases are proteins, but neither calcium ions nor cAMP are proteins. So choices (B) and (D) are incorrect.

7. **(C)** Protein phosphatases remove phosphate groups from other molecules. Choice (A) is incorrect because cyclic AMP functions as a secondary messenger. Protein kinases add phosphate groups to molecules, so choice (B) is incorrect. Choice (D) is incorrect because adenylyl cyclase is the enzyme that catalyzes the formation of cAMP.

8. **(D)** Insulin travels long distances through the circulatory system, so this is an example of endocrine signaling. Choice (A) is incorrect because autocrine signaling involves a cell that produces a ligand, and the ligand then binds to a receptor on the same cell that produced the ligand. Juxtacrine signaling is between cells that are in direct contact, so choice (B) is incorrect. Choice (C) is incorrect because paracrine signaling occurs over short distances.

9. **(C)** Signaling over short distances is paracrine signaling. Choice (A) is incorrect because autocrine signaling involves a cell that produces a ligand and the ligand then binds to a receptor on the same cell that produced the ligand. Juxtacrine signaling is between cells that are in direct contact, so choice (B) is incorrect. Endocrine signaling occurs over long distances, so choice (D) is incorrect.

10. **(B)** If a cell lacks the receptor for a ligand, the signal transduction process cannot start. Choice (A) is incorrect because if cAMP was lacking, the

cell might still be able to mount an incomplete response if the ligand could bind to a receptor and trigger steps in the signal transduction process that do not depend on the presence of cAMP. Ligands are usually not produced in their target cells, so lacking a gene for the ligand would likely have no effect on the target cell, making choice (C) incorrect. Choice (D) is incorrect because the inability to metabolize a ligand would likely not affect the target cell's ability to respond to the ligand.

Short Free-Response

11. (A-i) Receptors for hydrophilic ligands are found on the cell membrane. Receptors for hydrophobic ligands are found in the cytosol.

    (B-i) The cell membrane is made of a phospholipid bilayer. Hydrophobic ligands can slide between the hydrophobic phospholipids of the bilayer and enter the cell.

    (C-i) A mutation in the enzyme adenylyl cyclase (that rendered it ineffective) would result in a cell that could not make the secondary messenger cyclic AMP. This would disrupt any cell signaling process that contains a step that requires cyclic AMP.

    (D-i) Adenylyl cyclase catalyzes the conversion of ATP into cAMP, so an ineffective adenylyl cyclase would result in no cAMP production.

12. (A-i) Cell types C and D did not respond to the presence of the hormone because they most likely did not have the receptor for the hormone.

    (B-i) Petri dishes 1, 3, 5, and 7 are the experimental controls for each of the four cell types in the experiment.

    (C-i) If a molecule that irreversibly binds to this hormone was added to Petri dish 4 before the hormone was added, the level of glucose in the Petri dish would most likely remain unchanged after 30 minutes.

    (D-i) The irreversible binding of the molecule to the hormone would prevent the hormone from binding to its receptor in the cell. Without the binding of the hormone/ligand, the signal to absorb glucose from the Petri dish would not occur, and the glucose level in the Petri dish would remain unchanged.

Long Free-Response

13. (A-i) Step 1 represents reception of the signal. Steps 2, 3, and 4 represent transduction. Step 5 represents the response.

    (B-i) The control is the cell in which adrenaline alone has been added.

    (B-ii) The independent variable is the presence or absence of molecules X, Y, or Z.

    (B-iii) The dependent variables are the activation of the G-protein, the production of cAMP, and the activation of transcription of target gene A.

    (C-i and ii) Molecule X and molecule Y most likely interfere with adenylyl cyclase because no cAMP was produced when those molecules were added to the cells. Molecule Y also most likely interferes with the GDP on the G-protein receptor since it was never activated.

    (D-i and ii) If cAMP-dependent phosphodiesterase was added to the cell, the conversion of the inactive protein kinase to activated protein kinase (step 4) would most likely be affected since that step is dependent on the presence of cAMP.

11

The Cell Cycle

> **Learning Objectives**
>
> In this chapter, you will learn:
> → Phases of the Cell Cycle
> → Regulation of the Cell Cycle, Cancer, and Apoptosis

Overview

The cell cycle is important in the growth, repair, and reproduction of cells in living organisms. Controlling the rate of the cell cycle ensures that these processes occur in a timely manner while also preventing the development of uncontrolled cell growth or tumors. This chapter will review the phases of the cell cycle and the factors that control the rate of the cell cycle.

Phases of the Cell Cycle

Figure 11.1 shows the phases of the cell cycle: **interphase** (G1, S, and G2), **mitosis** (also known as the M phase), and **cytokinesis**. Nondividing cells will leave the cell cycle and enter a stage called G0.

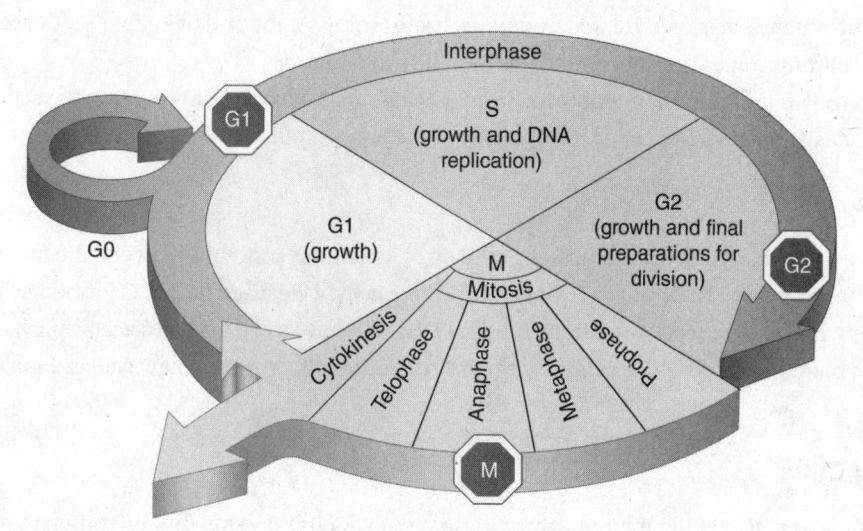

Figure 11.1 The Cell Cycle

Interphase

Interphase is the longest phase of the cell cycle. During interphase, the cell grows so that it has enough material to divide between two daughter cells. The cell also replicates its genetic material (DNA) during interphase. Interphase consists of the following three sequential stages:

1. **G1**—During this stage, the cell grows and prepares for the replication of DNA, and some cellular organelles (such as centrioles) are replicated.
2. **S**—During the S (synthesis) stage, DNA is replicated. When the S stage begins, each chromosome consists of one chromatid. After DNA replication is completed, each chromosome has two identical chromatids held together by one centromere. At the end of the S stage, the cell contains twice the amount of DNA it had at the end of G1, but the same number of chromosomes.
3. **G2**—During the G2 stage, the cell continues to grow and prepares the materials needed for mitosis, such as the proteins that will make up the spindle fibers.

Mitosis (M Phase)

The goal of mitosis is to make sure there is an accurate transfer of a parent cell's complete genome to each of the two resulting daughter cells. Mitosis consists of four stages: prophase, metaphase, anaphase, and telophase.

- In **prophase**, the nuclear membrane dissolves and the chromosomes condense and become visible. Spindle fibers also begin to form, and centrosomes move to opposite poles of the cell.
- In **metaphase**, the spindle fibers have fully attached to the centromeres of each chromosome. Chromosomes are then aligned along the "equator" of the cell in a single column. The center of the mitotic spindle is called the metaphase plate.
- During **anaphase**, each chromosome splits at its centromere as opposing spindle fibers begin to shorten. The identical chromatids are pulled toward opposite ends of the cell. At this point, each chromatid now has its own centromere and is considered a separate chromosome. At the end of anaphase, the cell has twice the number of chromosomes that it had at the start of the cell cycle.
- In **telophase**, two new nuclear membranes form. Each of the two nuclei now contain the same number of chromosomes and the same genetic information as the parent cell.

Cytokinesis

Cytokinesis is the division of the cytoplasm, along with all of its cellular contents, between the two daughter cells. Cytokinesis occurs after mitosis. During cytokinesis in animal cells, a cleavage furrow is formed, which partitions the cytosol and its contents between the two new cells. Plant cells have a cell wall and accomplish cytokinesis differently. In plant cells, a cell plate is built within the dividing cell, providing new cell wall material for each daughter cell.

Nondividing Cells

Some cells may stop dividing either temporarily or permanently. Cells may stop dividing when they reach their mature, fully differentiated state or when environmental conditions are not favorable for continued growth. These nondividing cells have exited the cell cycle and are in **G0**. Cells may enter G0 at any point in the cell cycle and may reenter the cell cycle if stimulated to do so by appropriate molecular signals.

Regulation of the Cell Cycle, Cancer, and Apoptosis

Proper regulation of the cell cycle is critical to appropriate growth, repair, and reproduction of cells in living organisms. This regulation is achieved through the use of **checkpoints** during the cell cycle. Some of these checkpoints are controlled by the interactions between **cyclins** and **cyclin-dependent kinases**. Cyclin-dependent kinases are present at constant levels throughout the cell cycle. These kinases add phosphate groups to other molecules, activating those molecules. However, cyclin-dependent kinases themselves are inactive until they are bound to cyclin proteins. The levels of cyclin proteins vary during the cell cycle, reaching their maximum just before mitosis starts. When cyclins are bound to cyclin-dependent kinases, a complex called **mitosis-promoting factor (MPF)** is formed. MPF triggers the start of mitosis. See Figure 11.2.

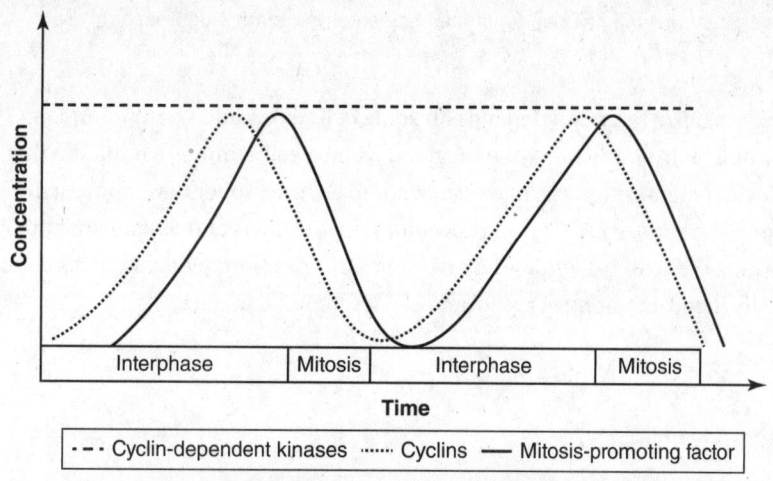

Figure 11.2 Levels of Cyclins, Cyclin-Dependent Kinases, and Mitosis-Promoting Factor in the Cell Cycle

All of the cells in an organism that are not involved with sexual reproduction are referred to as somatic body cells. The division of somatic cells, like those in tissues, can also be regulated by **density-dependent inhibition**. When cells in tissues, for example, become too crowded, they will stop dividing. Many types of somatic cells also exhibit **anchorage dependence**, which is when cells need to be attached to a surface in order to divide. Cancer cells are not regulated by density-dependent inhibition nor anchorage dependence and can continue to grow and divide under conditions that would cause normally functioning somatic cells to stop dividing.

Many genes are also involved in the regulation of the cell cycle. **Proto-oncogenes** propel cell division at a specific rate, much as an accelerator propels a car. Proto-oncogenes are necessary for regulated and controlled cell growth. If mutated, proto-oncogenes may become **oncogenes**, which promote abnormally high rates of cell division. An oncogene acts in a similar way to how an accelerator stuck in the down position would cause a car to go too fast. These oncogenes can cause tumors to form when cell division occurs too quickly and too often without regard for the neighboring cells. A mutation in a single allele of a proto-oncogene can cause a cell to grow out of control and can cause a tumor to form. Since a mutation in a single allele can cause a cell to grow out of control, proto-oncogenes are said to function in a dominant way.

Tumor suppressor genes code for proteins that detect mutations in cells that may cause tumors to develop. Tumor suppressor genes function much like the brakes on a car, preventing cell division from occurring at an abnormally fast rate. If a single mutation in a tumor suppressor gene allele occurs, the cell will still possess one remaining unmutated tumor suppressor allele that is functional. The tumor suppressor allele that is not mutated will help the organism identify cells that are dividing at a rate that is too fast. However, if both alleles of a tumor suppressor gene are mutated, the growth of a tumor may occur. Tumor suppressor genes are said to function in a recessive way because both alleles of a tumor suppressor gene must be nonfunctional for a cell to grow out of control.

> **TIP**
> While it IS important to know that cyclins and cyclin-dependent kinases work together to regulate the cell cycle, you do NOT need to know the names of specific cyclin/cyclin-dependent kinase pairs.

> BRCA (sometimes called the "breast cancer gene") is a mutation in a tumor suppressor gene. If a person has a BRCA mutation, he or she has a greatly increased risk of getting certain types of cancer compared to someone without the mutation. A person with the mutation in one allele would still have one functional tumor suppressor allele, so the occurrence of cancer would not be a certainty, but it would be a more likely occurrence than in a person without the mutation who has two functional copies of the tumor suppressor allele.

Sometimes a living organism's survival depends on some cells dying and not reproducing. This programmed cell death is called **apoptosis**. Apoptosis may be triggered when a cell acquires a mutation that could cause cancer. During embryonic development, apoptosis may also occur to ensure proper development of various organs or structures. For example, during early embryonic development, the digits of the hand are initially attached with a weblike structure. During the sixth to eighth weeks of embryonic development, apoptosis eliminates the webbing between the digits of the hand, resulting in the formation of separated fingers.

Practice Questions

Multiple-Choice

1. A chemotherapy drug stops the replication of DNA. During which stage of the cell cycle would this drug have the greatest effect?

   (A) G0
   (B) G1
   (C) G2
   (D) S

2. A cell has reached full maturity, is fully differentiated, and no longer divides. Which of the following describes this stage of the cell cycle?

   (A) G0
   (B) G1
   (C) G2
   (D) S

3. Which stage of mitosis has twice as many chromosomes at its end as it had at its beginning?

   (A) prophase
   (B) metaphase
   (C) anaphase
   (D) telophase

4. During which stage of mitosis do the centromeres of chromosomes attach to spindle fibers and line up in a single column in the center of the cell?

   (A) prophase
   (B) metaphase
   (C) anaphase
   (D) telophase

5. A researcher measures the number of cells in each phase of the cell cycle. The data are shown in the following table. Approximately, what percentage of these cells are in interphase?

Phase of Cell Cycle	Number of Cells
G1	73
S	89
G2	68
M	43

   (A) 16%
   (B) 33%
   (C) 52%
   (D) 84%

6. Which process best describes what occurs when soft tissue between the fingers dies during embryonic development?

   (A) apoptosis
   (B) DNA replication
   (C) mitosis
   (D) cytokinesis

7. During which process are the cytoplasm (and its cellular contents) divided between daughter cells?

   (A) apoptosis
   (B) DNA replication
   (C) mitosis
   (D) cytokinesis

8. Which of the following is a correct statement?

   (A) Cancer cells exhibit anchorage dependence.
   (B) Somatic cells exhibit density-dependent inhibition.
   (C) Somatic cells spend most of their time in mitosis.
   (D) Cancer cells exhibit density-dependent inhibition.

9. Which of the following best describes a role of mitosis?

   (A) Mitosis distributes cytosol between the daughter cells.
   (B) Replication of DNA occurs during mitosis.
   (C) Replication of cell organelles occurs during mitosis.
   (D) Mitosis distributes genetic material between the daughter cells.

10. During which phase of the cell cycle is a cleavage furrow or cell plate formed?

    (A) interphase
    (B) mitosis
    (C) cytokinesis
    (D) G0

Short Free-Response

11. Lectin is a molecule commonly found in legumes. Colchicine is a molecule that inhibits the formation of spindle fibers. An experiment was performed to observe the effects of lectin and colchicine on dividing cells. A total of 300 cells were observed for each treatment. Partial results of the experiment are shown in the following table.

	Control Cells	Lectin Added	Colchicine Added	Lectin and Colchicine Added
Number of Cells in Mitosis	75	195	?	80
Number of Cells in Interphase	225	105	?	220

Part A
  (i) **Describe** the effect of adding lectin to the cells.

Part B
  (i) **Identify** the independent variable in this experiment.
  (ii) **Identify** the dependent variable in this experiment.

Part C
  (i) Based on the data presented, **predict** the most likely effects of adding only colchicine to the cells.

Part D
  (i) **Justify** your prediction from Part C.

12. The following figure represents the concentrations of two different molecules (cyclin and cyclin-dependent kinase) during the cell cycle, and the arrow indicates the start of mitosis.

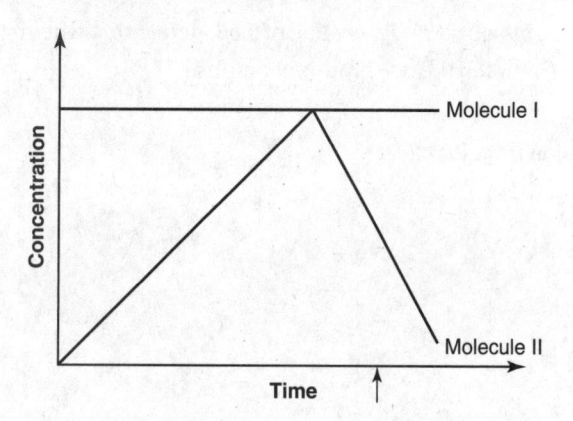

Part A
Cell division can be described as having three major events: replication of chromosomes, alignment of chromosomes, and separation of chromosomes.
  (i) **Describe** the stages of the cell cycle during which each of these three major events occurs.

Part B
  (i) **Identify** which molecule (cyclin or cyclin-dependent kinase) is represented by the line labeled "Molecule I" in the figure.
  (ii) **Identify** which molecule (cyclin or cyclin-dependent kinase) is represented by the line labeled "Molecule II" in the figure.

Part C

(i) **Explain** why mitosis starts at the point indicated by the arrow in the figure.

Part D

A cancer cell has a mutation that results in the constant high production of Molecule II.

(i) **Explain** how the process of mitosis might be affected by this mutation.

Long Free-Response

13. A student wants to test whether treating onion root cells with caffeine will significantly increase the number of cells in mitosis. Two groups of onions are grown: one group received distilled water, and the other received distilled water with caffeine. The root tips of both groups are harvested and stained, and the number of cells in interphase and mitosis in each group is counted. The expected and observed data are shown in the following table.

	Expected		Observed	
	Interphase	Mitosis	Interphase	Mitosis
Distilled Water	156	69	175	50
Distilled Water with Caffeine	169	76	150	95

Part A

(i) **State** the null hypothesis for this experiment.

Part B

(i) **Identify** the independent variable in this experiment.

(ii) **Identify** the dependent variable in this experiment.

Part C

(i) Using the formula for chi-square and a p-value of 0.05, **determine** whether there is a statistically significant difference between the two groups of onions.

Part D

(i) **Justify** your conclusion from Part C.

Answer Explanations

Multiple-Choice

1. **(D)** DNA is replicated during the S stage of the cell cycle, so a drug that blocks the replication of DNA would have the greatest effect during the S stage. DNA does not replicate during G0, G1, or G2, so choices (A), (B), and (C) are incorrect.

2. **(A)** During G0, the cell is not dividing and has exited the cell cycle. G1, G2, and S are all stages that prepare the cell to divide, so choices (B), (C), and (D) are incorrect.

3. **(C)** At the beginning of anaphase, each chromosome consists of two sister chromatids. During anaphase, these chromatids separate, and at the end of anaphase, each of these chromatids is considered a separate chromosome. Thus, anaphase ends with twice as many chromosomes as it had at its beginning. During prophase, chromosomes condense and become visible, but the number of chromosomes does not change. So choice (A) is incorrect. Choice (B) is incorrect because in metaphase, chromosomes line up along the center of the cell but the number of chromosomes does not change. In telophase, two new nuclei are formed and the number of chromosomes per cell at the end of telophase is less than it was at the beginning of telophase. Thus, choice (D) is also incorrect.

4. **(B)** During metaphase, chromosomes line up in the center of the cell on the metaphase plate. Prophase is when chromosomes condense and become visible, so choice (A) is incorrect. Anaphase is when sister chromatids separate and begin to move to opposite ends of the cell, so choice (C) is incorrect. Telophase is when two new nuclei are formed, so choice (D) is also incorrect.

5. **(D)** Interphase consists of G1, S, and G2. Therefore, the number of cells in interphase is $73 + 89 + 68 = 230$ cells. The total number of cells is $73 + 89 + 68 + 43 = 273$. So the percentage of cells in interphase is $\frac{230}{273} \approx 84\%$. Choice (A) is incorrect because 16% is the approximate percentage of cells that are in mitosis. Choice (B) is incorrect because 33% is the approximate percentage of cells that are in the S stage. Choice (C) is incorrect because 52% is the approximate percentage of cells in G1 and G2.

6. **(A)** Apoptosis is programmed cell death, and it can occur during embryonic development when the soft tissue between the fingers dies to allow the fingers to separate. While choices (B), (C), and (D) also occur during embryonic development, choice (A) is the best answer because it is the most specific to the formation of the fingers during embryonic development.

7. **(D)** Cytokinesis is the division of the cytoplasm. Choice (A) is incorrect because apoptosis is programmed cell death. DNA replication occurs during the S stage of the cell cycle, so choice (B) is incorrect. Choice (C) is also incorrect because mitosis is the process of evenly distributing replicated chromosomes between the daughter nuclei.

8. **(B)** Many somatic cells in organs and tissues exhibit density-dependent inhibition. Cancer cells do not exhibit anchorage dependence nor do they exhibit density-dependent inhibition, so choices (A) and (D) are incorrect. Choice (C) is incorrect because noncancerous cells spend most of their time in interphase, not mitosis.

9. **(D)** Mitosis is the distribution of genetic material between the daughter cells. Cytokinesis, not mitosis, is the distribution of cytosol between the daughter cells, so choice (A) is incorrect. Choice (B) is incorrect because the replication of DNA occurs during the S stage, not mitosis. Replication of cell organelles occurs during G1 and G2, not mitosis, so choice (C) is incorrect.

10. **(C)** Cytokinesis is the division of the cytoplasm. In animal cytokinesis, a cleavage furrow is formed. In plant cytokinesis, a cell plate is formed. DNA and cellular organelles are replicated during interphase, but the cytoplasm is not yet divided. So choice (A) is incorrect. Choice (B) is incorrect because mitosis is the distribution of genetic material between the daughter cells. G0 is the nondividing phase of the cell cycle, so choice (D) is incorrect.

Short Free-Response

11. **(A-i)** Lectin stimulates cell division, as shown by the increased number of cells in mitosis for the treatment with only lectin added (compared to the number of control cells in mitosis).

    (B-i) The independent variable is the presence or absence of lectin and/or colchicine.

    (B-ii) The dependent variable is the number of cells in mitosis or interphase.

    (C-i) The most likely effects of adding only colchicine to the cells would be that the number of cells in mitosis would be less than that of the control group and the number of cells in interphase would be more than that of the control group.

    (D-i) The addition of lectin alone increased the number of cells in mitosis (compared to that of the control group). The addition of both lectin and colchicine resulted in a number of cells in mitosis that was much closer to that of the control group than it was to that of the group with only lectin added. So colchicine most likely reduces the number of cells in mitosis, and thus a group with only colchicine added would have even fewer cells in mitosis than the control group had.

12. **(A-i)** Replication of chromosomes occurs during the S stage of interphase. Alignment of chromosomes occurs during metaphase of mitosis. Separation of chromosomes occurs during anaphase of mitosis.

    (B-i) Molecule I represents the concentration of cyclin-dependent kinases because it is at a constant level throughout the cell cycle.

    (B-ii) Molecule II represents the concentration of cyclins because its level peaks just before the start of mitosis and then falls off rapidly during mitosis.

    (C-i) Mitosis-promoting factor (MPF) triggers the start of mitosis. Mitosis-promoting factor forms when cyclins bind to cyclin-dependent kinases. The arrow shows the point at which there is enough cyclin to form the MPF needed to trigger the start of mitosis.

    (D-i) If a cancer cell had a mutation that resulted in the constant high production of cyclins (Molecule II), there would be constantly high levels of MPF in the cell. Mitosis would constantly be stimulated, and the cell would divide uncontrollably.

Long Free-Response

13. **(A-i)** The null hypothesis is that there is no statistically significant difference between the number of cells in interphase and the number of cells in mitosis in the untreated cells compared to the number of cells in interphase and the number of cells in mitosis in the cells treated with caffeine.

    (B-i) The independent variable is the presence of caffeine.

    (B-ii) The dependent variable is the number of cells in interphase or mitosis.

    (C-i) Chi-square $= \chi^2 =$

    $$\sum \frac{\text{observed} - \text{expected}^2}{\text{expected}}$$

    $$\text{Chi-square} = \frac{(175 - 156)^2}{156} + \frac{(50 - 69)^2}{69} + \frac{(150 - 169)^2}{169} + \frac{(95 - 76)^2}{76}$$

    Chi-square $= 2.314 + 5.232 + 2.136 + 4.750$

    Chi-square $= 14.432$

    There are four possible outcomes to the experiment, so there are 3 degrees of freedom (df = number of possible outcomes $- 1$).

    Using a *p*-value of 0.05, the critical value from the chi-square table is 7.81. Based on these calculations, there is likely a statistically significant difference between the two groups of onions.

    (D-i) Since the calculated chi-square value of 14.432 is greater than the critical value of 7.81, the null hypothesis is rejected, and it is possible to say there is likely a statistically significant difference between the treated and untreated groups. The data support the alternative hypothesis that the addition of caffeine increases the mitotic rate in onion root cells.

UNIT 5
Heredity

12

Meiosis and Genetic Diversity

> **Learning Objectives**
>
> In this chapter, you will learn:
> → How Meiosis Works
> → How Meiosis Generates Genetic Diversity

Overview

Meiosis allows parents to pass on genetic information to their offspring. Meiosis helps accomplish this task while also maintaining a consistent number of chromosomes in the offspring *and* creating genetic diversity in the species. This chapter reviews the process of meiosis and how meiosis accomplishes these tasks. This chapter also compares the similarities and differences between meiosis and mitosis.

How Meiosis Works

Meiosis is the process of cell division that is used in gamete formation. Meiosis forms **haploid** (n) gametes from **diploid** ($2n$) parent cells. This helps maintain the proper number of chromosomes in the offspring. When a haploid (n) egg is fertilized by a haploid (n) sperm, the resulting zygote has the correct diploid ($2n$) number of chromosomes.

Unlike mitosis, which results in the creation of two genetically identical diploid daughter cells, meiosis has two rounds of cell division (meiosis I and meiosis II) that result in the creation of four genetically different haploid gametes.

> When determining the number of chromosomes in a cell, count the number of centromeres—this will tell you the number of chromosomes. Be aware that before DNA replication, one chromosome will have one chromatid, and after DNA replication, one chromosome will have two chromatids.

Meiosis I

The first round of cell division in meiosis is called **meiosis I**, and it is sometimes referred to as a reduction division. Meiosis I consists of four stages: prophase I, metaphase I, anaphase I, and telophase I.

- In the prophase stage of mitosis, the nuclear membrane breaks down and the chromosomes condense and become visible. This also occurs in the **prophase I** stage of meiosis. However, unlike in mitosis, homologous chromosomes pair up and **genetic recombination** (also known as **crossing-over**) can occur in prophase I of meiosis. This has profound effects on genetic diversity, which will be discussed later in this chapter.
- In the metaphase stage of mitosis, chromosomes line up in a *single* column along the center of the cell. In the **metaphase I** stage of meiosis, however, chromosomes line up in homologous *pairs* along the center of the cell.

- During the anaphase stage of mitosis, the sister *chromatids* in each chromosome separate and move toward opposite ends of the cell. Each of these sister chromatids has its own centromere after this separation, so the number of chromosomes in the cell is doubled by the end of the anaphase stage of mitosis. In the **anaphase I** stage of meiosis, *pairs of homologous chromosomes* separate. There are the same number of chromosomes at the end of the anaphase I stage of meiosis as there are at the beginning of this stage.
- During the telophase stage of mitosis, two new nuclei are formed. This is followed by cytokinesis and results in two *diploid* cells. In the **telophase I** stage of meiosis, two new nuclei are formed. When followed by cytokinesis, this results in two *haploid* cells, as shown in Figure 12.1.

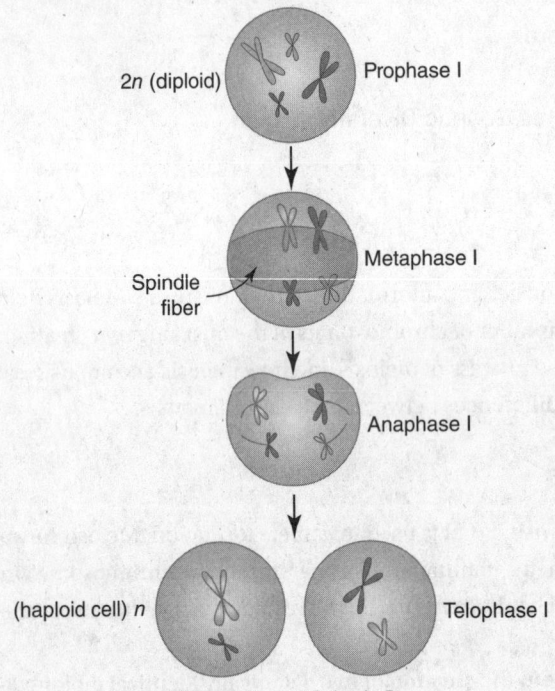

Figure 12.1 Meiosis I

Meiosis II

The second round of cell division in meiosis is called **meiosis II**, and it is very similar to mitosis. Meiosis II consists of four stages: prophase II, metaphase II, anaphase II, and telophase II. However, unlike in mitosis, the events of meiosis II are not preceded by the replication of DNA.

> DNA is replicated prior to the start of mitosis and is followed by one cell division, resulting in two diploid cells at the end of mitosis. However, while DNA is also replicated before the start of meiosis I, it is then followed by *two* cell divisions (meiosis I and meiosis II). This is the reason why the final result of meiosis is four haploid cells.

- In **prophase II** of meiosis II, chromosomes again condense and become visible.
- During **metaphase II**, chromosomes line up in a single line along the center of each cell, similar to how chromosomes line up in the metaphase stage of mitosis.
- Then, in **anaphase II**, sister chromatids separate and move to opposite ends of the cell. Each sister chromatid will have its own centromere once this separation has occurred. At the end of anaphase II, these separated sister chromatids are now considered separate chromosomes.
- **Telophase II** then splits each of the two cells in half, resulting in four cells. At the end of meiosis II, there are four haploid gametic cells, as shown in Figure 12.2.

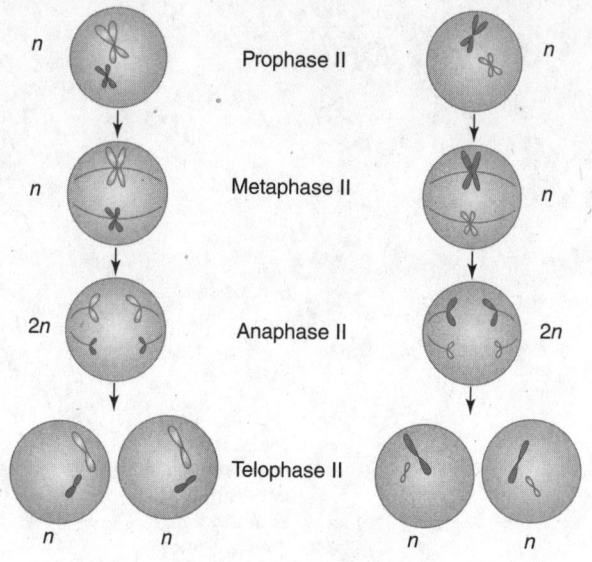

Figure 12.2 Meiosis II

The process of meiosis results in gametes with the haploid number of chromosomes. This is important because the joining of two haploid gametes in fertilization produces a diploid zygote. Together, meiosis and fertilization ensure consistency in the number of chromosomes from one generation to the next.

How Meiosis Compares to Mitosis

Table 12.1 summarizes the similarities and differences between mitosis and meiosis.

Table 12.1 Mitosis vs. Meiosis

	Mitosis	Meiosis
Purpose	Reproduction in asexually reproducing organisms; growth; repair	Reproduction in sexually reproducing organisms
DNA Replication	Occurs once (before the start of mitosis)	Occurs once (before the start of meiosis I)
Cell Division	One round	Two rounds
Result	Two genetically identical, diploid somatic cells	Four genetically different, haploid gametes

How Meiosis Generates Genetic Diversity

Meiosis generates genetic diversity in multiple ways. In the prophase I stage of meiosis, homologous chromosomes pair up in a process called **synapsis** to form **tetrads**.

Once the tetrads are formed, homologous chromosomes may exchange genetic information through the process of genetic recombination (or crossing-over). This generates new combinations of genetic material on each chromatid that may be passed on to the offspring. This increases the genetic diversity of the species, as shown in Figure 12.3.

> **TIP**
> To recall the meaning of *tetrad*, remember that the prefix *tetra-* means "four." During synapsis, two homologous chromosomes (with two chromatids each) pair up. This *tetrad* has *four* chromatids.

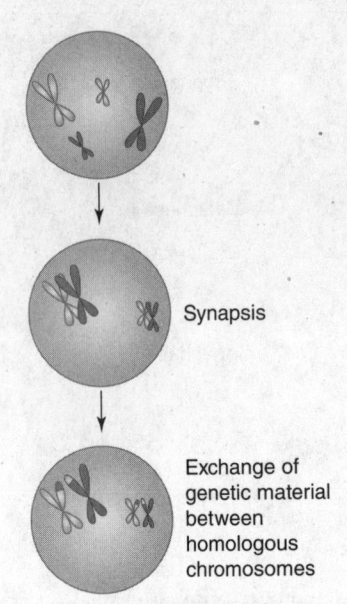

Figure 12.3 Genetic Recombination (Crossing-Over)

The frequency of genetic recombination events can be used to estimate the distance between genes that are on the same chromosome. Genes that are farther apart will have a higher recombination frequency, and genes that are closer to each other will have a lower recombination frequency. The frequency of genetic recombination between genes on the same chromosome can be used to generate genetic maps that show the relative positions of genes on chromosomes. Genes that are close together on the same chromosome tend to be inherited together more often (since the frequency of recombination between them is lower) and are referred to as **linked genes**.

> Sometimes, genetic recombination may occur between nonhomologous chromosomes. These events create mutations called **translocations**. While translocations can also generate new combinations of genetic material, if they occur in the middle of a gene, they may inactivate it, resulting in an unfavorable phenotype.

Genetic diversity can also be generated during the metaphase I stage of meiosis. Recall that during metaphase I, homologous pairs of chromosomes line up along the center of the cell. Each pair of chromosomes lines up and assorts independently, with different pairs having the paternal chromosome (from the male parent) on one side and the maternal chromosome (from the female parent) on the other side. For example, if an organism has two pairs of chromosomes, each time meiosis occurs, the two pairs may assort independently, resulting in a different combination of genetic material being passed on to the offspring, as shown in Figure 12.4.

NOTE
Closely linked genes will not demonstrate independent assortment.

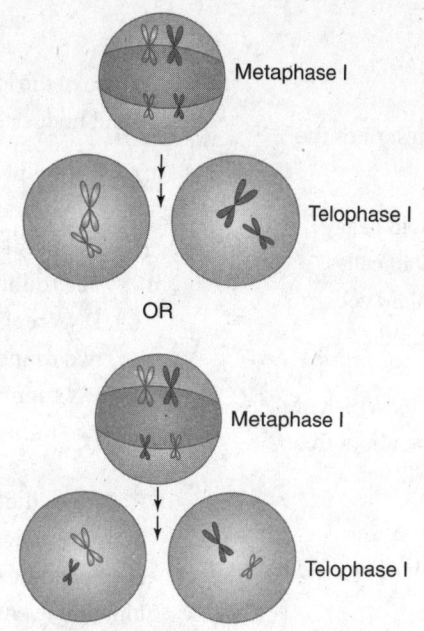

Figure 12.4 Independent Assortment of Chromosomes in Metaphase I of Meiosis

Occasionally during the anaphase I stage of meiosis, pairs of homologous chromosomes fail to separate and will instead move to the same side of the cell, eventually ending up in the same gamete. This is called **nondisjunction**. If the resulting gametes are fertilized, this can result in an **aneuploidy**, which is when there is an atypical number of chromosomes. One example of an aneuploidy is Down syndrome, also known as trisomy 21, which can result when an individual has three copies of chromosome 21 instead of two copies.

Practice Questions
Multiple-Choice

1. Which of the following correctly describes the products of meiosis?

    (A) two genetically identical diploid cells
    (B) two genetically different diploid cells
    (C) four genetically identical haploid cells
    (D) four genetically different haploid cells

2. An organism's diploid number is 28. Which of the following is a correct statement about this organism?

    (A) Its somatic cells would have 14 chromosomes, and its gametes would have 28 chromosomes.
    (B) Its somatic cells would have 28 chromosomes, and its gametes would have 14 chromosomes.
    (C) Both its somatic cells and its gametic cells would have 14 chromosomes.
    (D) Both its somatic cells and its gametic cells would have 28 chromosomes.

3. Which of the following stages of meiosis generates genetic diversity?

    (A) prophase I
    (B) prophase II
    (C) anaphase I
    (D) anaphase II

4. Independent assortment of chromosomes occurs during which stage of meiosis?

    (A) metaphase I
    (B) metaphase II
    (C) telophase I
    (D) telophase II

5. Which of the following is a correct statement about meiosis?

    (A) DNA replication occurs once, and there is one round of cell division.
    (B) DNA replication occurs twice, and there is one round of cell division.
    (C) DNA replication occurs once, and there are two rounds of cell division.
    (D) DNA replication occurs twice, and there are two rounds of cell division.

6. How does the frequency of recombination events between genes that are close together on the same chromosome compare to the frequency of recombination events between genes that are far apart on the same chromosome?

    (A) Genes that are close together on the same chromosome will recombine less frequently than genes that are far apart.
    (B) Genes that are close together on the same chromosome will recombine more frequently than genes that are far apart.
    (C) Genes that are close together on the same chromosome will recombine at the same rate as genes that are far apart.
    (D) Genes that are on the same chromosome never have recombination events between them.

7. Which of the following statements correctly describes a difference between mitosis and meiosis?

    (A) Mitosis produces four daughter cells, whereas meiosis produces two daughter cells.
    (B) Mitosis produces genetically different cells, whereas meiosis produces genetically identical cells.
    (C) Mitosis produces somatic cells, whereas meiosis produces gametes.
    (D) Mitosis produces haploid cells, whereas meiosis produces diploid cells.

8. Aneuploidies occasionally occur during which stage of meiosis?

   (A) prophase I
   (B) metaphase I
   (C) anaphase I
   (D) telophase I

Questions 9 and 10

The following figure is a karyotype, a picture that shows the chromosomes found in a somatic human cell.

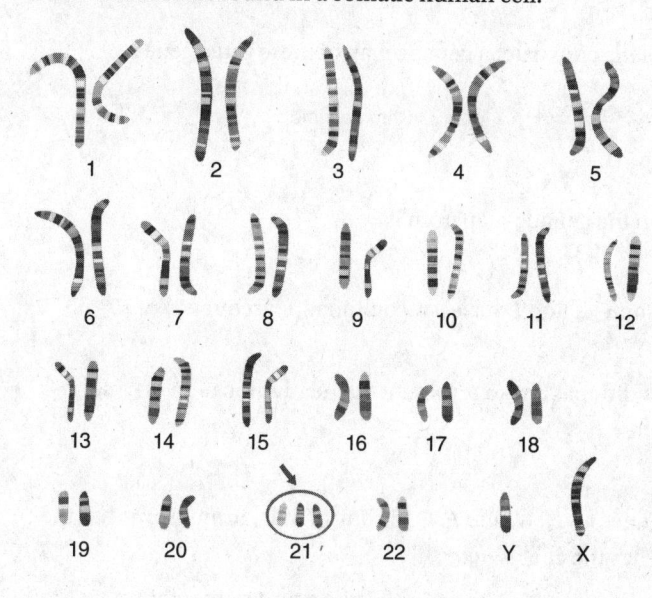

9. Which of the following best describes the condition shown in the circle?

   (A) trisomy
   (B) haploidy
   (C) monosomy
   (D) diploidy

10. Which of the following was the most likely cause of the condition shown in the circle?

    (A) translocation
    (B) independent assortment
    (C) nondisjunction
    (D) genetic recombination

Short Free-Response

11. Genes *A*, *B*, and *C* are on the same chromosome. The frequencies of genetic recombination events between the genes are shown in the following table.

Genes	Recombination Frequency
AB	45%
BC	15%
AC	30%

Part A

(i) Using the data above and the following template, **construct** a genetic map of these three genes.

⎯⎯⎯⎯⎯⎯⎯⎯⎯⎯⎯⎯⎯⎯⎯⎯⎯⎯⎯⎯ Chromosome

Part B

(i) **Explain** your placement of the three genes on the genetic map from Part A.

Part C

A new gene, *D*, is discovered on the same chromosome. *A* and *D* have a recombination frequency of 10%. *B* and *D* have a recombination frequency of 35%.

(i) On your genetic map from Part A, **draw** an additional line to represent the relative location of gene *D* on the chromosome.

Part D

A different chromosome in the organism has the genes *M*, *N*, *O*, and *P*. One copy of that chromosome has the alleles *MNOP*. The other copy of the chromosome has the alleles *mnop*.

(i) If genetic recombination occurred between genes *N* and *O*, **explain** the two new gametes that would be produced by this crossing-over event.

12. An organism has a diploid number of six chromosomes.

Part A

(i) **Determine** the number of chromosomes in each of the four cells that would be produced upon the completion of meiosis II.

Part B

(i) Based on your answer to Part A, **explain** why the cells would each have that number of chromosomes at the end of meiosis II.

Part C

A nondisjunction event occurred during anaphase II.

(i) **Predict** the number of chromosomes that would be found in each of the four cells after this nondisjunction event.

Part D

(i) **Justify** your prediction from Part C.

Long Free-Response

13. An organism has the two chromosomes shown in the following figure.

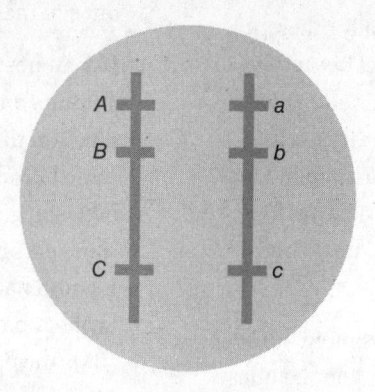

Gamete-producing cells from this organism are placed into two groups. One group is not exposed to UV light, and the other group is exposed to UV light prior to undergoing meiosis. The gametes produced at the end of meiosis and the frequencies of those gametes are shown in the following table.

Gamete	Frequency of Gametes in Cells Not Exposed to UV Light	Frequency of Gametes in Cells Exposed to UV Light
ABC	48%	35%
ABc	2%	15%
abC	2%	15%
abc	48%	35%

Part A

(i) **Identify** the stage of meiosis during which genetic recombination occurs.

Part B

(i) **Identify** the independent variable in this experiment.

(ii) **Identify** the dependent variable in this experiment.

Part C

A student claims that exposure to UV light prior to meiosis increases the frequency of genetic recombination during meiosis.

(i) Using the data, provide reasoning to **support the student's claim**.

Part D

The experiment is repeated with increased exposure to UV light (for cells that were exposed to UV light).

(i) **Predict** how this change will affect the recombination frequencies.

(ii) **Justify** your prediction.

Answer Explanations

Multiple-Choice

1. **(D)** Meiosis produces four genetically different haploid cells. Choice (A) is incorrect because it describes the products of mitosis. Choice (B) is incorrect because meiosis produces four cells, not two cells, and those four cells are haploid, not diploid. Choice (C) is incorrect because meiosis produces genetically different cells, not genetically identical cells.

2. **(B)** Somatic (body) cells have the diploid number of chromosomes (which is 28, based on the question), and gametes have the haploid number of chromosomes (which, in this case, is 14). Choice (A) is incorrect because it assigns the haploid number of chromosomes to the somatic cells and the diploid number of chromosomes to the gametic cells. Choices (C) and (D) are both incorrect because somatic cells and gametes have different numbers of chromosomes, not the same number of chromosomes.

3. **(A)** Genetic recombination (crossing-over), which generates genetic diversity, occurs during prophase I. Choices (B), (C), and (D) are incorrect because genetic diversity is not generated in prophase II, anaphase I, or anaphase II. (Note that genetic diversity is generated in metaphase I of meiosis, but that is not one of the choices in this question.)

4. **(A)** Pairs of homologous chromosomes assort independently during metaphase I. Choice (B) is incorrect because individual chromosomes line up on the metaphase plate in metaphase II, and no independent assortment of chromosomes occurs. In both telophase I and telophase II, new nuclear membranes are formed. No assortment of chromosomes occurs during those stages, so choices (C) and (D) are incorrect.

5. **(C)** DNA is replicated once before meiosis I starts, and meiosis is comprised of two rounds of cell division. Choice (A) is incorrect because it describes what happens during mitosis, not meiosis. Choice (B) is incorrect because DNA replication only occurs once in meiosis, and there are two rounds of cell division in meiosis. Choice (D) is incorrect because DNA replication only occurs once in meiosis.

6. **(A)** Genes that are close together on the same chromosome have fewer opportunities for recombination than genes that are farther apart on the same chromosome. Thus, recombination events between genes that are close together on the same chromosome occur less frequently than recombination events between genes that are farther apart on the same chromosome. Choice (B) is the opposite of the correct answer. The distance between genes does affect the frequency of recombination events, so choice (C) cannot be the answer. Typically, recombination events do occur between genes on the same chromosome, so choice (D) is not correct.

7. **(C)** Mitosis produces somatic (body) cells, whereas meiosis produces gametes. Choice (A) is incorrect because mitosis produces two cells, whereas meiosis produces four cells. Mitosis produces genetically identical cells, whereas meiosis produces genetically different cells, so choice (B) is incorrect. Haploid gametes are produced by meiosis, whereas diploid cells are produced by mitosis. So choice (D) is incorrect.

8. **(C)** Pairs of homologous chromosomes separate during anaphase I and ultimately end up in different cells at the end of telophase I. If a pair of homologous chromosomes did not separate in anaphase I, one of the two cells at the end of telophase I would have both members of the pair of homologous chromosomes and the other cell would have neither member of the pair of homologous chromosomes. The resulting gametes would have an incorrect number of chromosomes (either one too many or one too few), which results in an aneuploidy. Choices (A), (B), and (D) are incorrect because chromosomes do not separate in those stages, so those stages would not result in an atypical number of chromosomes.

9. **(A)** Trisomy describes the condition where there are three copies of a given chromosome when there would typically be two copies. Choice (B) is incorrect because haploidy describes the conditions where there is only one copy of each chromosome. Monosomy is when there is one copy of a given chromosome, so choice (C) is incorrect. Choice (D) is incorrect because diploidy describes the conditions where there are two copies of each chromosome.

10. **(C)** Nondisjunction is when two homologous chromosomes do not separate during meiosis. If a nondisjunction event occurs, the resulting gamete will have an extra copy of that chromosome. If that gamete is fertilized by a haploid gamete, the resulting cell will have three copies of the chromosome, as shown in the figure. Translocation describes mutations that are the result of genetic recombination between nonhomologous chromosomes, so choice (A) is incorrect. Choice (B) is incorrect because independent assortment is the independent alignment of pairs of homologous chromosomes in metaphase I; this does not affect the number of chromosomes found in a cell. Choice (D) is incorrect because genetic recombination describes the exchange of genetic material between chromosomes; this would not result in the presence of three copies of a chromosome.

162 AP BIOLOGY

Short Free-Response

11. (A-i) B ←— 15 —→ C ←——— 30 ———→ A Chromosome

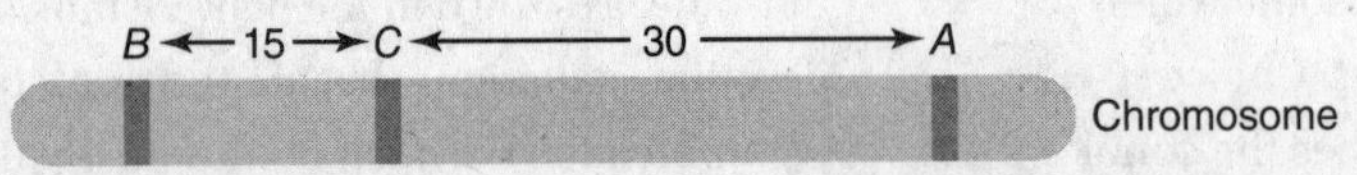

    (B-i) *B* and *C* have the lowest recombination frequency at 15%, so they must be the closest together of the three genes. *A* and *C* have a recombination frequency of 30%, and *A* and *B* have a recombination frequency of 45%, so *A* must be closer to *C* than *A* is to *B*. So the correct placement of the three genes on the chromosome must be as shown in Part A.

    (C-i) B ←— 15 —→ C ←——— 20 ———→ D ←— 10 —→ A Chromosome

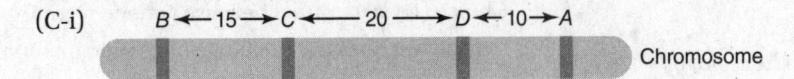

    (D-i) If a crossover event occurred between genes *N* and *O* as shown in the following figure, the resulting gametes would have the genotypes *MNop* and *mnOP*.

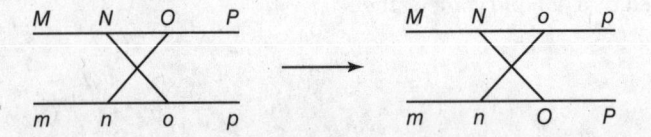

    M N   O   P   M N   o   p
         ╳   ╳
    m n   o   p   m n   O   P

12. (A-i) At the end of meiosis II, one would expect to find three chromosomes in each gamete.

    (B-i) Meiosis starts with a diploid cell and produces four haploid gametes. So if the original cell had a diploid number of six, the gametes would typically each have a haploid number of three.

    (C-i) Two cells would have three chromosomes each, one cell would have four chromosomes, and one cell would have two chromosomes.

    (D-i) If the nondisjunction event occurred during anaphase II, the result would be as shown in the following figure because the sister chromatids of the chromosome would not separate. Both chromatids would end up in the same cell instead of in two separate cells. This would result in an extra chromosome in the cell that received both chromatids and a missing chromosome in the cell that did not receive either chromatid.

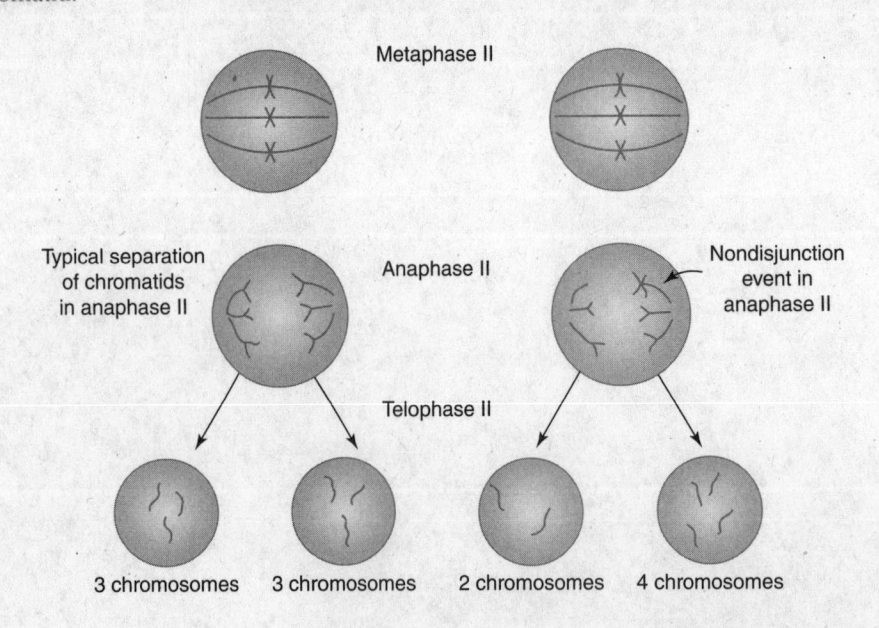

    Metaphase II

    Typical separation of chromatids in anaphase II   Anaphase II   Nondisjunction event in anaphase II

    Telophase II

    3 chromosomes   3 chromosomes   2 chromosomes   4 chromosomes

Long Free-Response

13. (A-i)　　Genetic recombination occurs during the prophase I stage of meiosis. During prophase I, homologous chromosomes synapse (pair up) to form tetrads. These tetrads can then exchange genetic information.

　　(B-i)　　The independent variable is exposure to UV light.

　　(B-ii)　　The dependent variable is the percentage of recombinant gametes produced.

　　(C-i)　　Cells that were exposed to UV light prior to undergoing meiosis produced more recombinant gametes. So this supports the student's claim that exposure to UV light prior to meiosis increases the frequency of genetic recombination.

　　(D-i and ii)　　More exposure to UV light would probably produce even more recombinant gametes. Since the initial exposure to UV light resulted in more recombinant gametes (than in those cells that were not exposed to UV light), more exposure to UV light would further increase the number of recombinant gametes.

13

Mendelian Genetics and Probability

Learning Objectives

In this chapter, you will learn:
- → Mendelian Genetics
- → Probability in Genetics Problems

Overview

The study of genetics started with Mendelian genetics. This chapter reviews the laws of Mendelian genetics and the laws of probability that relate to them.

Mendelian Genetics

The transfer of genetic information is a key process in living organisms. Information is transferred within an organism from the nucleus to the ribosome during the synthesis of proteins (which will be reviewed in Chapter 16). Information is transferred from generation to generation through inherited units of chemical information called genes. Information can be transferred between members of the same generation through the processes of conjugation, transformation, or transduction (all of which will be discussed in Chapter 17). In all cases, the carriers of this genetic information are DNA and RNA.

DNA, RNA, and ribosomes are found in all forms of life. DNA and RNA use a genetic code that is shared by all living organisms. This provides powerful evidence to support the common ancestry of all living organisms on Earth.

Mendel's laws of segregation and independent assortment explain how genetic information can be transferred from one generation to the next.

> It is important to note that Mendel's laws of segregation and independent assortment only apply to traits coded for by genes that are located on separate chromosomes. As discussed in Chapter 12, genes that are close together on the same chromosome are linked and tend to be inherited together more often. Mendel's laws of segregation and independent assortment do not apply to linked genes.

Mendel's law of segregation states that an organism carries two variations of every trait (called alleles), one from each parent, and these alleles segregate, or separate, independently into gametes. This segregation occurs during anaphase of meiosis. When homologous chromosomes separate during anaphase I, the alleles on those chromosomes will segregate into separate gametes. One variation (or allele) for a trait ends up in each gamete. When two gametes join during fertilization, the resulting offspring then again has two alleles for the trait.

Mendel's law of independent assortment states that genes for different traits segregate independently of one another. This independent assortment occurs during metaphase I of meiosis. For example, in pea plants, inheriting the allele for purple flower color is an independent event from inheriting the allele for wrinkled peas. This is because the gene for flower color in peas is on a separate chromosome than the gene for pea shape. These chromosomes assort independently in metaphase I of meiosis.

A **pedigree** is a chart that illustrates the inheritance of a trait through several generations. Horizontal lines between two individuals in a pedigree show that those individuals have had offspring together. Offspring of these individuals are indicated by vertical lines. Circles typically represent females, and squares represent males. Circles or squares that are shaded usually represent individuals who possess the trait shown in the pedigree.

Examining the pattern of inheritance of a trait shown in a pedigree can give clues to the trait's possible mode of inheritance. Dominant traits tend to be expressed in at least one parent and their offspring because only one allele for the trait is required for it to be expressed. Recessive traits often will be expressed in offspring, but not in either parent, because the parents are heterozygous carriers of the trait. Sex-linked recessive traits usually appear more often in males than in females. Chapter 14 will discuss the patterns of inheritance seen in non-Mendelian traits.

Before practicing some genetics problems, review the following terms:

- **Genotype**—the genetic makeup or alleles for the trait in an organism (i.e., *AA* or *Aa* or *aa*)
- **Phenotype**—the physical expression of the genotype in an organism (i.e., purple flowers or white flowers)
- **Homozygous**—having two copies of the same allele for a trait (i.e., *AA* or *aa*)
- **Heterozygous**—having two different alleles for a trait (i.e., *Aa*); Mendel referred to this as "hybrid"
- **Dominant**—requires only one copy of the allele for the trait to be expressed in the phenotype; dominant traits are those that are expressed in organisms that are heterozygous
- **Recessive**—requires two copies of the allele for the trait to be expressed in the phenotype; recessive traits are not expressed in organisms that are heterozygous

Probability in Genetics Problems

Problems that Involve a Monohybrid Cross

One of the simplest types of crosses is called a monohybrid cross. In a **monohybrid cross**, both parents are heterozygous (have two different alleles) for the trait. For example, purple flowers are dominant to white flowers in pea plants. Two heterozygous purple flowers are crossed. The results are shown in the Punnett square below.

> **TIP**
> Math is your friend when answering genetics problems! Using simple fractions and the laws of probability will help you solve genetics problems more accurately and efficiently.

	P	p
P	PP	Pp
p	Pp	pp

Figure 13.1 Cross Between Two Pea Plants that are Heterozygous for Flower Color

There are three possible genotypes that could result from this cross: *PP*, *Pp*, or *pp*. The probabilities of each are $\frac{1}{4}$ *PP*, $\frac{1}{2}$ *Pp*, and $\frac{1}{4}$ *pp*.

There are two possible phenotypes that could result from this cross: purple flowers or white flowers. Since both the *PP* and *Pp* genotypes would result in the purple flower phenotype, the probabilities of each phenotype are $\frac{3}{4}$ purple flowers and $\frac{1}{4}$ white flowers.

Consider another example. Two pea plants that are heterozygous for plant height are crossed. In pea plants, tall is dominant to short. The results of the cross are shown in Figure 13.2.

	T	t
T	TT	Tt
t	Tt	tt

Figure 13.2 Cross Between Two Pea Plants that are Heterozygous for Plant Height

There are three possible genotypes that could result from this cross: *TT*, *Tt*, or *tt*. The probabilities of each are $\frac{1}{4}$ *TT*, $\frac{1}{2}$ *Tt*, and $\frac{1}{4}$ *tt*.

There are two possible phenotypes that could result from this cross: tall plants or short plants. Since both the *TT* and *Tt* genotypes would result in the tall phenotype, the probabilities of each phenotype are $\frac{3}{4}$ tall plants and $\frac{1}{4}$ short plants.

> For the AP Biology exam, it is not necessary to memorize which specific traits are dominant or recessive. Information about which traits are dominant will either be given to you directly in the problem or you will be able to infer that information from the text or data given to you in the problem.

Problems that Involve a Dihybrid Cross

A **dihybrid cross** looks at the result when two organisms (that are both heterozygous for the same two traits) are crossed. When solving dihybrid crosses, math is your friend! Treat each trait separately, find the probabilities of each outcome for each trait, and use the laws of probability to solve the problem.

> Refer to Appendix A of this book for the relevant probability equations (and those same equations will be supplied to you on test day). The probability of two independent events occurring simultaneously is the product of their individual probabilities:
>
> $$P(A \text{ and } B) = P(A) \times P(B)$$

Try the following example. Two pea plants that are both heterozygous for purple flower color and plant height are crossed. The genotypes of the parents are *PpTt* × *PpTt*.

Recall from the monohybrid cross of *Pp* × *Pp* (Figure 13.1) that the probability of offspring with purple flowers is $\frac{3}{4}$ and the probability of offspring with white flowers is $\frac{1}{4}$.

Also recall from the monohybrid cross of *Tt* × *Tt* (Figure 13.2) that the probability of tall offspring is $\frac{3}{4}$ and the probability of short offspring is $\frac{1}{4}$.

So the probability of this cross producing a plant that is tall and has purple flowers is the product of the probability of producing a tall plant $\left(\frac{3}{4}\right)$ times the probability of producing a plant with purple flowers $\left(\frac{3}{4}\right)$. The result is $\frac{9}{16}$.

Using the same process, the probability of this cross producing a tall plant with white flowers is the product of the probability of producing a tall plant $\left(\frac{3}{4}\right)$ times the probability of producing a plant with white flowers $\left(\frac{1}{4}\right)$. The result is $\frac{3}{16}$.

Similarly, the probability of this cross producing a short plant with purple flowers is the product of the probability of producing a short plant $\left(\frac{1}{4}\right)$ times the probability of producing a plant with purple flowers $\left(\frac{3}{4}\right)$. The result is $\frac{3}{16}$.

Finally, the probability of this cross producing a short plant with white flowers is the product of the probability of producing a short plant $\left(\frac{1}{4}\right)$ times the probability of producing a plant with white flowers $\left(\frac{1}{4}\right)$. The result is $\frac{1}{16}$.

This same problem-solving process can be used for other types of crosses as well. Try this example. Imagine that there's a cross between a pea plant that is *PpTt* and a pea plant that is *Pptt*.

> **NOTE**
>
> Monohybrid crosses involve just ONE trait and produce a phenotypic ratio of 3:1, whereas *dihybrid* crosses involve TWO traits and produce a phenotypic ratio of 9:3:3:1.

First, find the probabilities of each possible outcome for the plant height trait, as shown in Figure 13.3.

	T	t
t	Tt	tt
t	Tt	tt

Figure 13.3 Results of a $Tt \times tt$ Cross

The probability of producing a tall offspring (Tt) is $\frac{1}{2}$. The probability of producing a short offspring (tt) is also $\frac{1}{2}$. Now, find the probabilities of each possible outcome for the flower color trait, as shown in Figure 13.4.

	P	p
P	PP	Pp
p	Pp	pp

Figure 13.4 Results of a $Pp \times Pp$ Cross

The probability of producing an offspring with purple flowers (PP or Pp) is $\frac{3}{4}$. The probability of producing an offspring with white flowers (pp) is $\frac{1}{4}$.

Now, using the laws of probability, solve for the following outcomes:

Offspring that are tall and have purple flowers $= \frac{1}{2} \times \frac{3}{4} = \frac{3}{8}$

Offspring that are tall and have white flowers $= \frac{1}{2} \times \frac{1}{4} = \frac{1}{8}$

Offspring that are short and have purple flowers $= \frac{1}{2} \times \frac{3}{4} = \frac{3}{8}$

Offspring that are short and have white flowers $= \frac{1}{2} \times \frac{1}{4} = \frac{1}{8}$

Problems that Involve a Test Cross

An organism that displays the dominant phenotype may have either the homozygous dominant genotype or the heterozygous genotype. One way to determine which genotype the organism has is by using a test cross. A **test cross** is a cross between an organism with a dominant phenotype (whose genotype is unknown) and an organism with the recessive phenotype. If the organism with the dominant phenotype is homozygous, all of the offspring of the test cross will show the dominant phenotype. If the organism with the dominant phenotype is heterozygous, $\frac{1}{2}$ of the offspring of the test cross will show the dominant phenotype and $\frac{1}{2}$ of the offspring of the test cross will show the recessive phenotype, as shown in Figure 13.5.

	T	T			T	t
t	Tt	Tt	or	t	Tt	tt
t	Tt	Tt		t	Tt	tt

Figure 13.5 Possible Outcomes of a Test Cross

Practice Questions

Multiple-Choice

1. Mendel's law of independent assortment is a result of
   (A) the random nature of fertilization of ova by sperm.
   (B) the random way homologous chromosomes undergo genetic recombination in meiosis.
   (C) the random way homologous chromosomes line up in metaphase I of meiosis.
   (D) the random nature of molecular movement at high temperatures.

2. The term *hybrid* can also mean _____.
   (A) homozygous
   (B) heterozygous
   (C) haploid
   (D) diploid

3. In peas, round (*R*) peas are dominant and wrinkled (*r*) peas are recessive. A plant with an unknown genotype for this trait has round peas. Which of the following results of a test cross would support the hypothesis that the genotype of the plant with the round peas is heterozygous?
   (A) 100% round offspring
   (B) 75% round offspring and 25% wrinkled offspring
   (C) 50% round offspring and 50% wrinkled offspring
   (D) 100% wrinkled offspring

4. In pea plants, smooth (*S*) pods are dominant and constricted (*s*) pods are recessive. Which of the following is a correct statement?
   (A) A genotype of *SS* would result in the constricted pod phenotype.
   (B) A genotype of *Ss* would result in the constricted pod phenotype.
   (C) Constricted pods would always be homozygous.
   (D) Smooth pods would always be homozygous.

5. Round peas are dominant to wrinkled peas. Smooth pea pods are dominant to constricted pea pods. Two pea plants that are heterozygous for both traits are crossed, and 800 offspring are produced. How many of the offspring are expected to have round peas and constricted pods?
   (A) 450
   (B) 150
   (C) 50
   (D) 0

6. A pea plant that is heterozygous for round peas and heterozygous for smooth pods is crossed with a pea plant that is heterozygous for round peas and has constricted pods. What fraction of the offspring are expected to have round peas and constricted pods? (Remember that round peas are dominant to wrinkled peas, and smooth pods are dominant to constricted pods.)
   (A) $\frac{3}{4}$
   (B) $\frac{3}{8}$
   (C) $\frac{1}{4}$
   (D) $\frac{1}{8}$

7. In humans, the gene for polydactyly (extra fingers or toes) is dominant over the gene for the typical number of fingers and toes. If a person who is heterozygous for polydactyly has children with a person with the typical number of fingers and toes, what is the probability that their first child will have polydactyly?
   (A) 100%
   (B) 50%
   (C) 25%
   (D) 0%

8. In the cross *AaBbCc* × *AaBbCc*, what is the probability of producing an offspring with the *AABBCc* genotype?

   (A) $\frac{1}{4}$
   (B) $\frac{1}{8}$
   (C) $\frac{1}{16}$
   (D) $\frac{1}{32}$

Questions 9 and 10

Rh positive (*R*) is a trait that is dominant over Rh negative (*r*). A woman, who is Rh positive, has a son, who is Rh negative.

9. Which of the following is a true statement about the mother?

   (A) She must have the *rr* genotype.
   (B) She must have the *RR* genotype.
   (C) She must be heterozygous for the trait.
   (D) She must have the same genotype as her son.

10. Which of the following is a true statement about the father of the woman's son?

    (A) He must have the *RR* genotype.
    (B) He must have the *Rr* genotype.
    (C) He must have the *rr* genotype.
    (D) He may have either the *Rr* or the *rr* genotype.

Short Free-Response

11. Sickle cell disease is an autosomal recessive disorder. Homozygous recessive individuals have red blood cells that can assume a sickle shape when oxygen levels are low. These sickled blood cells can cause blockages in blood vessels, leading to pain and possibly fatal complications. Heterozygous individuals do not suffer the pain or complications from sickle cell disease and are more resistant to malaria than individuals with two nonsickle alleles. The following figure shows the pattern of inheritance of sickle cell disease through three generations of a family. Squares indicate males, circles indicate females, and shaded shapes indicate individuals who have sickle cell disease.

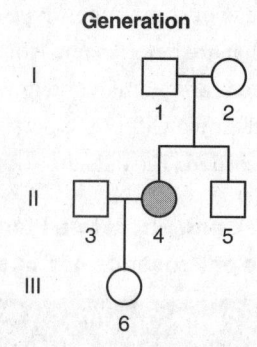

Generation

I: 1 (square) — 2 (circle)

II: 3 (square), 4 (shaded circle), 5 (square)

III: 6 (circle)

Part A

(i) **Identify** the genotype of individual I-2.

(ii) **Justify** your answer with evidence from the pedigree.

Part B

(i) **Identify** the genotype of individual III-6.

(ii) **Justify** your answer with evidence from the pedigree.

Part C

(i) Create a Punnett square that **represents** the possible genotypes of the offspring of individuals 1 and 2 from generation I.

Part D

(i) **Explain** which genotype from this pedigree would provide an advantage in San Francisco (where the mosquito that carries malaria does not live) and which genotype would be most advantageous to survival in Haiti (where the mosquito that carries malaria is found).

12. Fur color in hamsters is determined by a single gene, with yellow fur color being dominant to white fur color. Inheriting two copies of the allele for yellow fur color is fatal, so all individuals with yellow fur are heterozygous.

Part A

(i) **Determine** the probability that the offspring of two hamsters with yellow fur would also have yellow fur.

Part B

(i) **Construct** a Punnett square that represents the two hamster parents from Part A.

(ii) Use the Punnett square you constructed to **explain** your answer to Part A.

Part C

A hamster population lives in an area covered with golden-yellow grasses. An unusually cold winter results in the death of all the golden-yellow grasses, so in the spring, the area is covered only in white sand.

(i) **Predict** what would happen to the frequency of the allele for yellow fur color over time.

Part D

(i) **Justify** your prediction from Part C.

Long Free-Response

13. In pea plants, tall plant height is dominant over short. A new type of plant food (Supersize Food) is advertised to increase plant height. Two pea plants that are heterozygous for plant height are crossed, and the seeds that are produced are harvested. Half of the seeds are planted and grown in the presence of Supersize Food, and the other half of the seeds are planted and grown without Supersize Food. When the plants reach full maturity, the number of tall and short plants are recorded. The data are shown in the table.

Number of Tall and Short Pea Plants Grown in the Absence or Presence of Supersize Food

Height of Plant	Plants Grown Without Supersize Food	Plants Grown with Supersize Food
Tall	1,290	1,265
Short	430	455

Part A

(i) **Describe** the proportions of tall and short plants that were expected to result from the cross.

(ii) Use a Punnett square to **support your answer**.

Part B

(i) **Identify** the independent variable in this experiment.

(ii) **Identify** the dependent variable in this experiment.

(iii) **Identify** one possible appropriate experimental constant.

Part C

A student makes the claim that "there is no statistically significant difference between the plants grown in the presence or absence of Supersize Food."

(i) **Calculate** the chi-square value using the experimental data in the table.

(ii) **Evaluate** the student's claim based on your calculated chi-square value. Use a p-value of 0.05 for your analysis.

Part D

A tall plant with an unknown genotype is crossed with a short plant.

(i) **Predict** the expected proportions of tall and short offspring if the tall plant is heterozygous.

(ii) **Justify** your prediction using a Punnett square.

Answer Explanations

Multiple-Choice

1. **(C)** Each pair of chromosomes lines up independently in metaphase I of meiosis, leading to the independent assortment of the traits on those chromosomes. Choices (A) and (B) are incorrect because they do not lead to independent assortment (but they do contribute to increased genetic diversity). Choice (D) is incorrect because temperature does not affect independent assortment.

2. **(B)** *Hybrid* refers to heterozygous genotypes. Choice (A) is the opposite of the correct answer. Choices (C) and (D) are both incorrect because haploid and diploid are terms that refer to the number of chromosomes in a cell, not genotypes.

3. **(C)** If the plant with round peas is heterozygous, its genotype is *Rr*. The definition of a test cross is to cross an individual with the dominant phenotype (but an unknown genotype) with a homozygous recessive individual. Thus, if the *Rr* plant was crossed with an *rr* plant, 50% of the offspring would be expected to be round and 50% would be expected to be wrinkled. Choice (A) is incorrect because it describes the expected results if the original round plant was homozygous dominant (*RR*). Choice (B) is incorrect because in order to produce that proportion of offspring, both parent plants would have to be heterozygous (*Rr*), but the plant used to do the test cross must be *rr*. In order to produce 100% wrinkled offspring, both parent plants would have to be homozygous recessive (*rr*), so choice (D) is incorrect.

4. **(C)** A genotype of *ss* would result in the constricted pod phenotype. Choices (A) and (B) are incorrect because both *SS* and *Ss* would result in the smooth phenotype. Choice (D) is incorrect because smooth pods would not always have the *SS* genotype; smooth pods could also have the *Ss* genotype.

5. **(B)** The chances of this cross producing a plant with round peas is $\frac{3}{4}$. The chances of this cross producing a plant with constricted pods is $\frac{1}{4}$. So the chances of producing a plant with both round peas and constricted pods is the product of their individual probabilities: $\frac{3}{4} \times \frac{1}{4} = \frac{3}{16}$. Given a total number of 800 offspring, $800 \times \frac{3}{16} = 150$. Choice (A) is incorrect because it is the expected number of offspring with round peas and smooth pods. Choice (C) is incorrect because it is the expected number of offspring with wrinkled peas and constricted pods. Choice (D) is incorrect because $\frac{3}{4}$ of the offspring are expected to be round and $\frac{1}{4}$ are expected to be constricted, so some of the offspring must be both round and constricted and thus the answer cannot be 0.

6. **(B)** The genotypes of the plants described in the problem are *RrSs* and *Rrss*. First, calculate the probability for each trait separately. The probability of two plants that are heterozygous for round peas producing an offspring with round peas is $\frac{3}{4}$. The probability of a plant with heterozygous smooth pods crossed with a plant with constricted pods producing an offspring with constricted pods is $\frac{1}{2}$. Multiplying the two probabilities together gives the probability of a plant with both round peas and constricted pods: $\frac{3}{4} \times \frac{1}{2} = \frac{3}{8}$.

7. **(B)** A person who is heterozygous for polydactyly would have the *Ff* genotype. A person with the typical number of fingers and toes would have the *ff* genotype, so the chances their first child will have polydactyly is 50%. (Whether this is their first child, second child, etc., is irrelevant because each is an independent event.)

8. **(D)** Remember to break this down one trait at a time, find the probability of the outcomes for each trait, and then multiply the probabilities together. *Aa* × *Aa* would produce *AA* $\frac{1}{4}$ of the time. Similarly, *Bb* × *Bb* would produce *BB* $\frac{1}{4}$ of the time. *Cc* × *Cc* would produce *Cc* $\frac{1}{2}$ of the time. Multiply all three probabilities together: $\frac{1}{4} \times \frac{1}{4} \times \frac{1}{2} = \frac{1}{32}$.

9. **(C)** Since the mother is Rh positive, she must have at least one dominant *R* allele. Since she has a child who is Rh negative, she must have an *r* allele. Therefore, the mother must be heterozygous *Rr*. Choice (A) is incorrect because if the mother was *rr*, she would not be Rh positive. Choice (B) is incorrect because an *RR* mother could not produce an Rh negative child. Choice (D) is incorrect because since the mother is Rh positive and her son is Rh negative, they cannot have the same genotype.

10. **(D)** Since the son is Rh negative, his genotype must be *rr*. He must have inherited one *r* allele from each parent. So the father (whose Rh phenotype is unknown) could be either *Rr* or *rr*. Choice (A) is incorrect because an *RR* father would not be able to pass on an *r* allele to his offspring. Choices (B) and (C) are incorrect because while both are possible genotypes for the father, it is not possible to say with certainty which genotype the father has without knowing the father's phenotype.

Short Free-Response

11. (A-i) Individual I-2 is heterozygous for sickle cell.

    (A-ii) This is true because individual I-2 does not have sickle cell disease (so she cannot have the *ss* genotype), but she has a child who has sickle cell disease. So individual I-2 must be a carrier of the allele (*Ss*).

    (B-i) Individual III-6 must be heterozygous.

    (B-ii) This is true because individual III-6 does not have sickle cell disease, so she must have at least one *S* allele, and since one of her parents (individual II-4) does have sickle cell disease, she must have inherited a copy of the *s* allele from that parent.

    (C-i)

	Individual 1	
	S	s
S	SS	Ss
s	Ss	ss

    Individual 2

    (D-i) Since there is no malaria in San Francisco, the *SS* genotype would be most advantageous because those individuals would not have sickle cell disease. In Haiti, due to the presence of malaria, the heterozygous genotype (*Ss*) would be the most advantageous because individuals with that genotype would not have sickle cell disease but would also be resistant to malaria.

12. (A-i) The probability of producing a hamster offspring with yellow fur would be $\frac{2}{3}$.

    (B-i and ii) As shown in the following Punnett square, both parents must be heterozygous, so the possible genotypes are *YY*, *Yy*, or *yy*. However, since *YY* is always fatal as per the information stated in the question, there are only two possible outcomes for live offspring: *Yy* or *yy*. So $\frac{2}{3}$ would be heterozygous for yellow fur, and $\frac{1}{3}$ would have white fur.

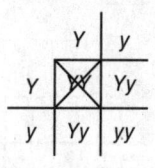

	Y	y
Y	✗	Yy
y	Yy	yy

    (C-i) The frequency of the allele for yellow fur color would probably decrease over time.

    (D-i) Hamsters with white fur color would be better able to blend in with the white sand environment and more likely to hide from potential predators than hamsters with yellow fur. Thus, the frequency of the allele for yellow fur color would likely decrease over time.

Long Free-Response

13. **(A-i and ii)** If both of the parental plants in the cross are heterozygous, one would expect $\frac{3}{4}$ of the progeny to be tall and $\frac{1}{4}$ of the progeny to be short. *TT* and *Tt* plants would be tall, and *tt* plants would be short.

	T	t
T	TT	Tt
t	Tt	tt

(B-i) The independent variable is the presence or absence of Supersize Food.

(B-ii) The dependent variable is the number of plants that are tall or short.

(B-iii) One appropriate constant might be the amount of light the plants were exposed to. Other appropriate constants include the amount of water the plants were given or the type of soil the plants were grown in.

(C-i) Using the formula for chi-square:

$$\chi^2 = \frac{(1,265 - 1,290)^2}{1,290} + \frac{(455 - 430)^2}{430} = \frac{625}{1,290} + \frac{625}{430} = 1.94$$

(C-ii) There are two possible outcomes in the experiment (tall or short), so there is one degree of freedom (df = number of possible outcomes − 1). Using a *p*-value of 0.05 and the chi-square table, the critical value is 3.84. The calculated chi-square value of 1.94 is less than the critical value. So it fails to reject the null hypothesis, and thus the student's claim is supported.

(D-i and ii) A test cross is when an organism with the dominant trait (but an unknown genotype) is crossed with an organism with the recessive phenotype. If the tall plant is heterozygous and is crossed with a short plant (the recessive phenotype), half of the offspring would be tall and half would be short, as shown in this Punnett square.

	T	t
t	Tt	tt
t	Tt	tt

14

Non-Mendelian Genetics

Learning Objectives

In this chapter, you will learn:

- → Linked Genes
- → Codominance, Incomplete Dominance, and Pleiotropy
- → Nonnuclear Inheritance
- → Phenotype = Genotype + Environment

Overview

The inheritance of many traits do not follow the laws of Mendelian genetics. When the observed ratios of phenotypes in the offspring do not follow the ratios predicted by the Punnett squares and Mendelian laws, it may be the result of linked genes, multiple genes coding for the phenotype, nonnuclear inheritance, or even the effects of the environment on the phenotype, all of which will be discussed in this chapter.

Linked Genes

As discussed in Chapter 12, genes that are close together on the same chromosome are said to be linked. Since **linked genes** are on the same chromosome (which is a long piece of DNA), they tend to be inherited together more often than unlinked genes (which are on separate chromosomes).

During prophase I of meiosis, genetic recombination may occur between linked genes. The farther apart two linked genes are on a chromosome, the more likely a genetic recombination event will occur because there is more space in which recombination can happen. The closer together two linked genes are on a chromosome, the less likely a genetic recombination event will occur. Recombination frequencies can be used to create genetic maps that show the distance between genes on a chromosome, as shown in Table 14.1 and Figure 14.1. The distance between genes is measured in **map units**. A recombination frequency of 10% between two genes would place those two genes 10 map units apart. A recombination frequency of 25% between two genes would place those two genes 25 map units apart.

Table 14.1 Genes and Recombination Frequencies

Genes	Recombination Frequency
A and *B*	10%
B and *C*	15%
A and *C*	25%

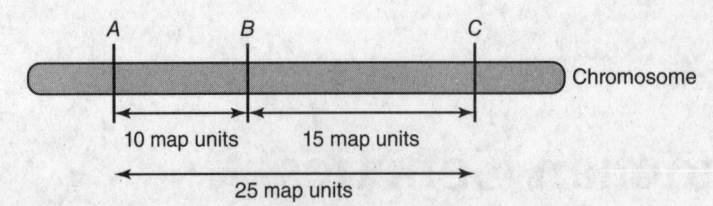

Figure 14.1 Genetic Map from Recombination Frequencies

Sex-Linked Genes

Most genes are located on **autosomes**, chromosomes that are not directly involved in sex determination. Males and females are equally likely to inherit genes located on autosomes. **Sex chromosomes** are involved in sex determination, and genes on sex chromosomes have different inheritance patterns than genes on autosomes.

A special case of linked genes are sex-linked genes. **Sex-linked genes** are genes that are located on sex chromosomes. Sex chromosomes are nonhomologous in humans; females typically have two X chromosomes, and males have one X and one Y chromosome. For this reason, traits that are coded for by sex-linked recessive alleles are more likely to be expressed in males since males have only one X chromosome. Females could also express a sex-linked recessive trait, but since females have two X chromosomes, a female would need to inherit the allele for the trait from both parents in order to express it. When looking at a pedigree that shows more males with the trait than females, it is likely the trait is coded for by a gene that is located on a sex chromosome, as shown in Figure 14.2.

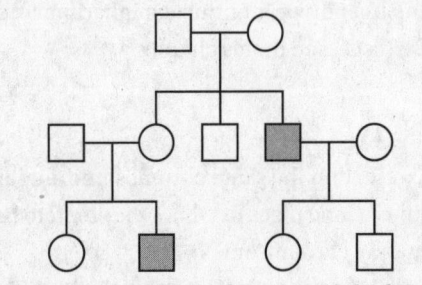

Figure 14.2 Pedigree for a Sex-Linked Trait

> **TIP**
> Remember, in pedigrees, squares typically represent males and circles represent females. Shaded figures usually represent individuals who possess the trait.

Examples of sex-linked recessive traits in humans include hemophilia and color blindness.

There are sex-linked dominant traits in humans, but these are very rare. If a male has a sex-linked dominant trait, all of his daughters will inherit the trait because all females inherit an X chromosome from their father. If a female has a sex-linked dominant trait, both her sons and daughters will have a 50% chance of inheriting the trait.

Codominance, Incomplete Dominance, and Pleiotropy

In **codominance**, both alleles in a heterozygote are expressed. An example of this is Type AB blood in humans. Individuals with the heterozygous $I^A I^B$ genotype express both the I^A and I^B alleles and have the Type AB blood phenotype. This contrasts with **incomplete dominance**, in which the heterozygote expresses a blended version of the alleles. An example of incomplete dominance is the pink flower phenotype in snapdragons. The R allele codes for red flower color, and the W allele codes for white flower color. Snapdragons with the heterozygous RW genotype have pink flowers.

Sometimes, expression of a single gene can have multiple effects—this is called **pleiotropy**. In chickens, a mutation in the alpha-keratin gene (*KRT75*) causes the frizzle phenotype, which results in curled feathers. This mutation also results in abnormal body temperatures and lower egg production in chickens.

Nonnuclear Inheritance

As discussed in Chapter 5, mitochondria and chloroplasts have their own DNA separate from nuclear DNA. Genes on mitochondrial or chloroplast DNA do not follow the inheritance patterns seen in genes located on nuclear DNA.

During gamete formation, the eggs produced in animals and the ovules produced in plants are much, much larger than the sperm (in animals) or pollen (in plants) that are produced. Since the eggs and ovules are larger, they contribute far more mitochondria and mitochondrial DNA (mtDNA) than the sperm or pollen do. In plants, the ovules also contribute more chloroplast DNA (cpDNA) than the pollen do. For this reason, traits on nonnuclear DNA in the mitochondria or chloroplast demonstrate maternal inheritance. Figure 14.3 shows a pedigree that illustrates an example of maternal inheritance. Note that the trait can be passed from mother to either her sons or her daughters. Males with the trait do not pass it on to their offspring.

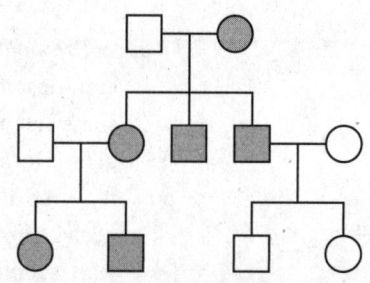

Figure 14.3 Pedigree Showing Nonnuclear Maternal Inheritance

> It is important to note the difference between nonnuclear maternal inheritance and sex-linked inheritance. In sex-linked inheritance, sex-linked genes are located on a sex chromosome (usually the X chromosome) in the nucleus and can be inherited from either fathers or mothers. In nonnuclear inheritance, genes are located in the mitochondria or the chloroplast and can only be inherited from the mother.

Phenotype = Genotype + Environment

The environment can affect gene expression and the resulting phenotype of an organism. For example, in the flowers of the hydrangea plant, a basic soil pH results in flowers with a pink color, while an acidic pH results in blue flowers. In humans, exposure to ultraviolet light can stimulate the expression of genes involved in the production of melanin. This ability of the same genotypes to produce different phenotypes in response to different environmental factors is called **phenotypic plasticity**.

Practice Questions
Multiple-Choice

1. Two genes that are close together on the same chromosome are said to be _____ and are _____ likely to be inherited together than _____ genes.

   (A) linked; less; unlinked
   (B) linked; more; unlinked
   (C) unlinked; less; linked
   (D) unlinked; more; linked

2. Sex-linked recessive traits in humans are _____ likely to be expressed in _____.

   (A) less; males
   (B) more; females
   (C) more; males
   (D) equally; both males and females

3. An example of a trait that involves multiple gene inheritance in plants is seed size. If seed size involves three genes (A, B, and C), in which each dominant allele contributes to increased seed size, which of the following genotypes would result in the smallest seed?

   (A) AABBCC
   (B) AaBBCc
   (C) AaBbCC
   (D) AaBbCc

4. Which of the following statements best explains the difference between nonnuclear inheritance and sex-linked inheritance?

   (A) Females may pass on nonnuclear and sex-linked traits to both their sons and daughters, but only their sons may pass on nonnuclear traits to the next generation.
   (B) Females may pass on nonnuclear and sex-linked traits to both their sons and daughters, but only their daughters may pass on nonnuclear traits to the next generation.
   (C) Females may pass on sex-linked traits to their offspring, but they cannot pass on nonnuclear traits to their offspring.
   (D) Females may pass on nonnuclear traits to their offspring, but they cannot pass on sex-linked traits to their offspring.

5. Two snowshoe hares (a type of rabbit) have the same genotype. During the winter, the hare that lives outdoors in the cold has white fur, but the hare that lives in a climate-controlled and warm environment has brown fur. This is an example of _____.

   (A) sex-linkage
   (B) pleiotropy
   (C) nonnuclear inheritance
   (D) phenotypic plasticity

6. In *Drosophila* (fruit flies), females have two X chromosomes, and males have one X chromosome and one Y chromosome. The eye color gene is located on the X chromosome, and the red-eyed allele is dominant to the white-eyed allele. A heterozygous, red-eyed female is mated with a white-eyed male. Which of the following is the most likely result in their offspring?

   (A) All females are red-eyed, and all males are white-eyed.
   (B) All females are red-eyed, 50% of males are red-eyed, and 50% of males are white-eyed.
   (C) 50% of females are red-eyed, 50% of females are white-eyed, and all males are white-eyed.
   (D) 50% of both sexes are red-eyed, and 50% of both sexes are white-eyed.

7. Sex determination in birds follows the ZW system, where males have two copies of the Z chromosome (ZZ) and females are heterozygous (ZW). A male bird (who is heterozygous for a trait on the Z chromosome) is mated with a female bird (who expresses the recessive phenotype for that trait). Which of the following best describes their most likely offspring?

    (A) 25% males with the dominant phenotype, 25% males with the recessive phenotype, 25% females with the dominant phenotype, and 25% females with the recessive phenotype
    (B) all males with the dominant phenotype and all females with the recessive phenotype
    (C) 25% males with the dominant phenotype, 25% males with the recessive phenotype, and 50% females with the recessive phenotype
    (D) 50% males with the dominant phenotype, 25% females with the dominant phenotype, and 25% females with the recessive phenotype

8. Cats with two X chromosomes are female, and cats with one X and one Y chromosome are male. The gene for fur color in cats is on the X chromosome: X^B is the black allele and X^o is the orange allele. Cats who are heterozygous express the calico phenotype, with patches of black fur and patches of orange fur. A female calico cat is mated with a black male cat. Which of the following best describes the predicted ratios of their offspring?

    (A) $\frac{1}{2}$ of the offspring would be female calico cats, and $\frac{1}{2}$ of the offspring would be male black cats.
    (B) $\frac{1}{4}$ of the offspring would be female black cats, $\frac{1}{4}$ of the offspring would be female calico cats, $\frac{1}{4}$ of the offspring would be male black cats, and $\frac{1}{4}$ of the offspring would be male orange cats.
    (C) $\frac{1}{2}$ of the offspring would be female black cats, and $\frac{1}{2}$ of the offspring would be male calico cats.
    (D) $\frac{1}{2}$ of the offspring would be female black cats, $\frac{1}{4}$ of the offspring would be male black cats, and $\frac{1}{4}$ of the offspring would be male orange cats.

Questions 9 and 10

Refer to the figure, which shows a pedigree of an inherited trait.

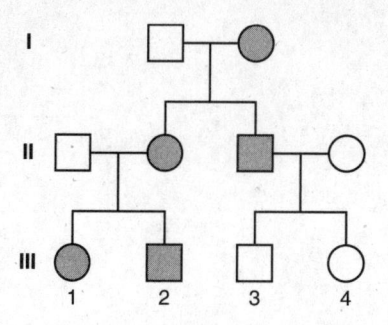

9. Based on this pedigree, which of the following most likely describes the inheritance of this trait?

    (A) autosomal dominant
    (B) autosomal recessive
    (C) sex-linked recessive
    (D) mitochondrial inheritance

10. If individual III-3 has a child with a woman who does not have the allele, what is the most likely probability that their child will have the trait?

    (A) 0%
    (B) 25%
    (C) 50%
    (D) 100%

Short Free-Response

11. Hemophilia is a sex-linked recessive disorder on the X chromosome. The following figure shows a portion of a pedigree of the descendants of Queen Victoria of England. Individuals with hemophilia are indicated by shading.

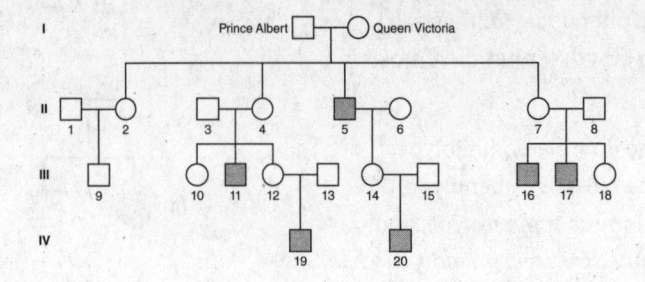

    Part A
    (i) **Describe** the genotype of individual 4 in generation II.
    (ii) **Justify** your answer with evidence from the pedigree.

    Part B
    (i) **Identify** the genotype of individual 14 in generation III.
    (ii) **Explain** how you know her genotype.

    Part C
    (i) Create a Punnett square that **represents** the possible genotypes of the offspring of individuals 7 and 8 in generation II.

    Part D
    (i) **Explain** why the offspring of two closely related individuals are more likely to express recessive traits than the offspring of unrelated individuals.

12. This pedigree shows the inheritance of a trait in three generations of a family.

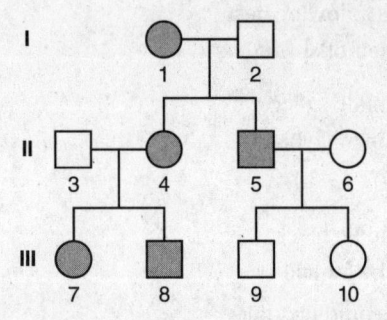

    Part A
    (i) **Describe** a likely mode of inheritance of this trait.
    (ii) Use evidence from the pedigree to **justify** your answer.

    Part B
    (i) **Explain** how a pedigree for a trait that is recessive would look different from a pedigree for a trait that is dominant.

Part C

A student makes a claim that this pedigree indicates the trait is inherited through mitochondrial DNA.

(i) **Evaluate** this claim using evidence from the pedigree.

Part D

(i) **Explain** why the pattern of inheritance of an autosomal dominant trait is different from the pattern of inheritance of a mitochondrial trait.

Long Free-Response

13. In *Drosophila* (fruit flies), the eye color gene is located on the X chromosome. Male flies have an X chromosome and a Y chromosome, and female flies have two X chromosomes.

    Part A

    A red-eyed male fly is mated with a white-eyed female fly. All the resulting female offspring have red eyes, and all the resulting male offspring have white eyes.

    (i) **Explain** which trait, red eyes or white eyes, is dominant and why.

    (ii) **Construct** a Punnett square to support your answer.

    Part B

    Scientists have discovered that white-eyed flies have a mutation in the enzyme TRY 2,3-dioxygenase (TDO2). TDO2 mutations prevent the accumulation of kynurenine, a compound associated with aging. White-eyed flies and red-eyed flies were raised under identical conditions, and the mean life span of both types of flies was recorded.

    (i) **Identify** the independent variable in that experiment.

    (ii) **Identify** the dependent variable in that experiment.

    Part C

    The data from the experiment described in Part B are shown in the following table.

Phenotype	Mean Life Span in Days ($\pm$ 2 SEM*)
White-eyed flies	42.8 ($\pm$ 5.8)
Red-eyed flies	29.1 ($\pm$ 4.9)

    *Standard Error of the Mean

    (i) **Evaluate** the data in the table above to determine whether or not there is likely a statistically significant difference between the life spans of white-eyed and red-eyed flies based on this experiment.

    Part D

    A nondisjunction event in meiosis leads to the production of a heterozygous red-eyed male fly with two X chromosomes and one Y chromosome.

    (i) **Predict** the possible offspring from a mating between this male and a white-eyed female.

    (ii) **Justify** your answer.

Answer Explanations

Multiple-Choice

1. **(B)** Genes that are close together on the same chromosome are *linked* and are *more* likely to be inherited together (since they are close together on the same piece of DNA) than *unlinked* genes. Choice (A) is incorrect because linked genes are more, not less, likely to be inherited together. Choices (C) and (D) are incorrect because unlinked genes may be either on separate chromosomes or far apart on the same chromosome.

2. **(C)** Sex-linked genes are *more* likely to be expressed in *males* since males have only one X chromosome. Choice (A) is incorrect because sex-linked genes are more likely, not less likely, to be expressed in males. Females have two X chromosomes, so a sex-linked recessive trait is less likely to be expressed in females because a female would have to inherit two copies of the allele while a male would only have to inherit one copy of the allele for it to be expressed. Thus, choices (B) and (D) are both incorrect.

3. **(D)** *AaBbCc* has the least number of dominant alleles (three) out of all the answer choices given. So the additive effect of the alleles in that genotype would be the least, and this genotype would result in the smallest seed. Since each dominant allele contributes to increased seed size, choice (A) is incorrect because it has the highest number of dominant alleles out of all the answer choices given. So it would produce the largest seeds. Choices (B) and (C) are incorrect because they both have four dominant alleles, so they both would produce larger seeds than choice (D).

4. **(B)** Females may pass on sex-linked traits to the next generation through their X chromosomes and nonnuclear traits through their mitochondrial DNA. Since mitochondrial DNA is passed on through eggs and not sperm, males cannot pass on nonnuclear traits to the next generation, but females may pass on nonnuclear traits to the next generation. Therefore, the only true statement is choice (B), and choices (A), (C), and (D) are incorrect.

5. **(D)** Phenotypic plasticity describes the ability of two individuals with the same genotype to produce different phenotypes, depending on the different environmental factors that affect each individual. Choice (A) is incorrect because sex-linkage involves traits found on the sex chromosomes. Pleiotropy is when a single gene locus affects two or more distinct phenotypic traits, so choice (B) is incorrect. Nonnuclear inheritance refers to traits coded for by genes outside of the nucleus (in the mitochondria or chloroplast), so choice (C) is incorrect.

6. **(D)** A heterozygous, red-eyed female fly would have the genotype $X^R X^r$, and a white-eyed male fly would have the $X^r Y$ genotype. The Punnett square for this cross is shown in the following figure.

	X^R	X^r
X^r	$X^R X^r$	$X^r X^r$
Y	$X^R Y$	$X^r Y$

$\frac{1}{4}$ red-eyed females
$\frac{1}{4}$ white-eyed females
$\frac{1}{4}$ red-eyed males
$\frac{1}{4}$ white-eyed males

The most likely result would be 50% of both sexes would have red eyes and 50% of both sexes would have white eyes.

7. **(A)** The figure that follows shows the Punnett square for the cross between a heterozygous male and a female with the recessive trait. So $\frac{1}{4}$ of the offspring would be males with the dominant phenotype, $\frac{1}{4}$ would be males with the recessive phenotype, $\frac{1}{4}$ would be females with the dominant phenotype, and $\frac{1}{4}$ would be females with the recessive phenotype.

	Z^A	Z^a
Z^a	$Z^A Z^a$	$Z^a Z^a$
W	$Z^A W$	$Z^a W$

$\frac{1}{4}$ males with dominant phenotype
$\frac{1}{4}$ males with recessive phenotype
$\frac{1}{4}$ females with dominant phenotype
$\frac{1}{4}$ females with recessive phenotype

8. **(B)** As shown in the figure, $\frac{1}{4}$ of the offspring would be black female cats, $\frac{1}{4}$ would be calico female cats, $\frac{1}{4}$ would be black male cats, and $\frac{1}{4}$ would be orange male cats.

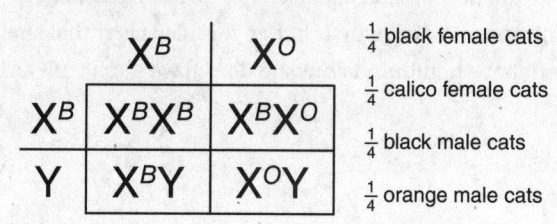

	X^B	X^O
X^B	$X^B X^B$	$X^B X^O$
Y	$X^B Y$	$X^O Y$

$\frac{1}{4}$ black female cats
$\frac{1}{4}$ calico female cats
$\frac{1}{4}$ black male cats
$\frac{1}{4}$ orange male cats

9. **(D)** In this pedigree, both females and males are affected, but only females appear to be able to pass the trait on to the next generation—these are hallmarks of mitochondrial inheritance. Autosomal traits are equally likely in both sexes, so choices (A) and (B) are incorrect. Sex-linked traits are more likely to appear in males than females, so choice (C) is incorrect.

10. **(A)** This is most likely a mitochondrial trait (see the explanation for Question 9). Individual III-3 does not have the trait, so he does not carry the allele. Even if he did carry the allele, males cannot pass on mitochondrial traits to the next generation. If he has a child with a woman who also does not have the allele, there is 0% chance of them having a child with the trait since neither of them have the allele for the trait. Choices (B), (C), and (D) are incorrect because if the woman does not have the allele for a mitochondrial trait, her offspring will not have the allele for the trait.

Short Free-Response

11. (A-i and ii) Individual II-4 is a heterozygous carrier of the hemophilia allele ($X^H X^h$) because she does not have hemophilia but she does have a male offspring who does have hemophilia.

    (B-i and ii) Individual III-14 is a heterozygous carrier of the hemophilia allele ($X^H X^h$) because her father has hemophilia and all females inherit an X chromosome from their father. It is also clear that she is a heterozygous carrier because she does not have hemophilia but she does have a male offspring who does have hemophilia.

    (C-i)

    Individual 7

	X^H	X^h
X^H	$X^H X^H$	$X^H X^h$
Y	$X^H Y$	$X^h Y$

    Individual 8

    (D-i) Closely related individuals are more likely to share the same alleles. If two closely related individuals share the same recessive alleles, their offspring would be more likely to express that recessive trait than two individuals who are not as closely related and do not have the same recessive alleles.

12. (A-i and ii) Because the trait is expressed in every generation, it could be an autosomal dominant trait. Since both sexes can express the trait but only females can pass the trait to the next generation, the mode of inheritance could be mitochondrial (nonnuclear).

    (B-i) If a trait is dominant, it will usually appear in every generation in a pedigree. Recessive traits will appear in offspring but do not necessarily appear in the parents because the parents may be unaffected heterozygous carriers of the trait.

    (C-i) Mitochondrial DNA is passed on from mother to offspring. The egg is thousands of times larger than the sperm, so during fertilization, the zygote inherits its mitochondria from the mother. Mothers will pass on mitochondrial DNA to all of their offspring, but their sons will not pass it on to the next generation. This is exactly what is shown in the three generations of this pedigree.

    (D-i) Autosomal dominant traits can be passed on to the next generation by either males or females. Mitochondrial traits can only be passed on to the next generation by females.

Long Free-Response

13. (A-i)   Females inherit one X chromosome from their male parent and one from their female parent. Since all of the females produced from the cross have the same phenotype as their male parent (red eyes), red eyes must be dominant to white eyes. Males inherit their only X chromosome from their female parent, so all of the males have the white-eyed phenotype.

    (A-ii)   The results of the cross are shown in the following figure.

	X^R	Y
X^r	$X^R X^r$	$X^r Y$
X^r	$X^R X^r$	$X^r Y$

    (B-i)   The independent variable is the color of the eyes (or the presence or absence of the TDO2 mutation).

    (B-ii) The dependent variable is the mean life span.

    (C-i)   There is likely a statistically significant difference if the 95% confidence intervals do NOT overlap. The lower limit of the 95% confidence interval for the mean life span of the white-eyed flies is 37 days (42.8 − 5.8). The upper limit of the 95% confidence interval for the mean life span of the red-eyed flies is 34 days (29.1 + 4.9). The 95% confidence intervals do not overlap, so there is likely a statistically significant difference in the life spans of the two groups of flies in this experiment.

    (D-i and ii)   In this mating, $\frac{2}{3}$ of the females would have red eyes, $\frac{1}{3}$ of the females would have white eyes, $\frac{2}{3}$ of the males would have white eyes, and $\frac{1}{3}$ of the males would have red eyes, as shown in the figure.

	Possible Gametes from $X^R X^r Y$					
Possible Gamete from $X^r X^r$	$X^R X^r$	$X^R Y$	$X^r Y$	X^R	X^r	Y
X^r	$X^R X^r X^r$	$X^R X^r Y$	$X^r X^r Y$	$X^R X^r$	$X^r X^r$	$X^r Y$
Phenotypes	Red-eyed female (with 3 X chromosomes)	Red-eyed male (with 2 X chromosomes)	White-eyed male (with 2 X chromosomes)	Red-eyed female	White-eyed female	White-eyed male

UNIT 6
Gene Expression and Regulation

15

DNA, RNA, and DNA Replication

Learning Objectives

In this chapter, you will learn:
- Structure of DNA and RNA
- DNA Replication

Overview

Nucleic acids, DNA and RNA, are the carriers of genetic information in living organisms. This genetic information is transferred from one generation to the next generation through these nucleic acids. While prokaryotes and eukaryotes typically use DNA as the carrier of genetic information between generations, some viruses use RNA as their primary genetic material. Prokaryotes package their DNA into circular chromosomes in a region called the nucleoid. Eukaryotes package their DNA into linear chromosomes in the nucleus. Both prokaryotes and eukaryotes can carry additional genetic information in the form of small, circular pieces of DNA called plasmids, which are referred to as extranuclear DNA.

DNA and RNA have some structural similarities and structural differences, which are important to their respective functions. Both will be discussed within this chapter followed by a review of DNA replication.

Structure of DNA and RNA

DNA and RNA are composed of nucleotides. Each nucleotide has three parts: a nitrogenous base, a five-carbon sugar, and a phosphate group. The nitrogenous bases in DNA are adenine, guanine, cytosine, and thymine. RNA contains the nitrogenous bases adenine, guanine, cytosine, and uracil. These nitrogenous bases are classified into two groups, as shown in Figure 15.1. Adenine and guanine are purines and consist of a two-ringed structure. Cytosine, thymine, and uracil are pyrimidines and have a one-ringed structure.

Figure 15.1 Nitrogenous Bases in DNA and RNA

DNA and RNA have consistent base-pairing rules that are seen in all living organisms. In DNA, adenine (A) pairs with thymine (T), and guanine (G) pairs with cytosine (C). RNA base-pairing rules are similar in that guanine pairs with cytosine but different in that adenine (A) pairs with uracil (U). When adenine pairs with another base, either thymine or uracil, it forms two hydrogen bonds. Guanine and cytosine pairs form three hydrogen bonds between them.

The five-carbon sugar also differs slightly in DNA and RNA. DNA contains the five-carbon sugar deoxyribose, while RNA has the five-carbon sugar ribose. An important difference between these sugars is at the 2′ carbon. Deoxyribose has a hydrogen atom attached to the 2′ carbon, while ribose has a hydroxyl (−OH) group attached to the 2′ carbon, as shown in Figure 15.2.

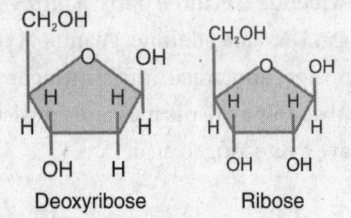

Figure 15.2 Deoxyribose and Ribose

This small difference makes DNA much more stable than RNA. This increased stability of DNA may explain why it is the carrier of genetic information between generations, while RNA usually has more temporary functions. For example, in transcription, mRNA carries the genetic information from the nucleus to the ribosome.

DNA is a double helix in which the two strands are antiparallel. One strand of the DNA is oriented with the 5′ phosphate group at the start of the strand, while the opposite strand has the 3′ hydroxyl (−OH) group at the start of the strand. A purine on one of the strands is always paired with a pyrimidine on the opposite strand. Since purines have a double-ringed structure and pyrimidines have a single-ringed structure, this keeps the width of the double helix consistent.

RNA is typically single-stranded (in mRNAs and microRNAs) but can fold to form three-dimensional structures in rRNAs in the ribosome and in tRNAs.

> **TIP**
>
> An easy way to remember which nitrogenous bases are purines and which are pyrimidines are the mnemonic devices "PureAsGold" for Purines are Adenine and Guanine, and "CUtThePy" for Cytosine, Uracil, and Thymine are Pyrimidines.

DNA Replication

The purpose of DNA replication is to ensure the continuity of genetic information between generations. Each of the original two strands in the double helix serves as a template for a new strand, as shown in Figure 15.3. Since each new double helix is composed of one strand from the original piece of DNA and one newly synthesized strand, DNA replication is described as being semiconservative.

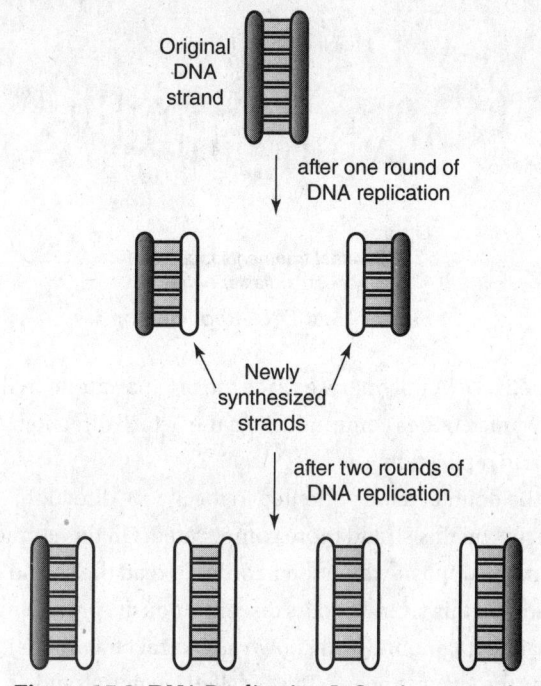

Figure 15.3 DNA Replication Is Semiconservative

To start DNA replication, the enzyme helicase first unwinds the two DNA strands in an area called the origin of replication, or "ori" site. As part of the double helix is unwound, other sections of the double helix become more tightly wound, and this results in supercoiling in those areas. Topoisomerase enzymes make temporary nicks in the sugar-phosphate backbone of DNA to relieve this supercoiling and then reseal these nicks. The enzyme RNA polymerase then synthesizes an RNA primer using a few complementary RNA nucleotides. New DNA nucleotides can then be added to this RNA primer. DNA polymerase is the enzyme that adds new nucleotides to the 3′ hydroxyl group at the end of this RNA primer. DNA polymerase adds new nucleotides in the 5′ to 3′ direction, always connecting the 5′ phosphate on the new nucleotide to the 3′ hydroxyl on the growing nucleotide strand.

As discussed previously, DNA molecules have directionality; the two strands of the DNA double helix are antiparallel (oriented in opposite directions). Because DNA polymerase can only add new nucleotides in the 5′ to 3′ direction, and because the two strands of DNA are antiparallel, DNA must proceed slightly differently on the two strands, as shown in Figure 15.4.

TIP

When you think of DNA polymerase adding new nucleotides in the 5′ to 3′ direction, think of adding new cars to a toy train. New cars can only be added to the back end of the train, not to the engine, just like new nucleotides can only be added to the 3′ hydroxyl, not the 5′ phosphate.

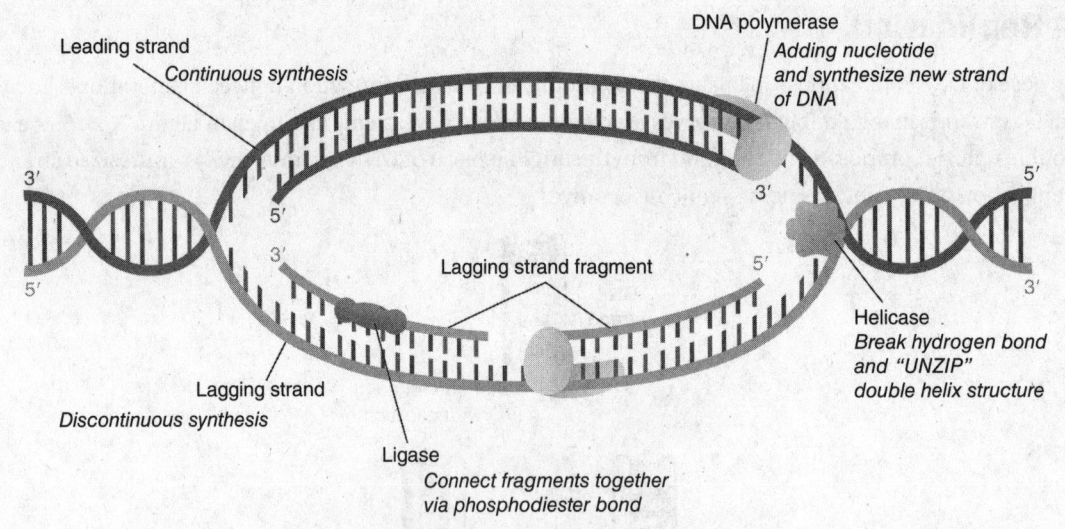

Figure 15.4 DNA Replication

On one strand of the double helix, DNA polymerase reads the original strand in the 3' to 5' direction and can add new nucleotides continuously in the 5' to 3' direction. This process is called leading strand replication.

However, the other strand of the double helix is oriented in the 5' to 3' direction, which makes the replication process on this strand more complicated. On this strand, DNA polymerase must proceed in the opposite direction in order to read the strand in the 3' to 5' direction. Replication on this strand occurs discontinuously, producing short fragments called lagging strand fragments (also known as Okazaki fragments), which are then joined together by the enzyme ligase. This is called lagging strand replication.

> DNA replication can be an intimidating process to learn, and many textbooks have far more detail than you need to know for the AP Biology exam. The most important things to know are why DNA replication is considered semiconservative and the differences between leading and lagging strand replication. Make sure you understand the functions of helicase, topoisomerase, DNA polymerase, RNA polymerase, and ligase in this process.

Practice Questions

Multiple-Choice

1. The genome of a newly discovered virus has the following nucleotide composition: 22% guanine, 16% cytosine, 34% adenine, and 28% uracil. Based on the nucleotide composition, the genome of this virus is most likely made of which of the following?

   (A) single-stranded DNA
   (B) double-stranded DNA
   (C) single-stranded RNA
   (D) double-stranded RNA

2. Prokaryotic genomes are packaged into _____ chromosomes in the _____.

   (A) circular; nucleoid region
   (B) circular; nucleus
   (C) linear; nucleoid region
   (D) linear; nucleus

3. Energy is required to separate the two strands of the DNA double helix because of the hydrogen bonds between the base pairs. Based on the base pair content, which of the following would require the least amount of energy to separate the strands of DNA?

   (A) 20% G, 20% C, 30% A, 30% T
   (B) 30% G, 30% C, 20% A, 20% T
   (C) 15% A, 15% T, 35% G, 35% C
   (D) 25% G, 25% C, 25% A, 25% T

4. Which of the following correctly describes a structural difference between DNA and RNA?

   (A) DNA has a five-carbon sugar; RNA has a four-carbon sugar.
   (B) DNA has thymine; RNA has uracil.
   (C) DNA has adenine; RNA has cytosine.
   (D) DNA is typically single-stranded; RNA is typically double-stranded.

Questions 5 and 6

N-15, also known as heavy nitrogen, is an isotope of nitrogen that is heavier than the isotope that is typically found in nature, N-14. Conducting chemical reactions in the presence of different isotopes of nitrogen allow a scientist to follow nitrogen atoms in a metabolic pathway. In a classic experiment, Meselson and Stahl allowed parent DNA (containing N-15) to replicate in the presence of N-14.

5. After **one** round of DNA replication, which of the following results would support the statement "DNA replication is semiconservative"?

   (A) All DNA molecules have one strand containing N-14 and one strand containing N-15.
   (B) 50% of the DNA molecules only contain N-14, and 50% of the DNA molecules only contain N-15.
   (C) All DNA molecules only contain N-15.
   (D) All DNA molecules only contain N-14.

6. After **two** rounds of DNA replication, which of the following results would support the statement "DNA replication is semiconservative"?

   (A) All DNA molecules have one strand containing N-14 and one strand containing N-15.
   (B) 50% of the DNA molecules only contain N-14, and 50% of the DNA molecules only contain N-15.
   (C) All DNA molecules only contain N-15.
   (D) 50% of the DNA molecules contain only N-14, and 50% of the DNA molecules have one strand containing only N-15 and one strand containing only N-14.

7. Why does DNA replication proceed continuously on one strand of the double helix but discontinuously on the other strand of the double helix?

   (A) One strand of the double helix contains thymine, and the other strand contains uracil.
   (B) Only G and C nucleotides appear on one strand of the double helix, and only A and T nucleotides appear on the other strand of the double helix.
   (C) RNA polymerase can only synthesize RNA primers on one strand of the double helix.
   (D) DNA polymerase can only add new nucleotides in the 5′ to 3′ direction.

8. The enzyme _____ unwinds the double helix of DNA, and the enzyme _____ relieves the supercoiling created by this unwinding.

   (A) helicase; topoisomerase
   (B) DNA polymerase; topoisomerase
   (C) DNA polymerase; ligase
   (D) helicase; ligase

9. The enzyme _____ adds new nucleotides to both the lagging and leading strand, and the enzyme _____ joins the discontinuous segments synthesized on the lagging strand.

   (A) helicase; topoisomerase
   (B) DNA polymerase; topoisomerase
   (C) DNA polymerase; ligase
   (D) helicase; ligase

10. Which of the following are small, circular pieces of extranuclear DNA that can be found in either prokaryotes or eukaryotes?

    (A) Okazaki fragments
    (B) RNA primers
    (C) plasmids
    (D) linear chromosomes

Short Free-Response

11. Reverse transcriptase is an enzyme that makes a complementary DNA copy of RNA in retroviruses. This DNA copy can then insert itself into the genome of the host cell. Reverse transcriptase has a higher error rate than DNA polymerase, which results in more mutations in the DNA copy of the RNA. Reverse transcriptase is not typically used by eukaryotic cells for any function.

    Part A
    (i) **Describe** which nucleotides you would expect to find in the genome of a virus that uses reverse transcriptase.

    Part B
    The human immunodeficiency virus (HIV) contains RNA as its genetic material. Reverse transcriptase inhibitors have been shown to be effective in slowing the replication of HIV.
    (i) **Explain** why reverse transcriptase inhibitors have few side effects in eukaryotic cells.

    Part C
    (i) **Predict** the mutation rate of a retrovirus compared to that of a DNA virus.

    Part D
    (i) **Justify** your prediction from Part C.

12. An experiment is conducted to study the effect of a ligase inhibitor on DNA replication.

    Part A
    (i) **Describe** the function of ligase in DNA replication.

    Part B
    (i) **Identify** an appropriate control for this experiment.

    Part C
    (i) **Predict** which strand of the DNA would be most affected by a ligase inhibitor.

    Part D
    (i) **Justify** your prediction from Part C.

Long Free-Response

13. Polymerase chain reaction (PCR) uses a heat-stable DNA polymerase and repeated cycles of DNA replication to amplify specific sequences of DNA. Primers specific to the desired DNA sequences are used to direct DNA polymerase to the beginning of the sequence to be amplified. The following table shows the number of copies of the DNA sequence that exist at the end of each cycle.

Number of Cycles of DNA Replication	Number of Copies of Desired DNA Sequence
0	1
1	2
2	4
3	8
4	16
5	32
6	64
7	128

Part A
(i) **Explain** why a primer is needed to direct DNA polymerase to copy the desired sequence.

Part B
(i) On the axes provided, **construct** a graph of the data from the table.

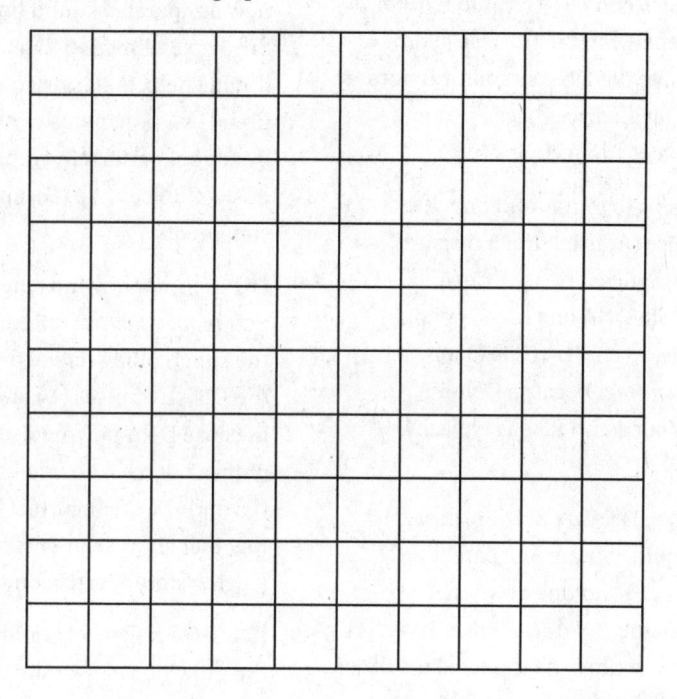

Part C
(i) Analyze the data and **state** the mathematical relationship between the number of cycles of PCR and the number of copies of the desired DNA sequence generated.

Part D
(i) **Predict** the minimum number of cycles required to generate 1,000 copies of the desired DNA sequence.

(ii) **Justify** your prediction.

Answer Explanations
Multiple-Choice

1. **(C)** Since the question states that the virus's genome contains uracil, choices (C) and (D) are possibilities since RNA contains uracil and DNA does not; thus, rule out choices (A) and (B). However, since the percentage of guanine does not equal the percentage of cytosine and the percentage of adenine does not equal the percentage of uracil, it cannot be a double-stranded virus, so choice (C) is the best answer.

2. **(A)** Prokaryotes have *circular* chromosomes in the *nucleoid region* of their cells. Choice (B) is incorrect because prokaryotes do not have a nucleus. Choices (C) and (D) are incorrect because prokaryotes do not have linear chromosomes; eukaryotes have linear chromosomes.

3. **(A)** Since it has the lowest G-C content and since each G-C pair has three hydrogen bonds between them (instead of the two hydrogen bonds that are between each A-T pair), choice (A) would require the least amount of energy to break those hydrogen bonds and separate the DNA strands. Choices (B), (C), and (D) are all incorrect because they have higher G-C contents than choice (A).

4. **(B)** Only DNA contains thymine, and only RNA contains uracil. Choice (A) is incorrect because RNA has a five-carbon sugar (ribose), not a four-carbon sugar. Both DNA and RNA contain adenine and cytosine, so choice (C) is incorrect. Choice (D) is incorrect because DNA is typically double-stranded and RNA is typically single-stranded.

5. **(A)** After one round of DNA replication, every molecule of DNA would contain one parent strand (containing N-15) and one newly synthesized strand (containing N-14). Choice (B) is incorrect because it describes the expected result from conservative DNA replication, not semiconservative replication. Choices (C) and (D) are both incorrect because after one round of DNA replication, no DNA molecules would contain solely N-15 or N-14.

6. **(D)** As shown in Figure 15.3, after two rounds of DNA replication, 50% of the DNA molecules would contain strands with only N-14. The other 50% of the DNA molecules would contain one strand with only N-15 and one strand with only N-14. Choice (A) describes the result of only one round of DNA replication if DNA replication was semiconservative, not the result of two rounds of DNA replication. So (A) is incorrect. Conservative DNA replication would produce the result described in choice (B), so (B) is incorrect. Choice (C) is incorrect because it describes the result if no replication was happening at all.

7. **(D)** Because the two strands of DNA are antiparallel and because DNA polymerase can only add new nucleotides in the 5' to 3' direction, replication must occur differently on the two strands of DNA. On the strand of the double helix that is oriented in the 3' to 5' direction, DNA polymerase can perform replication continuously, adding a new antiparallel strand one nucleotide at a time in the 5' to 3' direction. On the opposite strand of the double helix that is oriented in the 5' to 3' direction, DNA polymerase must work in the opposite direction, performing replication discontinuously. Choices (A), (B), and (C) are all incorrect statements.

8. **(A)** *Helicase* unwinds the double helix of DNA, and *topoisomerase* relieves the stress caused by the supercoiling (created when helicase unwinds the DNA). Choices (B) and (C) are incorrect because DNA polymerase adds new nucleotides; it does not have an unwinding function. Choices (C) and (D) are incorrect because ligase joins together short segments of DNA created on the lagging strand by discontinuous replication.

9. **(C)** *DNA polymerase* adds new nucleotides to the growing DNA strand, and *ligase* joins the discontinuous segments of DNA on the lagging strand together. Choice (A) is incorrect because helicase unwinds the DNA double helix and topoisomerase relieves the supercoiling created by this unwinding. Since topoisomerase relieves

supercoiling, choice (B) is also incorrect. Choice (D) is incorrect because helicase is responsible for unwinding the DNA double helix, not adding new nucleotides.

10. **(C)** Plasmids are small, circular pieces of DNA outside of the nucleus that can be found in prokaryotes and eukaryotes. Okazaki fragments are the short segments of DNA created by discontinuous replication on the lagging strand of DNA, so choice (A) is incorrect. Choice (B) is incorrect because RNA primers are used to give DNA polymerase a place to start adding nucleotides to on the growing DNA strand. Linear chromosomes are found in eukaryotes, not prokaryotes, and they are in the nucleus (they are not extranuclear). Thus, choice (D) is also incorrect.

Short Free-Response

11. (A-i) A virus that uses reverse transcriptase would have RNA as its genetic material, so its genome would contain the nucleotides adenine, cytosine, guanine, and uracil. (Thymine is not found in RNA.)

    (B-i) Eukaryotic cells contain DNA as their genetic material and do not need to use reverse transcriptase to make a DNA copy of their genetic material. Therefore, eukaryotic cells do not contain reverse transcriptase. A reverse transcriptase inhibitor would have few, if any, side effects on eukaryotic cells.

    (C-i) Retroviruses would be expected to have a higher mutation rate than that of DNA viruses.

    (D-i) One reason why a retrovirus would be expected to have a higher mutation rate than that of a DNA virus is because retroviruses use reverse transcriptase to copy their genome. Reverse transcriptase is less accurate and generates more mutations than DNA polymerase, which would lead to a higher mutation rate in retroviruses.

12. (A-i) The function of ligase in DNA replication is to join together the fragments created during replication of the lagging strand in DNA.

    (B-i) An appropriate control would be to conduct replication of the same DNA sequence without the presence of the ligase inhibitor.

    (C-i) The lagging strand of DNA would be most affected by a ligase inhibitor.

    (D-i) Replication of the lagging strand of DNA would produce multiple short fragments that would need to be joined by ligase. An inhibitor of ligase would prevent this from happening.

Long Free-Response

13. (A-i)   DNA polymerase needs a 3' hydroxyl group to which it can add new nucleotides, so a primer is necessary to allow DNA polymerase to add the first new nucleotide.

(B-i)

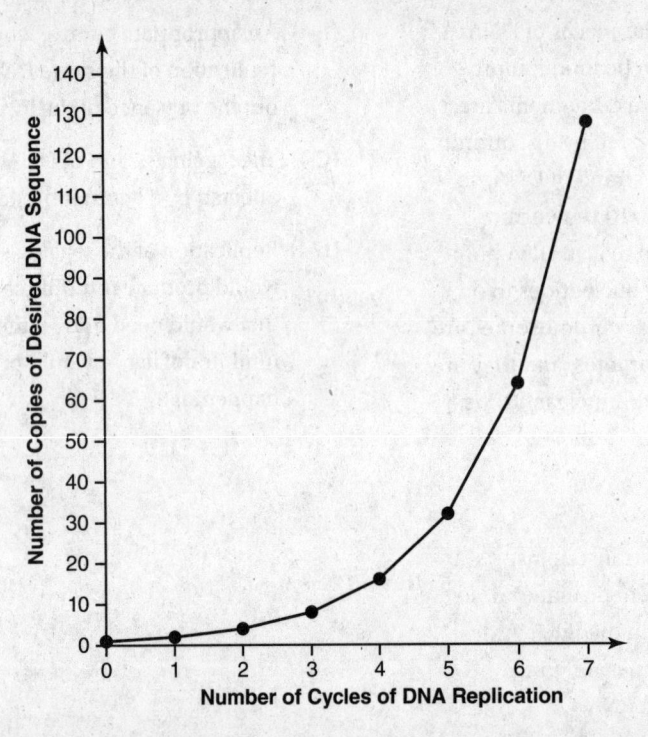

(C-i)   Each cycle of PCR doubles the number of copies of the desired DNA sequence. In other words, this is an exponential growth relationship.

(D-i and ii)   Since the number of copies of the desired DNA sequence doubles with each cycle, a minimum of 10 cycles would be required to produce at least 1,000 copies. After 8 cycles, 256 copies would be present. After 9 cycles, 512 copies would be present, and after 10 cycles, 1,024 copies would be present.

16

Transcription and Translation

Learning Objectives

In this chapter, you will learn:
- → Transcription in Prokaryotes vs. Transcription in Eukaryotes
- → Translation
- → Flow of Information from the Nucleus to the Cell Membrane

Overview

The term *central dogma* describes the typical flow of genetic information in a cell: DNA is transcribed into mRNA; mRNA is then translated into proteins by ribosomes. This chapter will review the processes of transcription and translation and explore how these processes are similar or different in prokaryotes and eukaryotes.

Transcription in Prokaryotes vs. Transcription in Eukaryotes

Transcription is the process in which the genetic information in a sequence of DNA nucleotides is copied into newly synthesized RNA molecules. **RNA polymerase** is the enzyme that transcribes a DNA sequence into RNA molecules. The function of an RNA molecule depends on its structure and sequence:

- **mRNA (messenger RNA)** is single-stranded and carries information from DNA to the ribosome. mRNA contains three base pair sequences called **codons**, which are complementary to the DNA base pair sequence. These codons will specify specific amino acids during translation.
- **tRNA (transfer RNA)** folds into a three-dimensional structure that acts as an adaptor molecule during translation. One end of the tRNA will bind to a specific amino acid, while the other end of tRNA contains an anticodon, which will pair with the appropriate mRNA codon at the ribosome during translation.
- **rRNA (ribosomal RNA)** also folds into a three-dimensional structure. rRNA and proteins form ribosomes that perform translation. The three-dimensional rRNA acts as a **ribozyme**, catalyzing the reactions needed in translation.

To start transcription, the enzyme RNA polymerase must bind to a noncoding DNA sequence called a **promoter**. The promoter sequence does not code for any amino acids and instead serves as a binding site for RNA polymerase upstream from the start of the coding region of a gene. Promoter sequences are highly conserved in living organisms. Most eukaryotic promoters contain a region called a **TATA box**, so named because it is rich in thymine and adenine nucleotides. Proteins called **transcription factors** help RNA polymerase bind to the promoter sequence and begin transcription.

During transcription, RNA polymerase adds new RNA nucleotides in the 5' to 3' direction, similar to how DNA polymerase adds new DNA nucleotides in the 5' to 3' direction during DNA replication. The strand of DNA being transcribed by RNA polymerase is called the template strand. The newly synthesized RNA must be antiparallel to the template DNA strand, so RNA polymerase reads the template DNA

> In music, *transcription* means the arrangement of a piece of music for different instruments. Think of transcription in biology as the arrangement of the "music" of genetic information in DNA for the different "instruments" of mRNA, tRNA, and rRNA.

strand in the 3' to 5' direction. The sequence of base pairs in the RNA strand that is transcribed is complementary to the template DNA sequence. For example, if the DNA sequence is 3' – ACG TAC GTA CGT – 5', the newly synthesized RNA sequence will be 5' – UGC AUG CAU GCA – 3'. The strand of the double helix that functions as the template strand can vary depending on the gene being transcribed.

Remember that one of the differences between prokaryotic and eukaryotic cells is that prokaryotic cells do not have a nucleus. Therefore, in most prokaryotic cells, the mRNA transcript formed in transcription is immediately accessible to ribosomes and can be translated without delay. In eukaryotic cells, the initial mRNA transcript, referred to as **pre-mRNA**, needs to be modified before it can leave the nucleus of the cell and travel to the ribosomes for translation.

In eukaryotic cells, three modifications must occur to the pre-mRNA before it can leave the nucleus: the removal of introns and the joining of exons, the addition of a guanosine triphosphate (GTP) cap to the 5' end of the RNA, and the addition of a poly-adenine (poly-A) tail to the 3' end of the RNA.

> **TIP**
> Remember that *RNA polymerase* is the enzyme responsible for *transcription* and forms *RNA* molecules while *DNA polymerase* is the enzyme responsible for *DNA replication* and forms *DNA* molecules. Be sure not to confuse these two enzymes on the AP exam.

1. Eukaryotic pre-mRNAs contain noncoding RNA sequences called **introns**. These introns are interspersed between coding sequences of eukaryotic RNAs called **exons**. Structures called **spliceosomes**, which are made of small nuclear RNAs (**snRNAs**) and small nuclear ribonucleoproteins (**snRNPs**), remove the introns from the pre-mRNA and then splice together the exons. These exons can be joined in different combinations to generate multiple RNA transcripts from the same gene, as shown in Figure 16.1. This **alternative splicing** gives eukaryotes the ability to generate a greater variety of RNA transcripts from just one gene than what can be generated from one gene in prokaryotes.

> In the human immune system, alternative splicing of exons in the antibody-producing B cells allows for the production of a wide variety of antibody molecules in response to evolving pathogens.

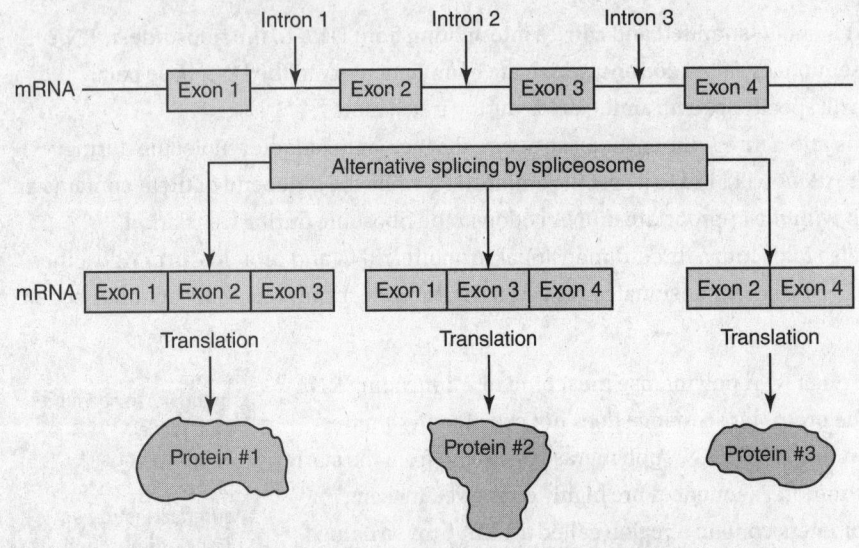

Figure 16.1 Alternative Splicing

While introns do not code for amino acids, some introns may function in the regulation of gene expression.

2. To protect the 5' end of the pre-mRNA transcript from degradation before it can be translated, a **5' GTP cap** is added. Nuclear pores recognize this 5' GTP cap and allow mRNAs with the cap to exit the nucleus. This GTP cap also helps in the initiation of translation when the RNA reaches the ribosome.
3. The enzyme poly-A polymerase adds a string of adenine nucleotides to the 3' end of the pre-mRNA transcript. This **3' poly-A tail** helps prevent degradation of the transcript. mRNAs with longer 3' poly-A tails tend to have longer durations in the cytosol, which allows more copies of the protein (that the mRNA codes for) to be generated.

After the excision of the introns and splicing of the exons, as well as the additions of the 5' GTP cap and the 3' poly-A tail, the transcript is now referred to as a **mature mRNA** and is ready to be translated by the ribosome.

Translation

Translation of mRNA molecules occurs at the ribosomes in both prokaryotes and eukaryotes. Ribosomes are found in the cytoplasm of both prokaryotes and eukaryotes. Ribosomes are also found on the rough endoplasmic reticulum of eukaryotes. In eukaryotes, cytoplasmic ribosomes usually translate proteins that will stay inside the cell, while ribosomes on the rough endoplasmic reticulum usually translate proteins that will be exported from the cell.

Because prokaryotes do not have a nucleus, translation of the mRNA can occur as the mRNA is being transcribed. Multiple ribosomes can be simultaneously translating a prokaryotic RNA, forming **polyribosomes** (also known as polysomes), as shown in Figure 16.2.

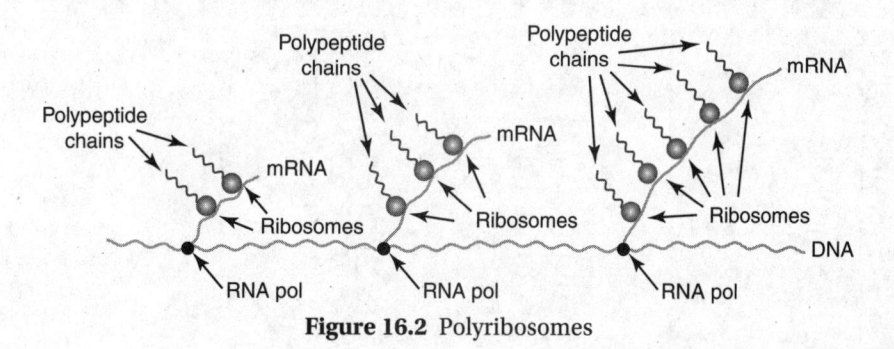

Figure 16.2 Polyribosomes

The process of translation requires energy and involves three main steps in both prokaryotes and eukaryotes: initiation of translation, elongation of the polypeptide chain, and termination of translation.

1. **Initiation:** The genetic code in mRNA is read in three base pair units called codons. Translation is initiated when the rRNA in the ribosome pairs with the **start codon** (AUG, which codes for the amino acid methionine). A tRNA with the complementary **anticodon** (in this example, UAC) brings the appropriate amino acid to the ribosome, and the anticodon on tRNA pairs with the codon on the mRNA. In this way, tRNA functions as an "adaptor" molecule, linking the correct amino acid with the correct codon on the mRNA. This process is called initiation and is shown in Figure 16.3.

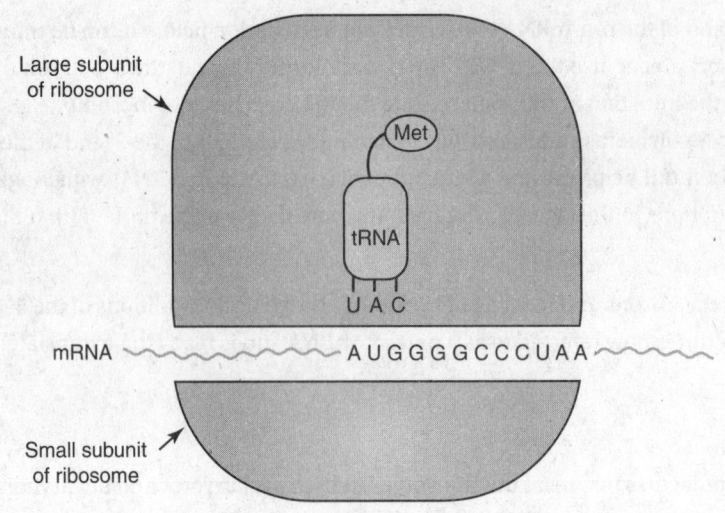

Figure 16.3 Initiation of Translation

As Figure 16.4 shows, 61 of the 64 possible codons code for amino acids. Three of the 64 codons do not code for amino acids and are stop codons.

First Letter	Second Letter: U	Second Letter: C	Second Letter: A	Second Letter: G	Third Letter
U	UUU, UUC – Phe; UUA, UUG – Leu	UCU, UCC, UCA, UCG – Ser	UAU, UAC – Tyr; UAA, UAG – Stop	UGU, UGC – Cys; UGA – Stop; UGG – Trp	U, C, A, G
C	CUU, CUC, CUA, CUG – Leu	CCU, CCC, CCA, CCG – Pro	CAU, CAC – His; CAA, CAG – Gln	CGU, CGC, CGA, CGG – Arg	U, C, A, G
A	AUU, AUC, AUA – Ile; AUG – Met	ACU, ACC, ACA, ACG – Thr	AAU, AAC – Asn; AAA, AAG – Lys	AGU, AGC – Ser; AGA, AGG – Arg	U, C, A, G
G	GUU, GUC, GUA, GUG – Val	GCU, GCC, GCA, GCG – Ala	GAU, GAC – Asp; GAA, GAG – Glu	GGU, GGC, GGA, GGG – Gly	U, C, A, G

Figure 16.4 The Genetic Code

Each codon on mRNA codes for one amino acid, but multiple codons can code for the same amino acid. For example, the codons UCA, UCC, UCG, and UCU all code for the amino acid serine. This gives redundancy to the genetic code. This redundancy in the genetic code may result in "silent" mutations that do not affect the amino acid sequence of the polypeptide chain. Almost all organisms on Earth use the same genetic code, which is evidence for the common ancestry of all organisms.

TIP

With the exception of AUG, the start codon, you do not need to memorize any codons of the genetic code. If needed, a codon chart will be provided on the AP Biology exam. Codons are specified here for illustrative purposes and clarity.

2. **Elongation:** After the first amino acid is placed in the ribosome by tRNA, the ribosome **translocates** (or moves) to the next codon. A new tRNA with the appropriate anticodon and amino acid then pairs with this codon. The ribosome then catalyzes the formation of a peptide bond between the amino acids brought to the ribosome by the first two tRNAs, forming the beginning of the polypeptide chain. (See Figure 16.5.)

Once the peptide bond is formed between the amino acids, the first tRNA releases its amino acid (which is now linked to the second amino acid), and the first tRNA is released from the ribosome. This translocation of the ribosome and the addition of new amino acids to the polypeptide chain is called elongation and is repeated one codon at a time until a **stop codon** is reached.

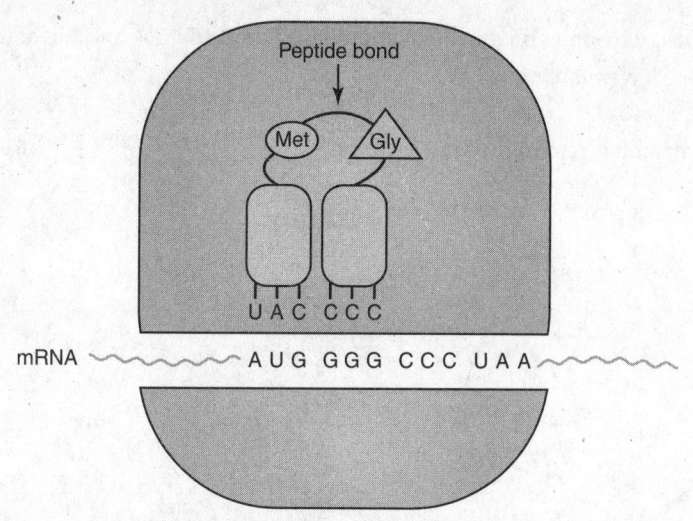

Figure 16.5 Elongation of Translation

3. **Termination:** Stop codons (also known as "nonsense" codons) do not code for any amino acid. As shown in Figure 16.6, when the ribosome reaches a stop codon, proteins called release factors bind to the ribosome. These release factors cause the ribosome to disassemble and release the polypeptide chain. This ends translation and is called termination.

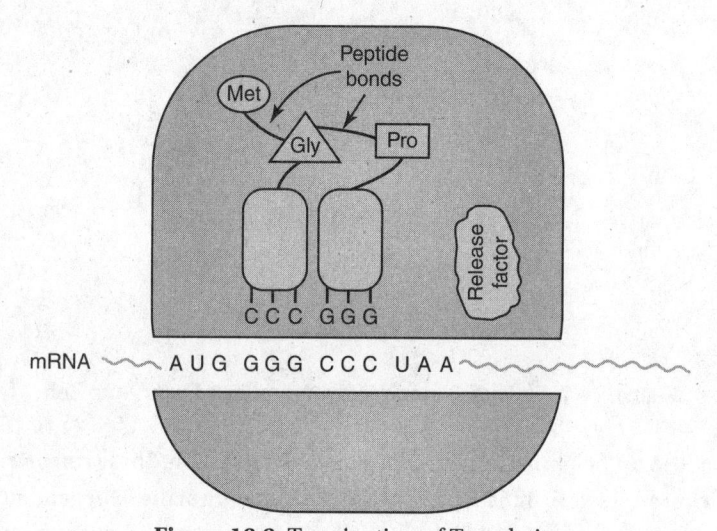

Figure 16.6 Termination of Translation

Flow of Information from the Nucleus to the Cell Membrane

DNA and RNA are the carriers of genetic information in organisms. In most organisms, genetic information starts with DNA, which provides the information for the transcription of mRNA. mRNA then provides the information for the sequence of amino acids in a protein.

In eukaryotes, the endomembrane system (see Chapter 5) works to modify, package, and transport proteins to their final destinations. Genetic information in DNA is transcribed into mRNA in the nucleus. Ribosomes on the rough endoplasmic reticulum use the information in mRNA to translate proteins. After the protein is translated, a vesicle containing the protein will bud off from the rough endoplasmic reticulum and travel to the Golgi. At the Golgi, the proteins will be modified and packaged into vesicles for export from the cell. These vesicles will bud off from the Golgi and travel to the cell membrane. The vesicles then fuse with the cell membrane and release their protein contents from the cell. So the flow of genetic information in a eukaryotic cell is as follows:

DNA → mRNA → protein at ribosomes on the rough endoplasmic reticulum → protein at the Golgi → protein in vesicle → protein outside of cell membrane

This flow of genetic information is represented in Figure 16.7.

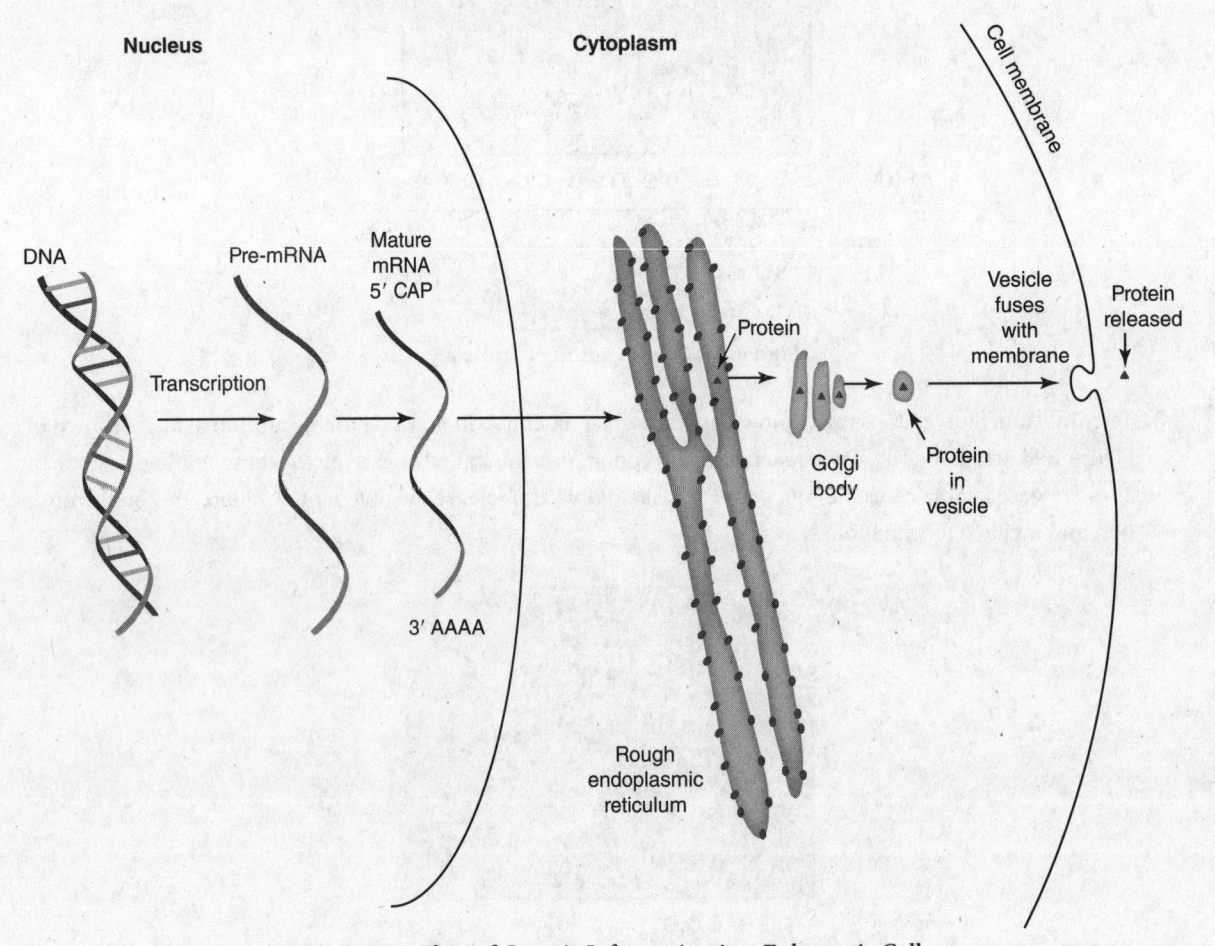

Figure 16.7 Flow of Genetic Information in a Eukaryotic Cell

Some viruses contain RNA as their primary carrier of genetic information. These **retroviruses** contain the enzyme **reverse transcriptase**. Reverse transcriptase makes a DNA copy of the RNA genome of the virus. This DNA copy is then inserted into the genome of the host cell that is infected by the virus. The host cell will then transcribe and translate the information in the viral DNA inserted into the host cell's genome. Reverse transcriptase is less accurate than RNA polymerase, so retroviruses have a relatively high mutation rate.

Practice Questions

Multiple-Choice

1. What would be the minimum number of nucleotides required to code for a protein made of 12 amino acids?

   (A) 6
   (B) 12
   (C) 36
   (D) 48

2. A scientist inserts a eukaryotic gene directly into a bacteria's genome. However, the protein produced by the bacteria from the eukaryotic gene does not have the same amino acid sequence as the protein produced from the gene in eukaryotic cells. Which of the following best explains this?

   (A) Prokaryotes and eukaryotes do not use the same genetic code, so the bacteria cannot decode the eukaryotic gene.
   (B) Eukaryotic genes contain introns, which must be removed before translation; prokaryotes cannot remove introns.
   (C) Prokaryotes do not have a nucleus and cannot perform transcription.
   (D) Prokaryotes do not have rough endoplasmic reticulum, so they cannot translate eukaryotic genes.

3. Which of the following correctly represents the mRNA sequence that would be transcribed from the DNA sequence 3' ACC GGT AAG TTC 5'?

   (A) 3' TGG CCA TTC AAG 5'
   (B) 3' UGG CCA UUC AAG 5'
   (C) 5' TGG CCA TTC AAG 3'
   (D) 5' UGG CCA UUC AAG 3'

4. What is the amino acid sequence that would be translated from the following gene? (Use Figure 16.4 to answer this question.)

   3' AAT CGT TTC AAT CAA 5'

   (A) Asn-Arg-Phe-Asn-Gln
   (B) Leu-Ala-Lys-Leu-Val
   (C) Phe-Arg-Asn-Phe-Gln
   (D) Gly-Ala-Lys-Gly-Val

5. Which of the following processes is most similar in prokaryotes and eukaryotes?

   (A) alternative splicing of exons
   (B) addition of a 3' poly-A tail to mRNA
   (C) addition of a 5' GTP cap to mRNA
   (D) transcription by RNA polymerase

6. The following figure depicts steps in the flow of genetic information in a eukaryotic cell.

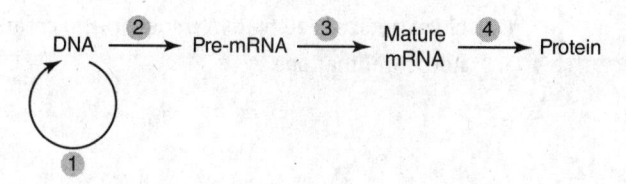

   DNA →2→ Pre-mRNA →3→ Mature mRNA →4→ Protein
   (with step 1 as a loop on DNA)

   During which step will spliceosomes remove introns?

   (A) 1
   (B) 2
   (C) 3
   (D) 4

7. The antibiotic tetracycline inhibits bacterial growth by blocking the binding of tRNAs to the ribosome. Which of the following processes is most directly affected by tetracycline?

   (A) transcription
   (B) exon splicing
   (C) initiation of translation
   (D) export of proteins from the cell

8. Why do retroviruses have a high mutation rate?

   (A) RNA polymerase has a high error rate when reading viral genomes.
   (B) DNA polymerase has a high error rate when reading viral genomes.
   (C) Viruses use a different genetic code than prokaryotes and eukaryotes use.
   (D) Reverse transcriptase has a high error rate.

9. Humans can generate over 10^{12} different antibody proteins, but humans have fewer than 25,000 genes. Which of the following best explains how this is possible?

   (A) Humans acquire new antibody genes when they are infected with pathogens.
   (B) Alternative splicing of exons can generate many different transcripts from the same gene.
   (C) The error rate in RNA polymerase generates new transcripts for antibody proteins.
   (D) Golgi bodies modify RNA transcripts to create new combinations.

10. Which of the following catalyzes the formation of peptide bonds during translation?

    (A) RNA polymerase
    (B) mRNA
    (C) rRNA
    (D) tRNA

Short Free-Response

11. Ehlers-Danlos syndrome (EDS) type IV is a result of a mutation in the *COL3A1* gene, which results in the deletion of one of the exons in the procollagen transcript.

    Part A

    (i) **Describe** the location in the cell where the splicing of exons occurs.

    Part B

    (i) **Explain** the difference between exons and introns.

    Part C

    (i) **Predict** the effect of the EDS mutation on the structure of the procollagen protein.

    Part D

    (i) **Justify** your prediction from Part C.

12. A mutation results in the deletion of the TATA box in the promoter of a eukaryotic gene.

    Part A

    (i) **Describe** the function of the promoter in gene expression.

    Part B

    (i) **Explain** why promoter sequences are highly conserved in living organisms.

    Part C

    The figure that follows shows the gene and the DNA surrounding it.

    (i) **Draw** an "X" on the most likely location of the promoter of the gene.

    ———————[GENE]———————

    Part D

    (i) **Explain** how the deletion of the TATA box in the promoter would most likely affect the levels of protein produced by the gene.

Long Free-Response

13. The levels of mRNA and protein produced by three different genes (*PDHA1*, *MBP*, and *INS*) were measured in different tissues of the human body. The levels of mRNA and protein expression from these genes were compared to mRNA and protein expression from the gene *POLR2A* (RNA polymerase II subunit A, which is expressed in all human tissues). The results are shown in the table. "0" indicates that mRNA or protein were not detected in that tissue, and "+," "++," and "+++" indicate relatively low, medium, or high levels of mRNA or protein detected, respectively.

	mRNA Levels				Protein Levels			
	Brain	Liver	Pancreas	Lung	Brain	Liver	Pancreas	Lung
POLR2A	++	++	++	++	++	++	++	++
PDHA1	+	+	+	+	+	+	+	+
MBP	+++	0	0	0	+++	0	0	0
INS	0	0	+++	0	0	0	+++	0

Part A

(i) **Identify** which gene was used as a control in this experiment.

Part B

(i) **Explain** why the gene you selected in Part A would be an appropriate control for this experiment.

Part C

(i) Based on the data in the table, **determine** which of the genes (*PDHA1*, *MBP*, or *INS*) is most likely involved in glycolysis.

Part D

INS is the insulin gene. Insulin is a hormone that is released when blood sugar levels are high.

(i) **Predict** the levels of *INS* mRNA and protein that would be found in cells in the pancreas when blood sugar levels are low.

(ii) **Justify** your prediction.

Answer Explanations
Multiple-Choice

1. **(C)** Each amino acid is coded by a three base pair codon, so a protein made of 12 amino acids would require a minimum of 12 × 3 = 36 nucleotides. Choice (A) is incorrect because 6 nucleotides would only contain two codons. Since 3 nucleotides are required to make a codon, 12 nucleotides would only be sufficient to code for four amino acids, so choice (B) is incorrect. While 48 nucleotides might code for 12 amino acids if there were introns to be removed, 48 is not the minimum number of nucleotides required, so choice (D) is incorrect.

2. **(B)** Eukaryotic genes contain introns; prokaryotic genes do not. Prokaryotes do not have the spliceosomes required to remove introns. So if a eukaryotic gene was directly inserted into a bacterial genome, the bacteria would likely try to translate the introns and produce a different protein. Choice (A) is incorrect because prokaryotes and eukaryotes *do* use the same genetic code. While prokaryotes do not have a nucleus, they do contain RNA polymerase and can perform transcription, so choice (C) is incorrect. Choice (D) is incorrect because rough endoplasmic reticulum is not required for the translation of all proteins.

3. **(D)** The transcribed mRNA would be antiparallel to the given DNA sequence, so it would start with the 5′ end. Also, in RNA, uracil replaces thymine, and the other base-pairing rules are the same as those found in DNA. Choices (A) and (B) are incorrect because the mRNA is not antiparallel to the DNA sequence. Choices (A) and (C) are incorrect because they contain thymine, which does not appear in RNA.

4. **(B)** The mRNA transcribed from the DNA sequence in the problem would be 5′ UUA GCA AAG UUA GUU 3′. Using Figure 16.4, the amino acid sequence that would be translated from that mRNA would be Leu-Ala-Lys-Leu-Val.

5. **(D)** Both prokaryotes and eukaryotes use RNA polymerase for transcription. Only eukaryotes have alternative splicing of exons and the addition of a 3′ poly-A tail and a 5′ GTP cap to mRNA, so choices (A), (B), and (C) are incorrect.

6. **(C)** Introns are removed from the pre-mRNA to help form the mature mRNA, which is represented by step 3. Step 1 represents DNA replication, so choice (A) is incorrect. Transcription is represented by step 2, so choice (B) is incorrect. Choice (D) is incorrect because step 4 represents translation.

7. **(C)** If tRNAs could not bind to the ribosome, the initiation of translation could not occur. Choices (A) and (B) are incorrect because transcription and exon splicing do not involve the ribosome, so an antibiotic that interferes with ribosomes would not affect those processes. The export of proteins from the cell involves the Golgi bodies and vesicles, so choice (D) is incorrect.

8. **(D)** Retroviruses use reverse transcriptase to make a DNA copy of their RNA genome. Reverse transcriptase has a very high error rate, so many mutations occur.

9. **(B)** Eukaryotes can use alternative splicing of exons to generate multiple mRNA transcripts from one gene, which can lead to the production of multiple proteins from one gene. Humans do not acquire new genes when infected with pathogens, so choice (A) is incorrect. The error rate of RNA polymerase is not responsible for generating new transcripts for antibody proteins, so choice (C) is incorrect. Choice (D) is incorrect because Golgi bodies modify proteins, not RNA transcripts.

10. **(C)** rRNA catalyzes the formation of peptide bonds during translation. Choice (A) is incorrect because RNA polymerase performs transcription, not translation. mRNA brings the genetic information from the nucleus to the ribosome but does not catalyze the formation of peptide bonds, so choice (B) is incorrect. Choice (D) is incorrect because tRNA serves as an "adaptor" molecule that brings the amino acids that correspond to each codon to the ribosome.

Short Free-Response

11. (A-i) The splicing of exons occurs in the nucleus.

    (B-i) The codons in exons are expressed in the protein and code for amino acids in the protein. Introns are noncoding sequences and do not code for amino acids in the protein.

    (C-i) The EDS mutation would result in a shorter, less functional procollagen protein.

    (D-i) If an exon was deleted, the resulting mRNA transcript would not contain the complete code for the protein. So the translated protein would be shorter and likely less functional.

12. (A-i) The promoter serves as a binding site for RNA polymerase at which transcription begins.

    (B-i) All living organisms use RNA polymerase, an enzyme that has a consistent three-dimensional shape. All living organisms need similar nucleotide sequences in their promoters that RNA polymerase can recognize.

    (C-i) ──X──[GENE]──

    (Note that any "X" to the left of the gene is acceptable.)

    (D-i) The deletion of the TATA box in the promoter would most likely make it more difficult for RNA polymerase to recognize the promoter. Less transcription would occur, and less protein would be produced.

Long Free-Response

13. (A-i) *POLR2A* was used as the control because the expression of all the other genes was compared to the expression of *POLR2A*.

    (B-i) *POLR2A* would be the best gene to use as a control because it is known to be expressed in all human tissues, as stated in the question.

    (C-i) *PDHA1* is most likely involved in glycolysis because glycolysis occurs in the cytoplasm of all living cells, and *PDHA1* is expressed in all the tissues in the experiment.

    (D-i and ii) Since insulin is released when blood sugar levels are high, it is likely that the levels of insulin would be low when blood sugar levels are low. Therefore, when blood sugar levels are low, it is expected that both the *INS* mRNA and the insulin protein levels would be low.

17

Regulation and Mutations

Learning Objectives

In this chapter, you will learn:

- → Regulation of Gene Expression in Prokaryotes
- → Regulation of Gene Expression in Eukaryotes
- → Gene Expression Helps Cells Specialize
- → Mutations

Overview

The phenotype of an organism is determined by the genes that are expressed and the levels at which those genes are expressed. Regulation of gene expression is important in determining an organism's phenotype. Genes can be regulated by the interaction of regulatory proteins with regulatory sequences in the genome. **Regulatory proteins** are proteins that can turn on or turn off genes by binding to specific nucleotide sequences. The nucleotide sequences to which these regulatory proteins bind are called **regulatory sequences**. Mutations in the genome may also affect gene expression. This chapter will review how the expression of genes is regulated (in prokaryotes and in eukaryotes) and the different types of mutations that can affect an organism's phenotype.

Regulation of Gene Expression in Prokaryotes

Prokaryotes use operons to regulate gene expression. An **operon** is a cluster of genes with a common function under the control of a common promoter. Operons contain regulatory sequences, genes for regulatory proteins, and genes for structural proteins (which are responsible for the function of the operon). An example of an operon is shown in Figure 17.1.

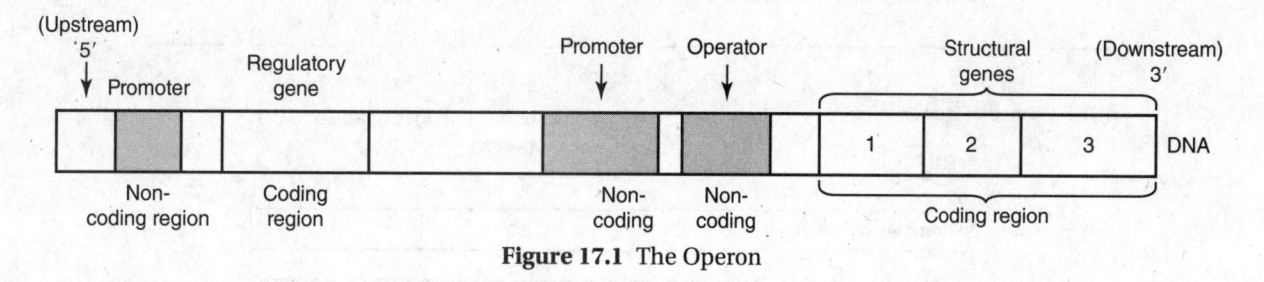

Figure 17.1 The Operon

Promoters are noncoding regulatory sequences that serve as binding sites for RNA polymerase. **Operators** are noncoding regulatory sequences that serve as binding sites for repressor proteins (a type of regulatory protein). **Structural genes** are coding sequences that contain the genetic code for the proteins required to perform the function of the operon.

The two types of prokaryotic operons you need to understand for the AP Biology exam are inducible operons and repressible operons.

When determining whether an operon is turned "on" or "off," consider what would be most advantageous to the bacteria in its environment. If an operon's function is to digest a molecule, it is advantageous to the bacteria to turn "off" the operon if the molecule is not present and to turn "on" the operon if the molecule is present. If an operon's function is to synthesize a molecule, then if the molecule is absent in the environment, the bacteria will need to turn the operon on, but if the molecule is present in the environment, then the bacteria will turn off the operon to conserve resources and energy.

Inducible Operons

Inducible operons usually have a catabolic function (digesting molecules) and are turned off unless the appropriate inducer molecule is present. The repressor protein in an inducible operon binds to the operator sequence, blocking transcription of the operon by RNA polymerase. However, when an inducer molecule is present, the **inducer** binds to the repressor protein, changing its shape so that it can no longer bind to the operator sequence. This allows RNA polymerase to begin transcribing the operon.

The classic example of an inducible operon is the *lac* **operon**, as shown in Figure 17.2. The function of the *lac* operon is to produce the proteins required to digest the sugar lactose. If no lactose is present, the *lac* repressor protein will bind to the operator, blocking transcription of the operon by RNA polymerase. Lactose serves as the inducer molecule for the *lac* operon. When lactose is present, it binds to the *lac* repressor protein, changing its shape so that it no longer can bind to the operator sequence. This allows RNA polymerase to transcribe the genes for the proteins that digest lactose. After all the lactose has been digested, the repressor can again bind to the operator sequence, shutting down the operon. This type of feedback mechanism allows the cell to manufacture the proteins needed to digest lactose only when they are needed, saving valuable resources in the cell.

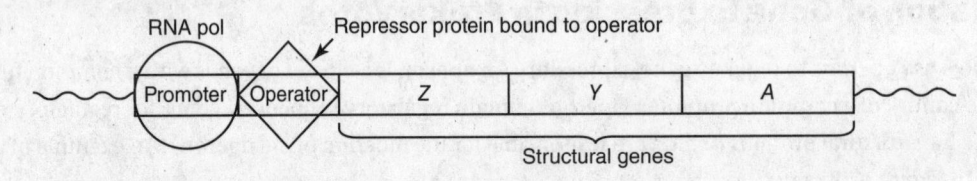

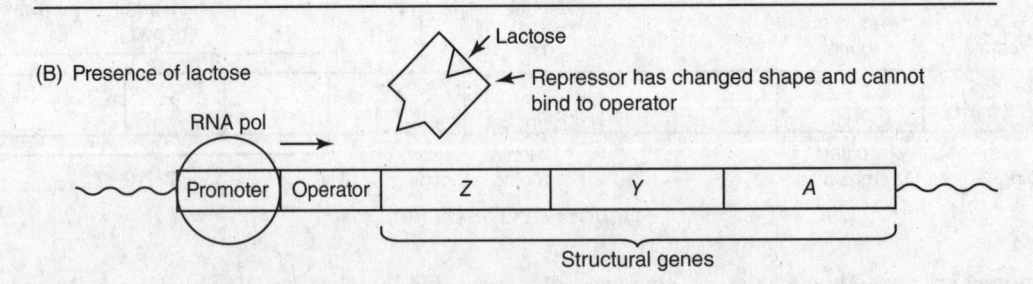

Figure 17.2 *Lac* Operon

While the binding of repressor proteins to the operator sequence can negatively regulate gene expression, there are also ways to positively "upregulate" gene expression. An example of this is the interaction between cyclic AMP (cAMP) and the catabolite activator protein (CAP) in the *lac* operon. When glucose levels are low, cAMP levels in the cell increase. cAMP binds to the catabolite activator protein (CAP), stimulating CAP to bind at a CAP binding site near the promoter. This increases the affinity of RNA polymerase for the promoter, stimulating transcription. So transcription of the *lac* operon is increased when glucose, another food source for the bacteria, is absent, allowing the cell to utilize the energy in lactose more efficiently.

Repressible Operons

Repressible operons usually have an anabolic function (synthesizing molecules) and are turned on unless the product of the operon is in abundance in the cell. The ***trp* operon** is an example of a repressible operon. (See Figure 17.3.) The function of the *trp* operon is to produce the enzymes needed to synthesize the amino acid tryptophan. In the *trp* operon, the amino acid tryptophan functions as a **corepressor**. The *trp* repressor protein cannot bind to the operator sequence on its own; the *trp* repressor must be bound to the amino acid tryptophan before it can bind to the operator. Therefore, if no tryptophan is present, the *trp* repressor will not be bound to the operator and RNA polymerase can transcribe the operon. When tryptophan is present, it will bind to the *trp* repressor, which will then bind to the operator, stopping transcription of the operon. This type of feedback mechanism allows the cell to make the enzymes needed to synthesize tryptophan only when the cell needs them, again saving valuable resources in the cell.

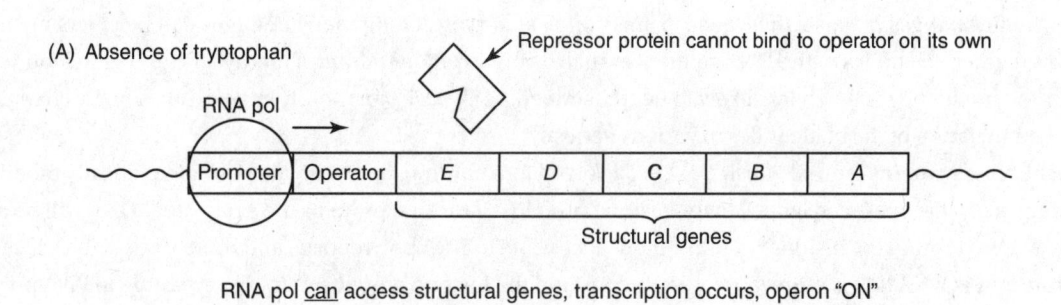

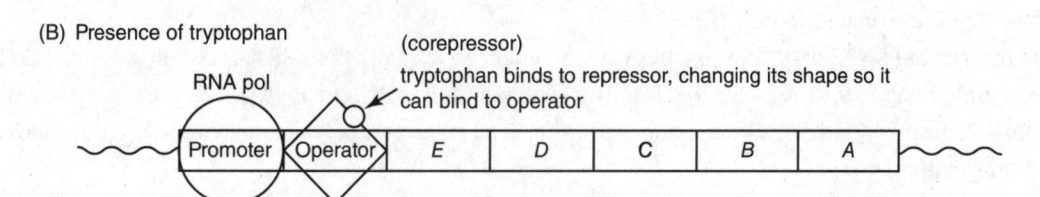

Figure 17.3 *Trp* Operon

In summary, there are some key differences between inducible operons and repressible operons:

- The function of the operon (catabolic or anabolic)
- Whether the repressor protein can bind to the operator on its own, or whether the repressor protein needs a corepressor to bind to the operator

> For the AP Biology exam, you are not required to memorize which operons are inducible or repressible, but you do need to understand how inducible and repressible operons work. You must be able to apply your knowledge about operons to novel scenarios. For example, if a question on the exam is about the arabinose operon, the question will tell you the type of operon it is (inducible). Then you should apply your knowledge of how the *lac* operon works to answer the question about the arabinose operon.

Regulation of Gene Expression in Eukaryotes

Eukaryotes can also use the interactions between regulatory sequences and regulatory proteins to turn genes on or off and to regulate the levels of gene expression.

The following are some examples of regulatory sequences used in eukaryotes:

- **Promoters**—sequences that serve as binding sites for RNA polymerase.
- **Regulatory switches**—sequences to which activator proteins or repressor proteins may bind. **Enhancers** are regulatory switches to which activator proteins or transcription factors bind. **Silencers** are regulatory switches to which repressor proteins bind.

Some examples of regulatory proteins include:

- **Repressors**—These bind to regulatory switches and turn off or suppress gene expression.
- **Activators**—These bind to regulatory switches and upregulate gene expression.
- **Transcription factors**—These help RNA polymerase bind to the promoter and start transcription.
- **Mediators**—These serve as "connectors" between other regulatory proteins and allow regulatory proteins to communicate.

Epigenetic changes can also affect gene expression in eukaryotes. **Epigenetic changes** can be reversible modifications to the nucleotides of the DNA sequence, such as methylation (adding a methyl group) of nucleotides. A methylated nucleotide is much less likely to be transcribed, so the cell can modify gene expression by changing the level of methylation of the nucleotides in various genes.

The **histone proteins** around which DNA is packaged into chromosomes can also be epigenetically modified by adding acetyl groups (acetylation) to the histone proteins. If histone proteins are acetylated, DNA will be more loosely wound around the histone proteins, more accessible to RNA polymerase, and more likely to be expressed. **Euchromatin** is DNA that is more loosely wound around the histone proteins, is more accessible to RNA polymerase, and usually results in more expression of the genes in the euchromatin. DNA in chromosomes may be tightly wound around the histone proteins, forming **heterochromatin**, which is less accessible to RNA polymerase and results in reduced gene expression.

Small interfering RNA (siRNA) molecules can also affect gene expression. siRNA is single-stranded and binds to complementary mRNA molecules, forming double-stranded RNA (dsRNA) molecules. Enzymes in the cell detect and destroy these dsRNA molecules, resulting in no translation of the targeted mRNA molecules and reduced gene expression.

Gene Expression Helps Cells Specialize

The regulation of gene expression results in different genes being expressed in different cells (also known as **differential gene expression**). This differential gene expression influences the functions of cells and the resulting phenotype of the organism. Different tissue types express different genes, which results in cell differentiation.

The phenotype of an organism is determined not only by which genes are expressed but also by the levels at which the genes are expressed. For example, a skin cell expressing higher levels of the melanin protein would appear darker in color than a skin cell expressing lower levels of the melanin protein.

The timing of the expression of different transcription factors during development is critical to the formation of specialized tissues and organs from a single-celled zygote. Errors in the timing of the expression of these genes can result in errors in the body plan or structures of an organism. The *Hox* genes, a family of genes that codes for transcription factors, were discovered in *Drosophila* flies. The *Hox* gene coding for antennapedia controls the formation of legs in *Drosophila* during development. Mutations in this gene or expression of this gene at the wrong time can result in legs not forming or legs being located in the wrong body segment.

Mutations

Mutations are changes in the genetic material of an organism. Mutations may result in changes to the organism's phenotype. Phenotypes can change if the mutation interferes with or changes the function of a protein. Mutations that change the amount of a protein produced can also change the phenotype of an organism. Mutations provide genetic variation in populations. These variations are acted upon by natural selection and can lead to the evolution of populations.

Point mutations occur when one nucleotide has been substituted for a different nucleotide. If the nucleotide substitution results in no change to the amino acid sequence, it is a **silent mutation**. A point mutation that results in a premature stop codon is a **nonsense mutation**. Insertions or deletions of one or more nucleotides cause the reading frame to be shifted and result in a **frameshift mutation**. Typically, but not always, frameshift mutations result in bigger changes in the amino acid sequence than point mutations; this is because frameshift mutations affect the genetic code at the site of the mutation *and* also affect all of the codons that follow it, whereas point mutations typically affect only the codon at the site of the point mutation.

Not all mutations are harmful. The organism's environment determines whether a mutation is beneficial, harmful, or has no effect on the survival of the organism. If the organism's environment changes, the benefit or harm a mutation confers on the organism can change. Some mutations do not change the amino acid sequence of a protein at all because of the redundancy of the genetic code.

Mutations can be caused by environmental factors, such as chemicals or radiation, or by random errors in DNA replication or DNA repair mechanisms.

Mistakes in mitosis or meiosis can also lead to mutations. Failure of homologous chromosomes to separate during meiosis can lead to an **aneuploidy** (an atypical number of chromosomes). Aneuploidies in animals can be fatal or may cause sterility or other disorders. However, in plants, aneuploidies that result in polyploids (an entire extra set of chromosomes) can confer an advantage to the plant, making it more likely to survive.

> Two of the most common aneuploidies in humans are Klinefelter syndrome and Down syndrome (also known as trisomy 21). Individuals with Klinefelter syndrome have three sex chromosomes: two X and one Y. Individuals with Down syndrome typically have three copies of chromosome 21.

Genetic mutations can also be transmitted horizontally between members of the same generation. Some methods of **horizontal transmission** of genetic information include:

- **Transformation**—the uptake of naked foreign DNA by a cell
- **Transduction**—the transmission of DNA from one organism to another by viruses; as the virus transfers the DNA, the DNA sequence may be recombined or otherwise changed, leading to new mutations and variations
- **Conjugation**—the transmission of DNA through cell-to-cell contact, usually through a connection called a pilus
- **Transposition**—the movement of DNA between chromosomes or within a chromosome; these are sometimes referred to as "jumping genes"

Practice Questions
Multiple-Choice

1. A repressible operon in bacteria codes for the genes required to produce the amino acid serine. Serine functions as a corepressor for the operon. If serine is present in the bacteria's environment, which of the following is most likely?

    (A) increased digestion of serine
    (B) increase in the levels of serine
    (C) increased production of the repressor protein
    (D) increased binding of the repressor protein to the operator

2. The arabinose operon is an inducible operon that codes for the genes required to digest the sugar arabinose. Arabinose functions as an inducer molecule for the operon. If arabinose is present in the bacteria's environment, which of the following is most likely?

    (A) increased digestion of arabinose
    (B) increase in the levels of arabinose
    (C) increased production of the repressor protein
    (D) increased binding of the repressor protein to the operator

3. Which sequences in bacterial operons are noncoding sequences?

    (A) operators and promoters only
    (B) operators, promoters, and genes for regulatory proteins
    (C) operators, promoters, and genes for structural proteins
    (D) promoters, genes for regulatory proteins, and genes for structural proteins

4. Which of the following is a key difference between inducible operons and repressible operons?

    (A) Inducible operons have promoter sequences, and repressible operons do not have promoter sequences.
    (B) Inducible operons have operator sequences, and repressible operons do not have operator sequences.
    (C) The repressor in an inducible operon can bind to the operator sequence on its own, but the repressor in a repressible operon requires a corepressor in order to bind to the operator.
    (D) In inducible operons, RNA polymerase requires the presence of a corepressor in order to bind to the promoter; in repressible operons, RNA polymerase does not require a corepressor.

5. Under which of the following conditions will transcription of the *lac* operon be at its highest level?

    (A) low glucose, low lactose
    (B) low glucose, high lactose
    (C) high glucose, low lactose
    (D) high glucose, high lactose

6. Adding a methyl group to a DNA nucleotide is a type of _____ and will make a DNA sequence much less likely to be transcribed.

    (A) mutation
    (B) epigenetic change
    (C) aneuploidy
    (D) transposition

7. Which of the following are proteins in eukaryotes that bind to regulatory switches and upregulate gene expression?

    (A) activators
    (B) repressors
    (C) transcription factors
    (D) mediators

8. Not every change in the DNA sequence results in a change in the amino acid sequence of a protein. Which of the following explains this?

    (A) Each organism lives in a different environment, which changes the expression of its genes.
    (B) The genetic code is redundant, and more than one codon codes for most amino acids.
    (C) The proofreading function of ribosomes corrects changes in the DNA sequence.
    (D) Differential gene expression adapts to these changes in the DNA sequence.

9. Which of the following does not affect the phenotype of an organism?

    (A) the genes that are expressed
    (B) the level at which genes are expressed
    (C) the timing of gene expression
    (D) the number of genes in the organism

10. Which of the following methods of horizontal transmission of genetic material would be most likely to lead to new variants of the COVID-19 virus?

    (A) transformation
    (B) transduction
    (C) transposition
    (D) conjugation

Short Free-Response

11. One way that organisms respond to changing environmental conditions is through the regulation of gene expression.

    Part A

    (i) **Describe** the function of operators and promoters in prokaryotes.

    Part B

    (i) **Explain** how the operator and repressor interact to control the expression of the inducible *lac* operon.

    Part C

    Bacteria are placed in an environment that has low levels of glucose and high levels of lactose.

    (i) **Predict** the level of expression of the *lac* operon in this environment.

    Part D

    (i) **Justify** your prediction from Part C.

12. A student conducts an experiment in an effort to determine whether a specific bacterial operon is inducible or repressible. The level of transcription of the operon was measured after the addition of different molecules to the bacteria's environment. Data are shown in the table.

Fructose	Lysine	Level of Transcription
Absent	Absent	High
Absent	Present	Low
Present	Absent	High
Present	Present	Low

    Part A

    (i) **Describe** what, if any, effect fructose has on the level of transcription of the operon.

    Part B

    (i) **Describe** what, if any, effect lysine has on the level of transcription of the operon.

    Part C

    (i) **Make a claim** about whether this operon is more likely inducible or repressible.

    Part D

    (i) **Justify** your claim from Part C using evidence from the experiment and your knowledge of inducible and repressible operons.

Long Free-Response

13. The human *THY-1* gene (Thy-1 cell surface antigen) codes for a cell surface glycoprotein that is thought to function as a tumor suppressor in certain types of cancer. In order to find possible locations of enhancer sequences for the *THY-1* gene, scientists performed a series of DNA deletion experiments. In the experiments, different deletions were made in areas that were suspected to function as possible enhancer sequences, and the levels of *THY-1* transcription were measured after each deletion. The following diagram shows the sequences deleted. A, B, C, D, and E represent different DNA sequences. 1, 2, and 3 represent suspected enhancer sequences. P is the promoter of the *THY-1* gene. An X represents the deletion of that portion of the DNA sequence.

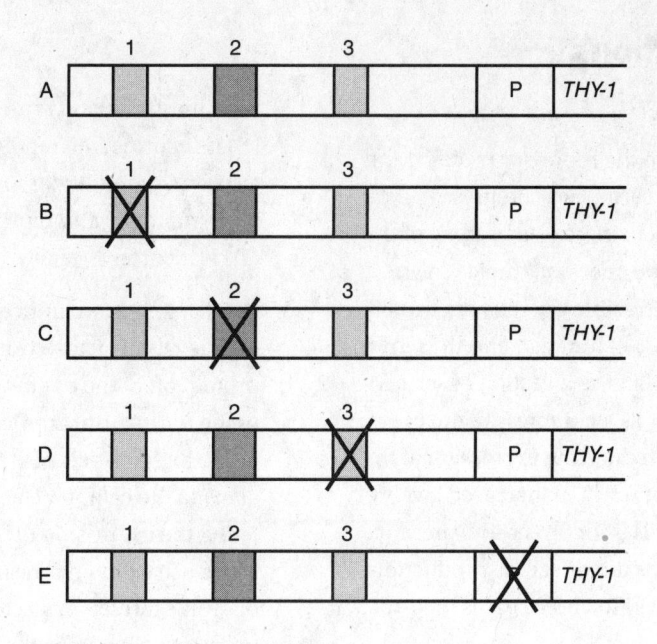

The levels of *THY-1* transcription in each part of the experiment are shown in the graph.

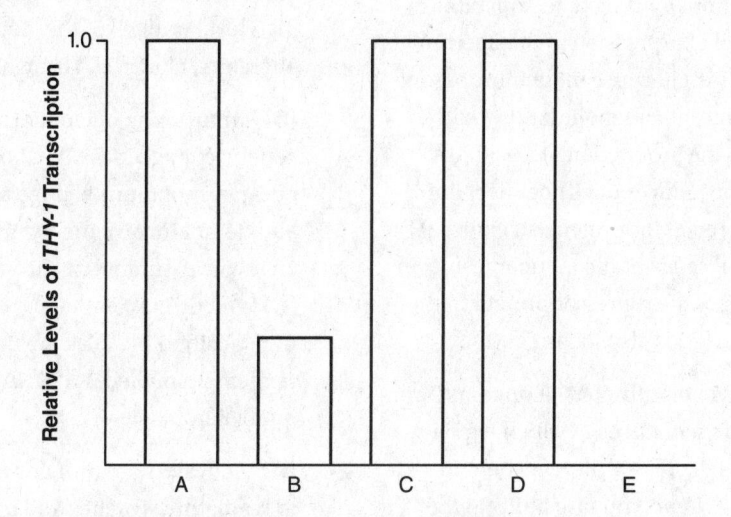

Part A

(i) **Explain** the function of enhancer sequences in eukaryotic gene expression.

Part B

(i) **Identify** which DNA sequence (A, B, C, D, or E) was the control in this experiment.

Part C

(i) **Identify** which sequence (1, 2, or 3) most likely functions as an enhancer sequence.

(ii) **Support your claim** using evidence from the data.

Part D

(i) **Predict** the effects of the deletion of P, as shown in sequence E, and what the relative level of *THY-1* transcription would be for E.

(ii) **Justify** your prediction with evidence from the data.

Answer Explanations

Multiple-Choice

1. **(D)** Corepressors help the repressor protein bind to the operator. Since serine is a corepressor, its presence would result in increased binding of the repressor protein to the operator. Choice (A) is incorrect because repressible operons are usually anabolic, not catabolic, in function and thus they would not be involved in the digestion of serine. Since serine functions as a corepressor, the presence of serine would reduce the expression of the operon and would not result in increased levels of serine. Thus, choice (B) is incorrect. The presence of the corepressor does not affect the production of the repressor protein, so choice (C) is incorrect.

2. **(A)** In inducible operons, the inducer molecule binds to the repressor protein, which prevents the repressor protein from binding to the operator, increasing the level of expression of the operon. Choice (B) is incorrect because inducible operons are usually catabolic, not anabolic, and would not be involved in the production of arabinose. The presence of the inducer does not affect the production of the repressor protein, so choice (C) is incorrect. The presence of the inducer results in decreased binding of the repressor protein to the operator, so choice (D) is also incorrect.

3. **(A)** The operators and promoters of operons both serve as noncoding sequences to which regulatory proteins bind. Repressor proteins bind to the operator, and RNA polymerase binds to the promoter. Genes for both regulatory proteins and structural proteins contain sequences that code for proteins, so choices (B), (C), and (D) are incorrect.

4. **(C)** In inducible operons, repressor proteins can bind to the operator without the assistance of a corepressor. In repressible operons, the presence of a corepressor is required in order for the repressor protein to bind to the operator sequence. Choice (A) is incorrect because both inducible and repressible operons have promoter sequences. Operator sequences are present in both inducible and repressible operons, so choice (B) is incorrect. RNA polymerase does not require a corepressor in order to bind to the promoter in both inducible and repressible operons, so choice (D) is incorrect.

5. **(B)** Lactose is an inducer for the *lac* operon, so high levels of lactose would pull the repressor protein off of the operator, increasing expression of the *lac* operon. Low levels of glucose would cause higher levels of cyclic AMP, which would activate the catabolite activator protein, upregulating the expression of the *lac* operon. Low levels of lactose would not result in high levels of *lac* operon expression, so choices (A) and (C) are incorrect. Glucose is the preferred food source for bacteria, so if high levels of glucose were present, the bacteria would not express the *lac* operon at a high level until all of the glucose was metabolized. Therefore, choice (D) is incorrect.

6. **(B)** Methylating nucleotides is a type of epigenetic change to DNA. Choice (A) is incorrect because mutations change the DNA sequences and do not involve methylation. An aneuploidy is an atypical number of chromosomes, so choice (C) is incorrect. Choice (D) is incorrect because transposition involves the movement or rearrangement of pieces of DNA and does not involve methylation.

7. **(A)** Activators are proteins in eukaryotes that bind to regulatory switches and upregulate gene expression. Choice (B) is incorrect because repressors downregulate gene expression. While transcription factors may bind to enhancers, transcription factors more often help RNA polymerase bind to the promoter. Thus, choice (A) is a better answer and therefore choice (C) is not the best answer. [Note that sometimes on the AP Biology exam, there is more than one plausible answer to a question, but you need to choose the *best* answer to the question.] Mediators are proteins that allow communication between the regulatory proteins involved in gene expression; they do not bind to regulatory switches, so choice (D) is incorrect.

8. **(B)** The genetic code contains more than one codon for most amino acids, so a change in a codon does not necessarily result in a change in the amino acid for which it codes. While changes in the environment can affect the expression of genes, the statement in choice (A) does not explain why changes in the DNA sequence may not result in changes in the amino acid found in the final protein. Thus, choice (A) is incorrect. Ribosomes do not have a proofreading function, so choice (C) is incorrect. Differential gene expression results in cell differentiation but not changes in the DNA sequence, so choice (D) is incorrect.

9. **(D)** The phenotype of an organism is determined by the genes that are expressed, the level of expression of those genes, and the timing of the expression of those genes. The phenotype is not affected by the number of genes in the organism, so choice (D) is the only choice that does not affect the phenotype of an organism.

10. **(B)** Transduction involves the transfer of genetic material by viral transmission. Transformation, transposition, and conjugation do not involve viruses, so choices (A), (C), and (D) are incorrect.

Short Free-Response

11. (A-i) Operators serve as binding sites for repressor proteins. Promoters are binding sites for RNA polymerase.

    (B-i) When the repressor is bound to the operator, it blocks RNA polymerase's access to the structural genes, and the operon is shut down.

    (C-i) The level of expression of the *lac* operon will be high.

    (D-i) Lactose is the inducer and will bind to the repressor, removing it from the operator and allowing the operon to be expressed. Low glucose levels will lead to high levels of cAMP, which will activate the catabolite activator protein, increasing the expression of the *lac* operon even more.

12. (A-i) Based on the data in the table, fructose appears to have no effect on the level of transcription.

    (B-i) The level of transcription is higher when lysine is absent and lower when lysine is present. Therefore, lysine has a negative effect on the level of transcription.

    (C-i) This operon is more likely repressible.

    (D-i) Repressible operons have a lower level of expression when the corepressor is present. Since the presence of lysine decreases the level of expression of the operon, it is more likely that the operon is repressible.

Long Free-Response

13. (A-i)   Enhancer sequences function as binding sites for activator proteins or transcription factors. When an activator or transcription factor binds to an enhancer site, more transcription of the gene occurs.

    (B-i)   DNA sequence A is the control because no part of that sequence is deleted.

    (C-i and ii)   Sequence 1 is most likely the enhancer sequence because the relative levels of *THY-1* gene transcription are lower when sequence 1 is deleted than when sequence 2 or sequence 3 is deleted.

    (D-i and ii)   Promoters serve as binding sites for RNA polymerase. If RNA polymerase cannot bind, transcription cannot occur. If P were deleted, no transcription of *THY-1* would occur, as shown in the following figure.

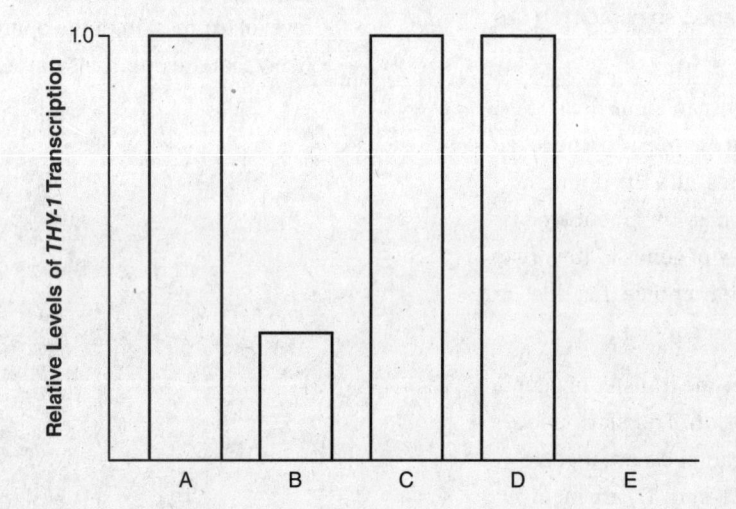

18

Biotechnology

Learning Objectives

In this chapter, you will learn:
- → Bacterial Transformation
- → Gel Electrophoresis
- → Polymerase Chain Reaction (PCR)
- → CRISPR-Cas9

Overview

The tools of biotechnology are being used for many purposes: cancer therapies, improvement in agriculture yields, gene therapies for genetic disorders, de-extinction projects (such as the one to bring back the woolly mammoth), and extinction projects (trying to eliminate pathogens that cause human disease), just to name a few. Understanding the basics of how these techniques work, and their possible uses and misuses, is very important. This chapter will briefly review a few basic biotechnology techniques to know for test day.

Bacterial Transformation

Bacterial transformation introduces foreign DNA into bacterial cells. The foreign DNA, usually a small, circular piece of DNA called a plasmid, may integrate into the host cell's chromosome or remain separate from the host cell DNA in the cell's cytoplasm. One method of transforming bacteria involves heat shock, in which bacterial cells are mixed with foreign DNA and then quickly exposed to a cold-hot-cold temperature transition. This heat shock creates temporary microscopic pores through which the foreign DNA can enter some of the bacterial cells.

The foreign plasmid DNA needs a selectable marker so that cells that have incorporated the plasmid DNA can be detected. The selectable marker is usually an antibiotic resistance gene that is not present in nontransformed bacterial cells. By growing the transformed bacteria on an agar plate that contains the antibiotic, one can select for the bacteria that absorbed and are now expressing the genes from the plasmid.

If the plasmid contains a gene from another organism, it is called recombinant DNA. **Recombinant DNA** is simply DNA that has been recombined from different source organisms. DNA can be cut at specific sequences using **restriction endonucleases** (also known as restriction enzymes), and these pieces of DNA can be recombined and connected with **DNA ligases**. Some recombinant plasmids may contain the selectable marker as well as a gene that codes for a desired protein product. Bacteria that take in the recombinant plasmid would be capable of producing the product that is coded for by the gene on the plasmid. Many pharmaceutical products (such as insulin) are now produced in large amounts by bacteria that contain recombinant plasmids. Before the advent of biotechnology, the insulin needed for patients with diabetes had to be isolated from the pancreas of animals, which was an expensive process and had the potential of transmitting diseases from other organisms. Now, recombinant insulin made by transformed bacteria is safer and less expensive than insulin was 50 years ago.

Gel Electrophoresis

Gel electrophoresis is a technique that is used to separate DNA fragments by size and by charge. DNA fragments can be created by treating a DNA sample with restriction enzymes, which cut DNA at specific base pair sequences. Due to the redundancy of the genetic code, even organisms with the same protein sequences will have slightly different DNA sequences. Treatment of DNA with restriction enzymes will cut these different DNA sequences at different locations and result in different fragment sizes.

The backbone of the DNA double helix consists of five-carbon deoxyribose sugars and phosphate groups. The phosphate groups have a slight negative charge. DNA samples are loaded into wells at the top of the gel, and an electric current is applied to the gel. A positive cathode is attached to the bottom of the gel and the negative anode to the top of the gel. DNA molecules have an overall negative charge (due to their abundance of the phosphate groups), causing them to migrate toward the bottom (positive end) of the gel.

The gel itself is usually made of agarose and contains microscopic pores through which the DNA fragments migrate. Shorter fragments will be able to travel more quickly through these pores; longer fragments will take more time. This leads to shorter fragments being found at the bottom of the gel, farthest from the well into which the DNA sample was loaded; longer fragments will be closer to the top of the gel. The addition of a DNA marker with fragments of known lengths runs alongside the other DNA samples, providing a "ruler" by which to estimate the size of each fragment produced. Each organism's pattern of fragments on the gel will be different and can be used to create a unique DNA fingerprint.

Polymerase Chain Reaction (PCR)

Polymerase chain reaction (PCR) is used to amplify specific DNA fragments. PCR can be used to create millions of copies of a specific fragment of DNA. PCR involves cycles of DNA replication using primers that are specific to the beginning and end of the fragment of the DNA sequence that is to be amplified. Each cycle of PCR consists of three stages:

1. **Denaturing the DNA**—This separates the two strands of the double helix. It is usually done at a high temperature.
2. **Annealing of the primers**—Once the DNA strands have been separated, the temperature is lowered slightly, and primers that are complementary to the desired DNA sequence are allowed to anneal (form hydrogen bonds with) the beginning and end of the fragment of the DNA sequence that is to be amplified.
3. **Extension of the primers**—DNA polymerase then adds new nucleotides to the primers, creating two copies of the desired DNA sequence. A heat-stable DNA polymerase is used for this step so that it will not be denatured during the rise in temperature with each cycle of PCR.

Each cycle of PCR doubles the number of copies of the desired DNA sequence. The number of copies of the DNA sequence produced at the end of PCR is equal to 2^n, where n is the number of cycles of PCR completed. After just 20 cycles of PCR, over one million copies of the DNA sequence can be generated.

PCR can also be used to rapidly sequence segments of DNA. DNA sequencing is important in forensics, bioinformatics, medicine, and evolutionary biology.

CRISPR-Cas9

> While CRISPR-Cas9 is not part of the latest AP Biology Course and Exam Description and likely will not be tested on the AP Biology exam, it is an important and timely topic, so a brief discussion of CRISPR is included here.

The clustered regularly interspaced short palindromic repeats (**CRISPR**)-Cas9 system functions in nature as an adaptive immune system in bacteria. Bacteria can be infected by viruses called bacteriophages. When a bacteriophage infects a bacterium, it cuts up the viral DNA and inserts those pieces between short palindromic repeats in the bacteria's DNA. This allows the bacteria to store information on which viruses have previously infected its cell. When another virus infects the bacteria, the bacteria compare the DNA from the new virus to these stored DNA sequences. If it finds a match, it cuts the DNA of the virus with the Cas9 enzyme.

CRISPR can be used to edit DNA sequences. The Cas9 enzyme uses a piece of guide RNA to find where to cut DNA. By using synthetic guide RNA pieces that correspond to the desired location of the cut to be made, one can direct the Cas9 enzyme to cut at a specific DNA sequence. If this cut is in the middle of a gene or one of its regulatory regions, this can create a "knockout" of the gene in which the gene is no longer functional. Observing the effects of this knockout can help scientists understand the function of that gene.

Sometimes instead of a knockout, a different DNA sequence at the site cut by the Cas9 enzyme is desired. Inserting multiple copies of a donor DNA sequence into the cell allows the cell's own DNA repair mechanisms to use the donor DNA as a template. The cell can then replace the DNA sequence at the cut site with a copy of the donor sequence.

Practice Questions

Multiple-Choice

1. Which of the following is used to cut DNA at specific base pair sequences?

    (A) DNA ligase
    (B) gel electrophoresis
    (C) polymerase chain reaction
    (D) restriction enzymes

2. A forensic scientist recovers a very small amount of evidence at a crime scene. The scientist would like to amplify the number of copies of DNA in the evidence sample. Which of the following techniques should the scientist use?

    (A) bacterial transformation
    (B) CRISPR-Cas9
    (C) gel electrophoresis
    (D) polymerase chain reaction

3. A forensic scientist cuts DNA from a crime scene and DNA from a suspect with the same enzyme. Which tool should the scientist use to separate the DNA fragments according to their size?

    (A) bacterial transformation
    (B) CRISPR-Cas9
    (C) gel electrophoresis
    (D) polymerase chain reaction

Questions 4 and 5

A scientist at a pharmaceutical company wants to create bacteria that will synthesize human growth hormone.

4. What should the scientist use to add the DNA code for human growth hormone to a plasmid?

    (A) bacterial transformation
    (B) DNA ligase
    (C) gel electrophoresis
    (D) polymerase chain reaction

5. The scientist has successfully created a plasmid that contains the DNA code for the human growth hormone gene. Which technique should the scientist use to insert the plasmid into a cell?

    (A) bacterial transformation
    (B) CRISPR-Cas9
    (C) gel electrophoresis
    (D) polymerase chain reaction

Questions 6–8

Huntington's disease is caused by a short tandem CAG repeat in the *HTT* gene. Individuals with fewer than 35 CAG repeats in the *HTT* gene do not develop Huntington's disease. Individuals with 40 or more CAG repeats will develop Huntington's disease.

6. Which of the following tools would be most useful in amplifying the number of copies of the *HTT* gene so that more DNA would be available for analysis?

    (A) CRISPR-Cas9
    (B) gel electrophoresis
    (C) polymerase chain reaction
    (D) restriction enzymes

7. Which tool would be most useful for estimating the size of the *HTT* gene isolated?

    (A) DNA ligase
    (B) gel electrophoresis
    (C) polymerase chain reaction
    (D) restriction enzymes

8. A transgenic mouse has the *HTT* gene. A scientist wants to investigate whether inserting a stop codon into the *HTT* gene might prevent the mouse from developing the disease. Which of the following tools would be most useful in inserting a stop codon into the *HTT* gene?

    (A) CRISPR-Cas9
    (B) gel electrophoresis
    (C) polymerase chain reaction
    (D) restriction enzymes

9. What property of DNA causes it to move toward the positive electrode in gel electrophoresis?

   (A) the nitrogen atoms in the nitrogenous bases
   (B) the hydrogen bonds that form between the bases in the two strands of the double helix
   (C) the charges on the oxygen atoms in the deoxyribose sugars
   (D) the charges on the phosphates in the backbone of the helix

10. Why is a heat-stable DNA polymerase required for the polymerase chain reaction (PCR)?

    (A) Heat-stable DNA polymerases add DNA nucleotides at a faster rate than other DNA polymerases.
    (B) The high temperatures that are required for each cycle of PCR to separate the DNA strands would denature DNA polymerases that are not heat stable.
    (C) Heat-stable DNA polymerases anneal more readily to the PCR primers.
    (D) Heat-stable DNA polymerases are more accurate than other DNA polymerases.

Short Free-Response

11. The pAMP-pKAN plasmid contains both the ampicillin resistance gene and a kanamycin resistance gene.

    Part A

    (i) **Describe** a method for inserting this plasmid into a bacterial cell.

    Part B

    (i) **Explain** how one could detect which bacterial cells had successfully absorbed the pAMP-pKAN plasmid.

    Part C

    A restriction enzyme makes a cut in the kanamycin resistance gene.

    (i) **Predict** the result if a bacterium absorbs this piece of DNA that has been cut by the restriction enzyme.

    Part D

    (i) **Justify** your prediction from Part C.

12. The comet assay technique is one way of measuring the relative amount of DNA damage in a cell. The nucleus of a cell is placed on a slide coated with an agarose gel. An electric current is applied to the gel, and smaller pieces of DNA will move farther away from the location where the nucleus was placed. These smaller pieces of DNA form the "tail," and the nucleus forms the "head" of the "comet." Longer tails indicate greater amounts of DNA damage.

    Part A

    (i) **Explain** why DNA fragments will move through an agarose gel when an electric current is applied.

    Part B

    (i) List and **describe** two things that could damage DNA and result in longer tails on the comet assay.

    Part C

    Two cells are treated with a known mutagen of DNA. Cell A has two functional p53 alleles (p53 is involved in DNA repair). Cell B has one functional p53 allele, and its second p53 allele is nonfunctional.

    (i) On the two templates provided, **construct** visual representations of the comet assay results for these two cells.

    Part D

    (i) **Explain** the visual representations you created on the two templates in Part C.

Long Free-Response

13. A student conducts a bacterial transformation experiment using a strain of *E. coli* that is not resistant to antibiotics. The student also uses a plasmid (pAMP-pTET) that carries genes for both ampicillin and tetracycline resistance. The four different plates used, and the results of this experiment, are shown in the following figure.

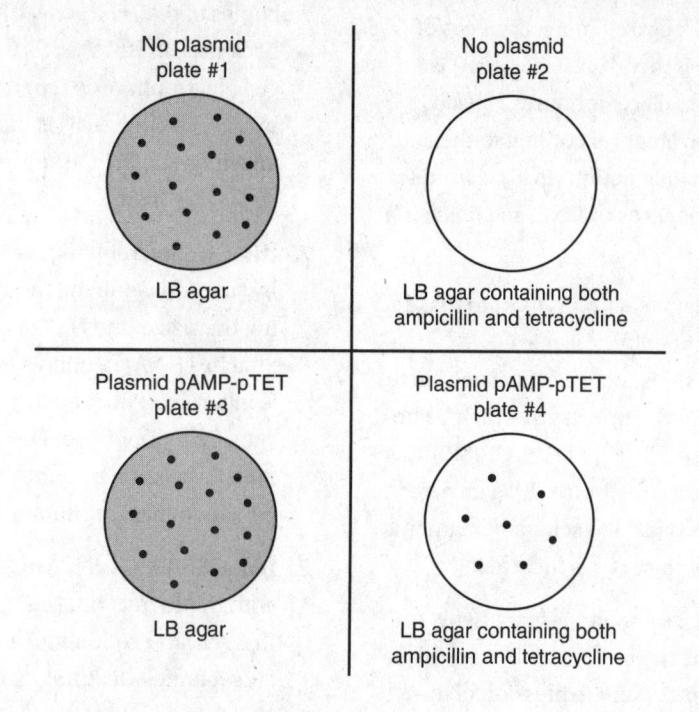

Part A

(i) **Describe** the two enzymes that would be used to create a recombinant plasmid that contained both the ampicillin resistance gene and the tetracycline resistance gene.

Part B

(i) **Identify** the plate that acts as a control to show that the untransformed bacteria were not originally resistant to ampicillin and tetracycline.

Part C

(i) **Determine** whether the relative numbers of bacteria colonies (seen on agar plate #3 and agar plate #4) are as expected if the bacterial transformation procedure was successful.

Part D

Bacteria are removed from plate #4 and are allowed to grow in nutrient medium. Bacteria that contain the F+ (fertility) plasmid can undergo the process of conjugation and transfer some of their DNA to other bacterial cells.

(i) **Predict** the result if bacteria (that were resistant to streptomycin and contained the F+ plasmid) were used to transform the bacteria from plate #4.

(ii) **Justify** your prediction.

Answer Explanations

Multiple-Choice

1. **(D)** Restriction enzymes (also known as restriction endonucleases) cut DNA at specific base pair sequences. DNA ligase connects pieces of DNA, so choice (A) is incorrect. Choice (B) is incorrect because gel electrophoresis is used to separate DNA fragments according to their size. Polymerase chain reaction (PCR) is used to amplify specific sequences of DNA, so choice (C) is incorrect.

2. **(D)** Polymerase chain reaction is used to make multiple copies of (amplify) specific DNA sequences. Choice (A) is incorrect because bacterial transformation is the process in which bacteria absorb and express foreign DNA. CRISPR-Cas9 is used in gene editing, so choice (B) is incorrect. Gel electrophoresis is used to separate fragments of DNA by size, so choice (C) is incorrect.

3. **(C)** Gel electrophoresis separates fragments of DNA according to their size. Bacterial transformation involves the uptake and expression of foreign DNA by bacteria, so choice (A) is incorrect. CRISPR-Cas9 is used for gene editing, so choice (B) is incorrect. PCR amplifies specific sequences of DNA, so choice (D) is also incorrect.

4. **(B)** DNA ligase attaches DNA fragments together, joining them with phosphodiester bonds. Choice (A) is incorrect because bacterial transformation is used to insert naked, foreign DNA into a cell. Gel electrophoresis separates DNA fragments by size, so choice (C) is incorrect. Polymerase chain reaction is used to amplify the number of copies of a specific sequence of DNA, so choice (D) is incorrect.

5. **(A)** Bacterial transformation inserts naked, foreign DNA into a bacterial cell. CRISPR-Cas9 is used for gene editing, not inserting plasmids into a cell, so choice (B) is incorrect. Choice (C) is incorrect because gel electrophoresis is used to separate DNA fragments according to their size. Polymerase chain reaction is used to amplify the number of copies of a specific DNA sequence, so choice (D) is incorrect.

6. **(C)** Polymerase chain reaction makes multiple copies of a specific DNA sequence and would be the best choice for amplifying the number of copies of the *HTT* gene. Choice (A) is incorrect because CRISPR-Cas9 is used for gene editing. Gel electrophoresis separates fragments of DNA by size, so choice (B) is incorrect. Restriction enzymes cut DNA at specific sequences, so choice (D) is incorrect.

7. **(B)** Gel electrophoresis separates DNA fragments by size and would be most useful for estimating the size of the *HTT* gene isolated. DNA ligase attaches DNA fragments together, so choice (A) is incorrect. Polymerase chain reaction amplifies specific sequences of DNA, so choice (C) is incorrect. Restriction enzymes cut DNA at specific base pair sequences, so choice (D) is also incorrect.

8. **(A)** CRISPR-Cas9 is used for gene editing, and with the correct donor DNA, it could be used to insert a stop codon into a gene. Gel electrophoresis separates DNA fragments by size, so choice (B) is incorrect. Choice (C) is incorrect because PCR amplifies the number of copies of a specific DNA sequence. Restriction enzymes cut DNA at specific sequences, so choice (D) is incorrect.

9. **(D)** The phosphates in the backbone of the DNA double helix have a slight negative charge and are attracted to the positive electrode. The nitrogen atoms in the nitrogenous bases are not charged, so choice (A) is incorrect. Hydrogen bonds between the base pairs do not result in a net negative charge, so choice (B) is incorrect. The oxygen atoms in the deoxyribose sugars are not charged, and thus choice (C) is incorrect.

10. **(B)** The denaturation stage that occurs during each cycle of PCR requires high temperatures that would denature most enzymes, so a heat-stable DNA polymerase is required for PCR. Heat-stable DNA polymerases do not add DNA nucleotides at a faster rate than other DNA polymerases, so choice (A) is incorrect. PCR primers anneal to the target DNA sequence, not to the DNA polymerase, so choice (C) is incorrect. Choice (D) is incorrect because heat-stable DNA polymerases are not more accurate than other DNA polymerases.

Short Free-Response

11. (A-i) Bacterial transformation through heat shock could be used to insert this plasmid into the cell. The transition from cold-hot-cold will create temporary small pores through which the plasmids may enter the cell.

    (B-i) Plating the transformed bacteria on an agar plate (that contains both ampicillin and kanamycin) would kill any bacteria that did not absorb the pAMP-pKAN plasmid.

    (C-i) If a bacterium absorbs the piece of DNA that has a cut in the kanamycin resistance gene, the bacterium would still be resistant to ampicillin but not resistant to kanamycin.

    (D-i) The ampicillin resistance gene would still be intact and functional, but the kanamycin resistance gene would not be functional due to the cut made by the restriction enzyme.

12. (A-i) The phosphates in the sugar-phosphate backbone of DNA have a slight negative charge. They would be attracted to a positively charged electrode.

    (B-i) Any mutagen could damage DNA and result in longer tails. Some examples are ionizing radiation, UV light, and carcinogens (all of which can cause breaks in the DNA sequence).

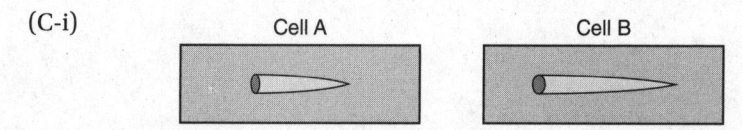

    (C-i)

    Cell A   Cell B

    (D-i) Cell A has two functional p53 alleles and thus would have a greater capacity to repair the DNA damage. Therefore, the tail in the comet assay would be shorter for cell A than the tail for cell B, which only has one functional p53 allele.

Long Free-Response

13. (A-i) Restriction endonucleases would be used to cut the desired genes from their original plasmids, and DNA ligase would be used to connect them together in a recombinant plasmid.

    (B-i) The plate that acts as this control is plate #2, which contains both ampicillin and tetracycline. This plate shows that the bacteria used in the transformation procedure were not originally resistant to the antibiotics. If no growth occurred on that plate after bacteria that did not receive the plasmid were plated upon it, that would confirm there was no previous antibiotic resistance.

    (C-i) Not every bacterium is expected to absorb the plasmid. So far fewer bacteria colonies would be expected on plate #4 (which contains ampicillin and tetracycline) than on plate #3 (which does not contain any antibiotic). All bacteria, regardless of whether they absorbed the plasmid, would grow on the LB agar in plate #3, while only bacteria that absorbed the plasmid could grow on plate #4. Based on what is shown in the figure, the relative numbers of bacteria colonies on these two plates meets expectations, and thus this bacterial transformation procedure was successful.

    (D-i and ii) The bacteria would now be resistant to three antibiotics: ampicillin, tetracycline, and streptomycin. The bacteria taken from plate #4 would be resistant to ampicillin and tetracycline, and after successful conjugation with the bacteria that have streptomycin resistance, the bacteria would be resistant to streptomycin as well.

UNIT 7
Natural Selection

19

Types of Selection

Learning Objectives

In this chapter, you will learn:
- → Evidence of Evolution
- → Natural Selection
- → Artificial Selection
- → Sexual Selection

Overview

Mutations generate genetic variation in a population. These genetic variations lead to different phenotypes in a population. Different types of selection act upon these variations in phenotypes, leading to differential reproductive success. Populations evolve when some members of the population have greater reproductive success than other members of the population. This chapter will review the evidence of evolution and the different types of selection that may affect a population.

Evidence of Evolution

Evidence of evolution can be found in both extant (living) and extinct species. Some categories of evidence of evolution include:

- **Molecular evidence**—Comparing DNA sequences and amino acid sequences in proteins from different organisms provides evidence of evolution. When comparing the DNA sequence of a gene that is shared by different organisms, the more recently the organisms share a common ancestor, the more similar their DNA sequences will be. For example, the *GAPDH* (glyceraldehyde-3-phosphate dehydrogenase) gene in humans and in chimpanzees is over 99% similar in sequence, but the similarity in the *GAPDH* gene in humans and in dogs is only about 91% similar. This indicates that humans and chimpanzees share a more recent common ancestor than humans and dogs. Molecular evidence is considered very strong evidence since environmental factors do not usually change an organism's DNA sequence.
- **Morphology**—Homologous structures, which have common ancestry but different functions, also provide evidence of evolution. For example, the number and arrangement of bones in human hands, bat wings, and whale fins are very similar, indicating common ancestry and evidence of evolution.
- **Fossils**—The existence of fossils from organisms that no longer live on Earth also provide evidence of evolution. Transitional fossils show intermediate states between ancestral and modern species. Fossils can be dated by studying the age of the rock layers in which they are found or by using radioactive isotopes to date the fossils.
- **Vestigial structures**—Some organisms contain anatomical features that no longer seem to have a purpose in the modern organism but may have had a function in an ancestral organism. An example of this in humans

> **TIP**
> The details of how the radioactive dating of fossils works will not be tested on the AP Biology exam.

is the tailbone, which currently serves no function in humans but may have helped our tree-dwelling ancestors balance upon or travel between branches.
- **Convergent evolution**—Species that live in similar environments may evolve similar adaptations even though they may not have a recent common ancestor. Sharks (cartilaginous fish) and dolphins (mammals) do not share a recent common ancestor, but they have evolved similar body shapes due to their similar environments.
- **Biogeographical evidence**—Biogeography is the study of the distribution of species. Species on islands off the coast of South America are more similar to species found in South America than to species found in North America.
- **Observations of evolution in current species**—When repeatedly exposed to antibiotics, bacteria populations evolve resistance over time. Mosquito populations have evolved resistance to pesticides like DDT.

Natural Selection

One of the most important things to remember about evolution is that individuals do not evolve; populations evolve. Charles Darwin was not the first person to propose that populations evolve, but he was the first to explain a mechanism for evolution, **natural selection**, that was supported by evidence for how populations evolve.

> Jean-Baptiste Lamarck was also an early proponent of the idea that biological evolution occurs. Lamarck's theory of inheritance of acquired characteristics emphasized changes to individual organisms during their lifetimes (acquired characteristics) and the inheritance of these acquired characteristics by their offspring. While modern epigenetics (discussed in Chapter 17) shows that some changes to an organism's DNA during its lifetime (such as methylation patterns in the genome) may be inherited by its offspring, most acquired characteristics are not inherited by the next generation.

Darwin's Theory of Natural Selection

Here are the core principles of Darwin's theory of evolution by natural selection that you should understand and be able to apply on the AP Biology exam:

- Variations in populations lead to different phenotypes in members of a population.
- Competition for limited resources, or predation, leads to some members of a population surviving while other members of a population do not.
- The environment determines which phenotypes are favorable.
- Individuals with phenotypes that give them a survival advantage are more likely to survive and reproduce (**differential reproductive success**).
- Over time, favorable phenotypes will become more prevalent in a population as members of the population without those favorable phenotypes do not survive.

This is how Darwin's theory of evolution by natural selection would explain the evolution of long necks in giraffes (see Figure 19.1):

- Variations in populations resulted in some giraffes having longer necks and others having shorter necks.
- Giraffes with longer necks can reach more food and therefore are more likely to survive and reproduce (differential reproductive success).
- Over time, only giraffes with longer necks will be present in a population because giraffes with shorter necks will not survive.

Figure 19.1 Natural Selection In Giraffes

Examples of Evolution by Natural Selection

There are many examples of evolution by natural selection. One example is antibiotic-resistant bacteria. In a bacteria population, some bacteria will have genotypes that cause them to be sensitive to an antibiotic; others have genotypes that confer resistance to an antibiotic. As antibiotics are applied to the environment of a population of bacteria, individual bacterium that are sensitive to the antibiotic will die out while bacteria that have resistance to the antibiotic will survive and reproduce. With continued exposure to the antibiotic, over time, only the antibiotic-resistant bacteria will survive and be present in the population. Note that individual bacterium do not evolve or "learn" to be resistant to the antibiotic; some bacteria already possess the variation that gives them resistance. Initially, these antibiotic-resistant bacteria are less prevalent in the population, but these bacteria are more likely to survive and reproduce in an environment that contains the antibiotic than those that are not antibiotic resistant. Over time, as these bacteria reproduce in the presence of the antibiotic, the frequency of bacteria with antibiotic resistance will increase.

> **TIP**
> Avoid using the term *fitness* on the AP Biology exam without explaining what it means (differential reproductive success).

Another example is the peppered moth found in England in the 1800s. Peppered moths have variations in wing color; some moths have darker wings, while others have lighter wings. The coloration of moth wings is an inherited trait. Birds eat the moths, and moths with wing coloration that blends in with their environment are more difficult for birds to see; therefore, those moths are less likely to be eaten by the birds. Prior to the Industrial Revolution in England, trees in the moths' habitat were covered with a light-colored lichen that allowed the lighter-colored moths to blend in with the trees. This made lighter-colored moths more difficult for birds to find and eat, and these lighter-colored moths were predominant in the moth population. During the Industrial Revolution in England in the 1800s, however, sulfur dioxide emissions from factories killed much of the light-colored lichen on the trees. Lighter-colored moths no longer had an advantage, and moths with darker wings were able to blend in with the darker tree bark. This made darker moths more difficult for birds to find and eat. By the 1950s, over 90% of the moths in the population had darker wings. As pollution controls were introduced during the 1960s, the light-colored lichen again covered the trees. Lighter-colored moths once again had a survival advantage and became more prevalent over time. (See Figure 19.2.)

Figure 19.2 Peppered Moths

> **TIP**
> A common mistake on the AP Biology exam is to describe evolution with Lamarckian statements, which describe *individuals* evolving in response to their environment. Remember that *populations* evolve; individuals do not.

Darwin's theory of evolution by natural selection would explain this by saying that there were natural variations in the moth population, with some moths having darker wings and some having lighter wings. The moths with darker wings had a survival advantage during the Industrial Revolution and became more prevalent in the population due to natural selection. When the environment changed in the second half of the 20th century, the moths with darker wings no longer had an advantage, so the moths with lighter wings then became more common.

According to the theory of natural selection, the environment selects for individuals with phenotypes that confer a survival advantage in that environment. If the environment changes, different phenotypes may confer an advantage, and the changing environment can change the direction of evolution, as was seen in the peppered moths example.

Directional Selection

The type of selection that was seen in the peppered moths example is directional selection. **Directional selection** occurs when one end of the range of phenotypes is favored by natural selection, causing the frequency of that phenotype to increase over time. Directional selection is represented by Figure 19.3.

> Remember, evolution is the result of changes in a population over multiple generations. Individuals with characteristics that provide a survival advantage are more likely to survive and reproduce than individuals without those characteristics. With each generation, the proportion of the population with characteristics that provide a survival advantage will increase, leading to the evolution of the population.

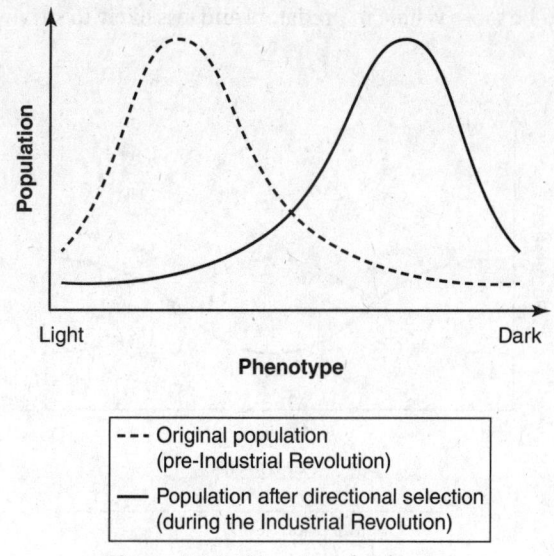

Figure 19.3 Directional Selection

Stabilizing Selection

Natural selection can also lead to **stabilizing selection**, where the intermediate phenotype is favored and extreme phenotypes are selected against. Clutch size (the number of eggs produced per reproductive cycle by birds) exhibits stabilizing selection. If a bird lays a large number of eggs, that bird may have too many offspring to care for and feed, leading to poor reproductive success. However, if the bird lays only one or two eggs per reproductive cycle, there is a risk that none of the offspring may survive. In robins, the average clutch size is five to six eggs per nest. Figure 19.4 represents stabilizing selection.

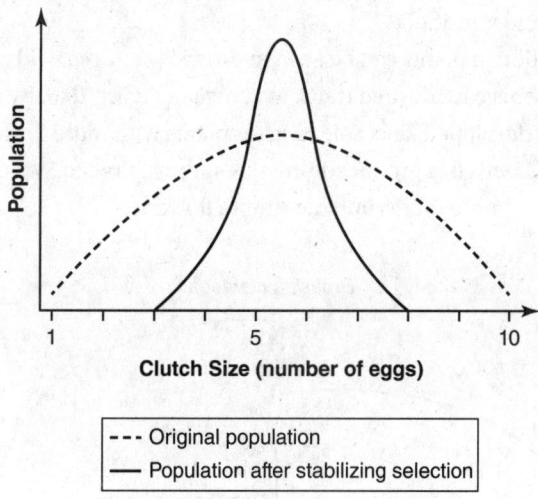

Figure 19.4 Stabilizing Selection

Disruptive Selection

Sometimes, natural selection can lead to **disruptive selection**, where individuals on both extremes of the phenotypic range are more likely to survive and reproduce than individuals with an intermediate phenotype. Consider a habitat with light-colored sandy soil that is interspersed with dark, rocky patches. Mice with light-colored fur in this habitat could blend in with the sandy soil and be less visible to predators. Mice with dark-colored fur could hide in the dark-colored, rocky patches. Mice with an intermediate fur color could not blend in with either the soil or the rocky patches; they would be more visible to predators and less likely to survive and reproduce. Figure 19.5 represents disruptive selection.

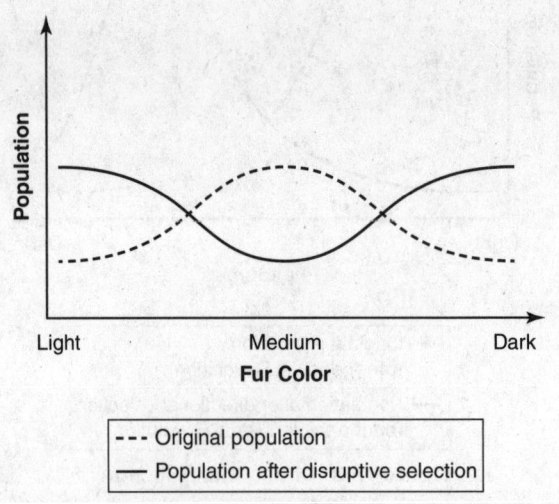

Figure 19.5 Disruptive Selection

Directional, stabilizing, and disruptive selection all rely on the mechanisms of natural selection. Differential reproductive success in different environments results in changes in populations.

Artificial Selection

Individuals in a population can also experience differential reproductive success through artificial selection. In **artificial selection**, humans selectively breed domesticated plants or animals to produce populations with desired traits. Instead of the environment selecting for individuals with favorable phenotypes, humans select which individuals in a population survive and reproduce.

An example of artificial selection in plants can be seen in *Brassica oleracea* wild cabbage. Over the years, farmers have selectively bred wild cabbage for desired traits, as shown in Figure 19.6. By selectively breeding plants with bigger leaves, farmers have developed kale. Selecting for plants with more flower clusters led to cauliflower. Crossbreeding plants with lateral buds has produced Brussels sprouts. Broccoli was produced by selectively breeding plants with both stems that are more robust and have more flowers.

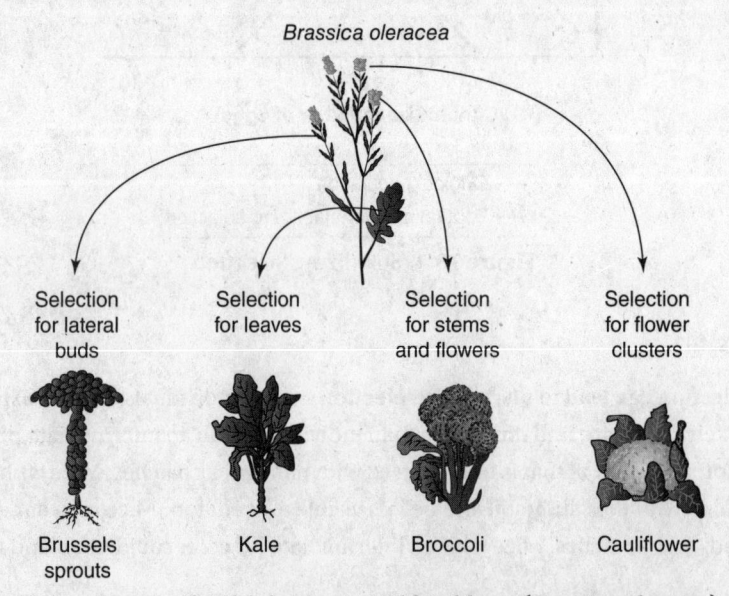

Figure 19.6 Artificial Selection in Wild Cabbage (*Brassica oleracea*)

Humans have used artificial selection to produce desired traits in animals as well. Dog breeds are also a result of artificial selection. Domesticated wolves were selectively bred for particular traits, leading to the wide variety of dog breeds seen today. Over the last 10,000 years, human populations have selectively bred members of the aurochs (the wild bovine related to oxen) to obtain the many breeds of domestic cattle seen today.

Sexual Selection

Sexual selection occurs when individuals with certain characteristics are more likely to attract mates than other individuals. Over time, individuals with traits that are more likely to attract mates become more prevalent in the population.

In intersexual selection, individuals of one sex are particular in selecting mates from the other sex. Mate choice may be based on perceived fitness of the members of the other sex, with members who seem stronger or healthier being more likely to produce offspring that will survive. Many bird behaviors involve mate choice. Birds may choose mates based on coloration, bird songs, mating dances, or nesting behaviors. An example of this is seen in *Pavo cristatus* (the peacock). Peacocks with brighter plumage are more likely to attract mates than their counterparts with duller plumage. Female members of *Sula nebouxii*, the blue-footed booby bird, select mates with the brightest blue feet. Younger males have brighter blue feet and are more likely to have higher fertility.

In intrasexual selection, members of one sex compete for mates of the other sex. This may involve asserting dominance to ward off competitors and gain better access to mates.

Practice Questions

Multiple-Choice

Questions 1 and 2

Drs. Peter and Rosemary Grant have studied the rate of evolutionary change in the finch populations of the Galápagos Islands. Beak size in these finches determines which types of seeds the finch populations feed on. Finches with larger beaks eat thick-walled seeds, while finches with smaller beaks eat thin-walled seeds. During a drought from 1981 to 1987, the number of plants that produced thin-walled seeds decreased. It was determined that the average beak size (both length and mass) of finches increased dramatically during the drought.

1. Which type of selection most likely led to the change in beak size during the drought?

   (A) directional selection
   (B) disruptive selection
   (C) stabilizing selection
   (D) sexual selection

2. During the years 1988 to 1995, the average beak size of the finches decreased. Which of the following is the most likely explanation for this change?

   (A) Female finches prefer to mate with males with smaller beaks.
   (B) The drought ended, and the average rainfall returned to predrought levels.
   (C) Predator populations increased.
   (D) Finches with larger beaks were more susceptible to diseases.

3. Which of the following would best determine whether two species of birds have a recent common ancestor?

   (A) DNA sequences
   (B) fossil record
   (C) habitat distribution
   (D) mating behaviors

4. The following figure shows the bone structures of the limbs of a whale, a horse, and a human.

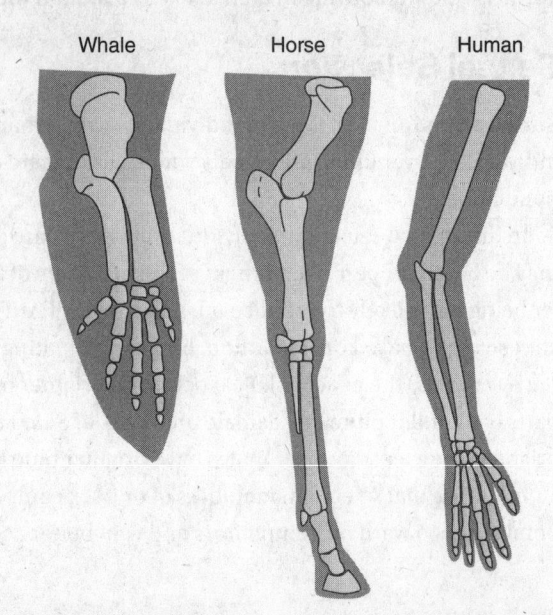

   These are examples of which types of structures?

   (A) analogous
   (B) convergent
   (C) homologous
   (D) vestigial

5. Triclosan is an antibacterial chemical that is used in some household products to reduce bacteria levels. Which of the following is the most likely result of increased use of products that contain triclosan over time?

   (A) All household bacteria species will be eliminated.
   (B) Household bacteria species will become resistant to triclosan.
   (C) Individual bacteria will learn how to resist the effects of triclosan.
   (D) Increased triclosan levels will increase the frequency of mutations in household bacteria.

6. Snakes feed on toads. Cane toads (*Rhinella marina*) excrete a toxic substance on their skin that is poisonous to many, but not all, snake species. If cane toads are introduced to a new environment, predict the most likely effect on the snake species in that environment.

   (A) All snake species in the environment will die out due to the cane toads' toxin.
   (B) Snake species that are resistant to the cane toads' toxin will increase in numbers.
   (C) Snakes that are susceptible to the cane toads' toxin will acquire resistance to the toxin.
   (D) All snake species will learn to avoid eating cane toads.

7. The human *TAS2R38* gene encodes a cell membrane protein that influences the ability to taste bitter compounds. Individuals who possess at least one *TAS2R38* allele have the "taster" phenotype and can taste certain types of bitter compounds. It is estimated that about 70% of humans have the taster phenotype. Which of the following best explains the frequency of the taster phenotype?

   (A) Many toxic compounds have a bitter taste, so the *TAS2R38* allele provided a survival advantage in ancestral humans.
   (B) Ancestral humans with the *TAS2R38* allele were more likely to consume bitter-tasting foods.
   (C) Bitter-tasting foods have a higher nutrient content and were more likely to be consumed by ancestral humans who did not have the *TAS2R38* allele.
   (D) A lack of the *TAS2R38* allele provided a survival advantage in ancestral humans.

8. The Aztecs were some of the first humans to slowly change teosinte, also known as wild corn, into the current form of corn eaten today. Which process did the Aztecs most likely use?

   (A) artificial selection
   (B) frequency-dependent selection
   (C) natural selection
   (D) sexual selection

9. Which of the following assertions (about how evolution by natural selection occurs) is incorrect?

   (A) There are variations among individuals of a species.
   (B) Some variations provide a survival advantage.
   (C) Variations acquired during an individual's lifetime are passed on to the individual's offspring.
   (D) Over time, the frequency of individuals with variations that provide a survival advantage will increase.

10. Rock pocket mice (*Chaetodipus intermedius*) are found in rocky outcrops in the desert of the southwestern United States. Some rock pocket mice have light-colored fur, while others have dark-colored fur. A population of rock pocket mice lives in a desert with both light-colored sand dunes and dark-colored rocks. Owls and hawks prey on the rock pocket mice they see. The initial ratio of light-colored to dark-colored mice in this population is approximately 1:1. A sandstorm occurs in the habitat of this population of rock pocket mice, and it covers all of the habitat in a thick layer of light-colored sand. Predict the most likely effect of this on the rock pocket mouse population.

    (A) The relative frequency of dark-colored rock pocket mice would increase.
    (B) The relative frequency of light-colored rock pocket mice would increase.
    (C) The numbers of both dark-colored and light-colored rock pocket mice would decrease.
    (D) The numbers of both dark-colored and light-colored rock pocket mice would increase.

Short Free-Response

11. Three new species (A, B, and C) of fossilized crocodile are discovered. The characteristics of these species are compared to those of the saltwater crocodile, *Crocodylus porosus*. The saltwater crocodile is found in Southeast Asia and Australia, and adults range in length from 5.5 to 5.8 meters and have 66 teeth. Data comparing the characteristics of the three fossilized crocodile species are shown in the table.

	A	B	C
Location of Fossils	Africa	Europe	Asia
Number of Teeth	48	56	72
Nose to Tail Length (m)	4.7	5.5	6.3
Percent of Homology of DNA with *Crocodylus porosus*	93%	90%	98%

Part A

(i) Based on the data given, **identify** the fossil species that has the most in common with *Crocodylus porosus*.

Part B

The number of teeth in a crocodile jaw correlates with increased predator efficiency.

(i) **Identify** the crocodile(s) that would be less efficient predators than *C. porosus*.

Part C

(i) **Evaluate** the claim that species B shares a more recent common ancestor with *C. porosus* than do species A or species C.

Part D

(i) **Explain** your reasoning for your response from Part C.

12. A student performs an experiment to study the effects of repeated exposure to antibiotics on bacteria. A strain of *E. coli* that is not antibiotic-resistant is grown on an antibiotic-free LB agar plate (plate 1), the starter plate. Some of the *E. coli* from the starter plate are then spread on a plate that contains LB agar and 50 μg/mL of the antibiotic ampicillin (plate 2), and some of the *E. coli* are spread on a plate that contains LB agar without antibiotic (plate 3). *E. coli* from plate 2 are then spread on a plate that contains LB agar and 100 μg/mL of the antibiotic ampicillin (plate 4). *E. coli* from plate 3 are then spread on a plate that contains LB agar without antibiotic (plate 5). A diagram of the experimental plates is shown below.

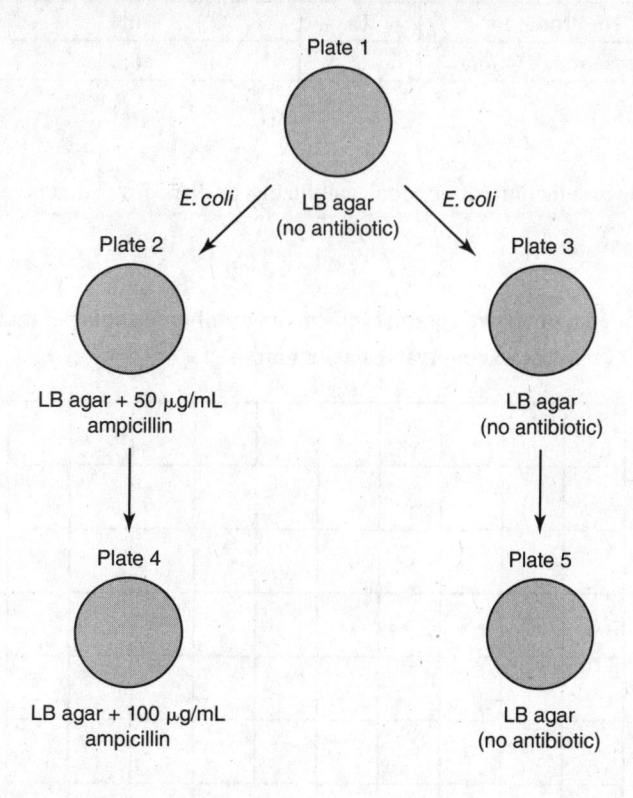

Part A

(i) **Describe** whether *E. coli* that are all susceptible to ampicillin, all resistant to ampicillin, or a mix of both will grow on the starter plate.

(ii) **Explain** your answer using the principles of natural selection.

Part B

(i) **Identify** the plates that serve as controls in this experiment.

(ii) **Identify** the independent variable in the experiment.

Part C

(i) **Predict** the relative amount of *E. coli* growth on plates 2, 3, 4, and 5.

Part D

(i) **Justify** your prediction from Part C.

Long Free-Response

13. Wild guppies (*Poecilia reticulata*) live in ponds on the island of Trinidad. Male guppies have great variation in the number and colors of spots, leading to a wide variety of color patterns among male guppies. Female guppies do not express these spots and are drably colored. Female guppies will more often choose to mate with males who possess bright color patterns. However, males with brighter color patterns are more visible to predators. An experiment was performed to measure the effect of the presence of a guppy predator

(*Cichlidae alta*) on the number of spots in male guppies. Guppies were placed into two different environments: one with no predators and the other in which *C. alta* was present. Guppies were allowed to reproduce in both environments for 20 generations. After 20 generations, the number of spots on each male guppy was counted. The mean number of spots on male guppies is shown in the table.

Total Number of Spots per Male Guppy		
	Mean	Standard Error of the Mean
No Predators	13	0.5
Presence of *C. alta*	9	0.5

Part A

(i) **Describe** the type of selection (directional, stabilizing, or disruptive) that is caused by the presence of *C. alta*.

Part B

(i) On the axes provided, **construct** a graph of the mean number of spots per male guppy for each group. Include 95% confidence intervals on your graph.

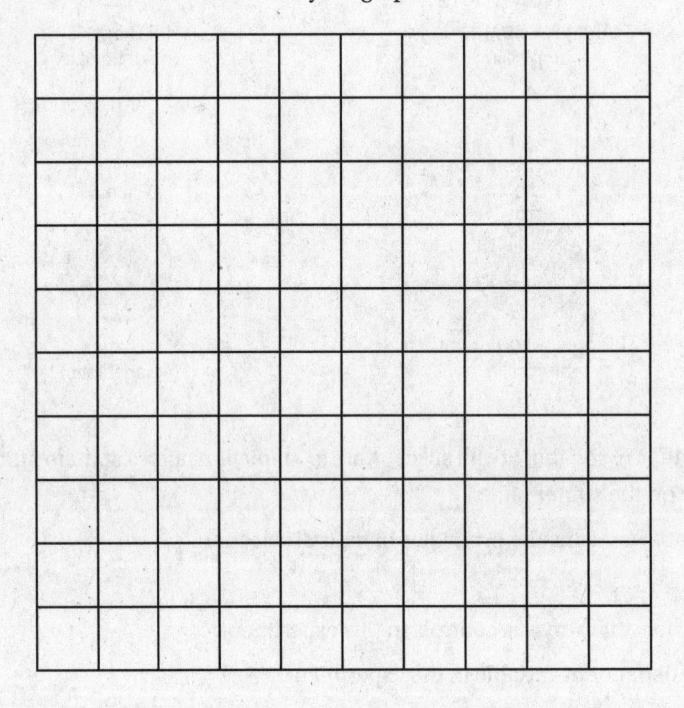

Part C

(i) Use the graph you constructed in Part B to **make a claim** about the mean number of spots per male guppy in the no-predator environment as compared to the mean number of spots per male guppy in the environment with *C. alta*.

(ii) **Support your claim** with evidence from the graph.

Part D

As a follow-up experiment, some of the guppies in the environment with *C. alta* were removed and placed in an environment with no predators. They were allowed to reproduce for 20 generations.

(i) **Predict** what you would expect to happen to the mean number of spots per male guppy in this new predator-free environment after the guppies were allowed to reproduce for 20 generations.

(ii) **Justify** your prediction.

Answer Explanations

Multiple-Choice

1. **(A)** Directional selection occurs when one end, or extreme, of the range of phenotypes has a survival advantage. In this example, birds with larger beaks were more likely to survive. Disruptive selection occurs when both extremes of the range of phenotypes increase in frequency; since only large beaks increased in frequency and small beaks decreased in frequency, choice (B) is incorrect. Choice (C) is incorrect because stabilizing selection favors the middle range of the phenotype, which did not occur in this case. There is no evidence in the question that suggests that mate preference is dependent on beak size, so choice (D) is incorrect.

2. **(B)** Before the drought, there were more thin-walled seeds and therefore more food available for finches with smaller beaks. Therefore, the end of the drought would increase food availability for finches with smaller beaks and allow more finches with smaller beaks to survive. There is no evidence in the question that suggests females have a mate preference that is dependent on beak size, so choice (A) is incorrect. Choice (C) is incorrect because there is no evidence for predators killing more finches of either beak size. The question does not give any information about disease susceptibility or resistance, so choice (D) is incorrect.

3. **(A)** DNA sequences are generally considered stronger pieces of evidence of evolution because environmental factors are far less likely to change the DNA sequence of a fossil. Choice (B) is incorrect because geological events can change the relative positions of fossils in rock layers and make the original location of the fossil position less certain. Birds can fly and move to new habitats, so choice (C) is incorrect. Choice (D) is incorrect because mating behaviors of birds may be influenced by their environments or by learning these behaviors from other birds.

4. **(C)** Homologous structures indicate common ancestry but may have different functions. Whales, horses, and humans are all mammals and share common ancestry, but their limbs have different functions. Analogous structures have similar functions. However, whale limbs are used for swimming, horse limbs are used for trotting or running, and human arms are used for reaching and grasping; they do not have similar functions, so choice (A) is incorrect. Choice (B) is incorrect because convergent evolution results in analogous structures and does not indicate common ancestry. Vestigial structures are structures in an organism that have no current function but may have had a function in an ancestral species, so choice (D) is incorrect.

5. **(B)** Increased use of products that contain triclosan will create an environment in which triclosan-resistant bacteria have a survival advantage, so over time the relative numbers of triclosan-resistant bacteria will increase. Choice (A) is incorrect because there is no evidence that triclosan kills all bacteria. Individuals do not evolve, populations evolve, so choice (C) is incorrect. Choice (D) is incorrect because there is no evidence in the question that the use of triclosan increases the frequency of mutations in bacteria.

6. **(B)** Snakes that are resistant to the cane toads' toxin would be more likely to survive and reproduce, so their relative numbers would be expected to increase over time. Choice (A) is incorrect because not all snakes are susceptible to the cane toads' toxin, so not all the snakes would die. Individual snakes cannot acquire resistance to the cane toads' toxin, so choice (C) is incorrect. Choice (D) is incorrect because there is no evidence in the question that snakes would learn to avoid eating cane toads.

7. **(A)** Since *TAS2R38* is present in the population at a higher frequency, it probably did provide a survival advantage and allowed individuals who possessed this allele to survive and reproduce at a greater rate than individuals who did not possess the allele. Choice (B) is incorrect because *TAS2R38* does not cause individuals to consume bitter foods; it just influences their ability to taste bitter foods. Bitter foods do not necessarily have a higher nutrient content than other foods, so

choice (C) is incorrect. There is no evidence in the question that the *TAS2R38* allele lowered an individual's chance of survival, so choice (D) is incorrect.

8. **(A)** Artificial selection occurs when humans selectively breed organisms for desired traits. Frequency-dependent selection occurs when the survival of an organism depends on its frequency in an environment, so choice (B) is incorrect. Choice (C) is incorrect because in natural selection, the environment selects for which individuals survive and reproduce. There is no evidence for mate choice in the question; plants usually depend on the wind or animal pollinators to exchange gametes, so choice (D) is incorrect.

9. **(C)** Variations that individuals acquire during their lifetime are called acquired characteristics and are not passed on to the individual's offspring. There are variations among individuals of a species, so choice (A) is not the answer. Choice (B) is not the answer because some variations do provide a survival advantage. Over time, natural selection will increase the frequency of individuals who possess variations that give them a survival advantage, so choice (D) is also not the answer.

10. **(B)** The light-colored mice will blend in with the sandy background and be less visible to predators, so their numbers would increase. Choice (A) is incorrect because dark-colored mice would be more visible to predators against the sandy background, and their numbers would decrease. Choices (C) and (D) are incorrect because the sandy background would increase the survival of light-colored mice and decrease the survival of dark-colored mice.

Short Free-Response

11. (A-i) Species C has the most in common with *Crocodylus porosus* because they are both found in Asia, the number of teeth in species C is closest to the number of teeth in *C. porosus*, and the DNA sequence of species C has the highest percentage of homology with *C. porosus*.

    (B-i) Species A and species B would be less efficient predators than *C. porosus* because they both have fewer teeth than *C. porosus*.

    (C-i) The data do not support the claim that species B shares a more recent common ancestor with *C. porosus* than do species A or species C. In fact, species C has the most in common with *C. porosus*.

    (D-i) The DNA from species B only has 90% homology with the DNA from *C. porosus*, while species C's DNA has 98% homology with the DNA from *C. porosus*. Thus, species C likely has a more recent common ancestor with *C. porosus* than species B does.

12. (A-i) It is likely that the *E. coli* on the starter plate (plate 1) contain mostly bacteria that are susceptible to ampicillin and some bacteria that are resistant to ampicillin.

    (A-ii) Natural selection acts upon variations in populations, and it is likely that there are varying degrees of resistance to ampicillin in the bacteria population.

    (B-i) The controls are plates 3 and 5 (the plates without ampicillin).

    (B-ii) The independent variable is the concentration of ampicillin on each plate.

    (C-i) Plates 3 and 5 will have many bacteria, perhaps a solid "lawn" of bacteria on each of those plates. Plates 2 and 4 will have very few bacteria growing on them.

    (D-i) Since the starter plate did not contain ampicillin, there was no selection for ampicillin resistance on that plate, and so the majority of the bacteria on the starter plate are expected to not have resistance to ampicillin. Therefore, very few bacteria are expected to grow on plates 2 and 4, which contain ampicillin. Many more bacteria will grow on plates 3 and 5 since those plates do not contain ampicillin.

Long Free-Response

13. (A-i) This is an example of directional selection because in the presence of the predator *C. alta*, the mean number of spots per male guppy decreases.

(B-i)

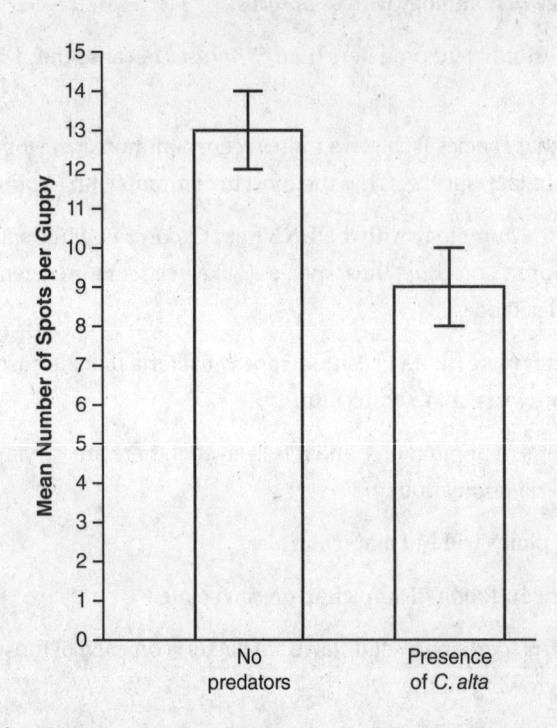

(C-i and ii) There is a statistically significant difference between the mean number of spots per guppy in the environment without predators and in the environment with *C. alta*. This claim is supported by the data because the 95% confidence intervals of the two groups do not overlap.

(D-i and ii) If the guppies were moved out of the environment that had the predator *C. alta* and were placed into an environment without predators, over time, it would be expected that the mean number of spots per male guppy would increase. This is because there would no longer be a disadvantage to having spots, and spots would attract mates, increasing the likelihood that the guppies with spots would reproduce.

20

Population Genetics

Learning Objectives

In this chapter, you will learn:
- → Population Genetics and Genetic Drift
- → Hardy-Weinberg Equilibrium

Overview

Evolution is driven by both random and nonrandom events. Mutations are random, but natural selection, which acts upon the phenotypes that result from those mutations, is not a random process. This chapter will review some of the processes that cause populations to evolve and some of the equations that can be used to describe changes in populations.

Population Genetics and Genetic Drift

Population genetics is the study of genetic variation within populations and the processes that can cause changes in allele frequencies within a population. The three major processes that drive changes in allele frequencies in a population are natural selection (which was discussed in Chapter 19), gene flow, and genetic drift.

Gene flow is the transfer of alleles from one population to another. Gene flow can be caused by the migration of individuals into a population. If these individuals carry different alleles than the receiving population, the allele frequency in the receiving population will change. In plants, gene flow can occur through the transfer of pollen (by wind or animals) into new plant populations.

Genetic drift is the random loss of alleles in a population. Genetic drift is more likely to occur in smaller populations. For example, assume an allele is found in 10% of a population. If there are 1,000 individuals in the population, 100 of those individuals would have that allele. It would be likely that at least one of those 100 individuals would survive and reproduce to pass on that allele to the next generation. Now consider a population of 10 individuals. In that small population, only one individual would have that allele. If that one individual failed to reproduce, that allele would not be passed on to the next generation and would be lost, thereby decreasing the genetic diversity of the population. Considerations around genetic drift are especially important to conservation biologists, who are trying to preserve the genetic diversity of endangered species.

Bottleneck Effect

One possible cause of genetic drift is the **bottleneck effect**. A population bottleneck occurs when the size of a population is greatly reduced for one or more generations. Natural disasters, like fires, floods, or volcanic eruptions, can cause population bottlenecks. Human-made events, such as overhunting or rapid habitat destruction, can also cause population bottlenecks. Because the population size is smaller after these bottlenecks, the surviving population is much less likely to possess all of the alleles the larger population had before the bottleneck and thus will likely have less genetic diversity. A model of the bottleneck effect is shown in Figure 20.1.

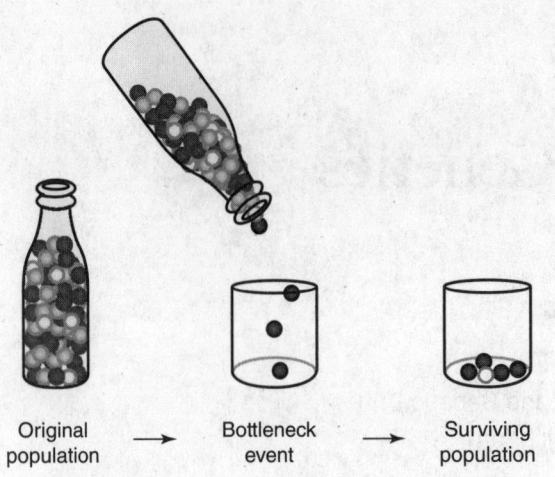

Figure 20.1 Model of the Bottleneck Effect

In Figure 20.1, each ball represents an allele in the population. In the original population, there are three different types of alleles, represented as black, gray, and white balls. When the population experienced a bottleneck event, none of the individuals that possessed the allele represented by the gray ball survived. Only two alleles, black and white, survived the bottleneck, reducing the genetic diversity of the surviving population.

The northern elephant seal population experienced a population bottleneck during the 1800s. By the 1890s, hunting had reduced the number of northern elephant seals to less than 30 individuals. Protection of this species in the 20th century resulted in an increase in their numbers, but they are not as diverse as the southern elephant seals, which were not subjected to intense hunting in the 1800s.

Founder Effect

Another cause of genetic drift is the **founder effect**. The founder effect occurs when a few members of a larger population start a new population. These few members of the larger population often have less genetic diversity compared to the larger population or may be a nonrandom sample of the larger population. A model of the founder effect is shown in Figure 20.2.

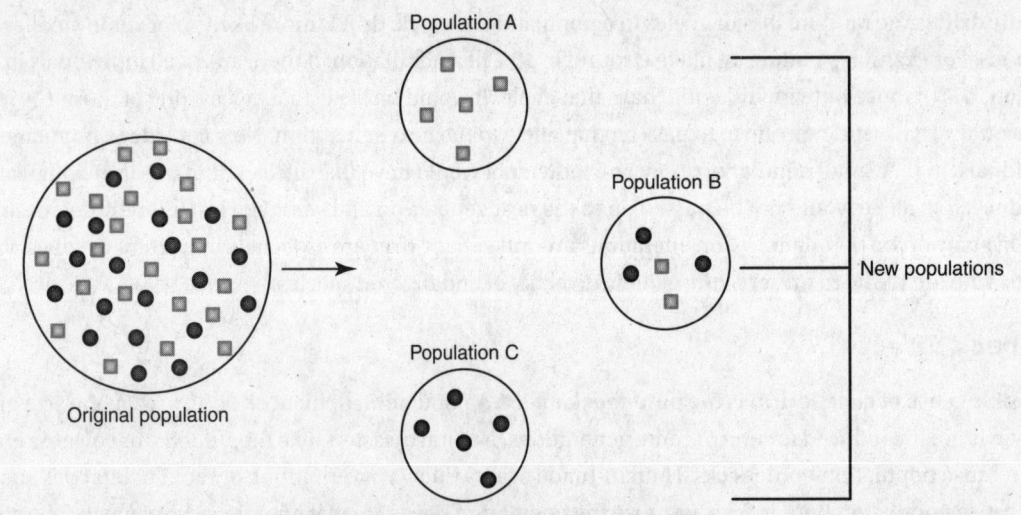

Figure 20.2 Model of the Founder Effect

In Figure 20.2, there are two different alleles present (represented by black circles and gray squares) in roughly equal frequencies in the original population. Only individuals who possess the allele represented by gray squares founded population A, leading to less genetic diversity in population A than existed in the original population.

Similarly, the founding members of population C only possessed the allele represented by black circles, again leading to less genetic diversity in population C than existed in the original population; this is also a very different allele frequency than that seen in population A.

An example of the founder effect can be seen in the Amish population in Pennsylvania. The founding members of this population came from Europe in the mid-1700s. The approximately 200 founding members of this population had a higher frequency of the allele that causes a rare form of skeletal dysplasia, Ellis-van Creveld syndrome. This has led to the frequency of this allele being relatively high in this population.

Hardy-Weinberg Equilibrium

Some populations have stable allele frequencies and are not evolving. Populations in an unchanging environment may not experience the selective pressures that lead to evolution. Scientists Godfrey Hardy and Wilhelm Weinberg each developed equations that describe these stable populations. For a population to have stable allele frequencies, five conditions must be met:

1. **Large population size**—A large population size reduces the chances of genetic drift occurring. In small populations, a change in allele frequencies (caused by genetic drift) is much more likely.
2. **Random mating**—Random mating eliminates the possibility of changing allele frequencies caused by sexual selection.
3. **No gene flow**—For allele frequencies to be stable, individuals must not be entering or leaving the population. The introduction of new individuals, or the movement of individuals out of a population, could change the allele frequencies.
4. **No selection**—All phenotypes in the population need to have equal reproductive success in order to keep allele frequencies stable. If one phenotype has a survival advantage, the alleles in that phenotype will become more prevalent in the population.
5. **No mutations**—Mutations are very rare and random occurrences. A mutation would change the allele frequencies in the population.

Populations that meet all five of these conditions are said to be in **Hardy-Weinberg equilibrium**. If a population is in Hardy-Weinberg equilibrium, the following equations apply:

$$p + q = 1$$

where p represents the frequency of the dominant allele (A) and q represents the frequency of the recessive allele (a)—note that this equation is used to describe *allele* frequencies—and

$$p^2 + 2pq + q^2 = 1$$

where p^2 represents the frequency of the homozygous dominant genotype (AA), $2pq$ represents the frequency of the heterozygous genotype (Aa), and q^2 represents the frequency of the homozygous recessive genotype (aa)—note that this equation is used to describe *genotype* frequencies.

> **TIP**
> Remember, if a question asks for an allele frequency, you need to find p or q. If a question asks for a genotype frequency or the number of individuals, you need to find p^2, $2pq$, or q^2.

> It is useful to review some laws of probability here. If you flip one coin, the probability of getting heads is $\frac{1}{2}$. If you flip two coins at the same time, the probability of both coins coming up heads is the product of their individual probabilities: $\frac{1}{2} \times \frac{1}{2} = \frac{1}{4}$. Similarly, if p is the probability of inheriting one copy of the dominant allele (A), then the probability of inheriting two copies of the dominant allele is $p \times p = p^2$. The probability of inheriting two copies of the recessive allele (a) is $q \times q = q^2$.
>
> The probability of inheriting one dominant allele and one recessive allele is $p \times q = pq$. However, there are two ways to create a heterozygous individual. The dominant allele could come from the father and the recessive from the mother, or the dominant allele could come from the mother and the recessive from the father. If there is more than one way that an event can occur, the probability of the event is the sum of the individual probabilities. So the frequency of heterozygotes would be the sum of these two events: $pq + pq = 2pq$.

Practice with Hardy-Weinberg Equilibrium

Now try some practice problems.

1. In cats, long hair is recessive to short hair. A large, randomly mating population of cats is in Hardy-Weinberg equilibrium. In this population, 32% of the cats have long hair. What percentage of the population are heterozygous for short hair?

   Start with the recessive phenotype, long hair, whose frequency is equal to q^2: $q^2 = 0.32$. Taking the square root of both sides of the equation gives you $q = 0.57$. Since the sum of p and q equals 1, $p = 1 - q = 1 - 0.57 = 0.43$. The frequency of heterozygotes equals $2pq$, so the heterozygotes $= 2(0.43)(0.57) = 0.49 = 49\%$.

2. In pigs, erect ears are dominant to drooping ears. In a pig population that is in Hardy-Weinberg equilibrium, 82% of the individuals have erect ears. What is the frequency of heterozygotes in this population?

> Since there are two genotypes that can produce the dominant phenotype, the frequency of the dominant phenotype is equal to $p^2 + 2pq$, which would give you an equation with two unknowns and is not solvable. In this situation, subtract the frequency of the dominant phenotype from 1 to obtain the frequency of the recessive phenotype. Since only one genotype can produce the recessive phenotype, the frequency of the recessive phenotype equals q^2, which gives you an equation with just one unknown and is solvable.

   Since 82% of the population has erect ears, $100 - 82 = 18\%$ of the population must have drooping ears (the recessive phenotype). Therefore, $q^2 = 0.18$, and $q = 0.42$. Subtracting q from 1 gives the value of p: $p = 1 - 0.42 = 0.58$. The frequency of heterozygotes then becomes $2pq = 2(0.58)(0.42) = 0.49$, or 49%.

3. In dogs, black fur is dominant to brown fur. In a population of dogs that is in Hardy-Weinberg equilibrium, 10% of the individuals have brown fur. What is the frequency of the allele for black fur, and what percentage of the population possesses at least one copy of the allele for black fur?

   In this problem, you are asked to find the frequency of an allele (p). First, start with the recessive phenotype, brown fur: $q^2 = 0.10$, so $q = 0.32$. Since $p + q = 1$, $p = 1 - q = 1 - 0.32 = 0.68$. So the frequency of the allele for black fur (p) is 0.68. The second part of the question asks for the percentage of the population that possesses at least one copy of the allele for black fur. Since black fur is a dominant trait, alleles for black fur will be found in both homozygous dominant individuals (p^2) and heterozygous individuals ($2pq$). The frequency of homozygous dominant individuals is

TIP

When solving Hardy-Weinberg problems, it is especially important to read the question carefully to understand what you are being asked to solve for. Pay attention to words like *individuals* versus *allele*.

$p^2 = (0.68)^2 = 0.46$. The frequency of heterozygous individuals is $2pq = 2(0.68)(0.32) = 0.44$. Adding the frequency of homozygous dominant individuals and heterozygous individuals gives you $0.46 + 0.44 = 0.90$, or 90% of the population has at least one copy of the dominant allele.

4. What do you do if a population is not in Hardy-Weinberg equilibrium? In these cases, you need to know the number of individuals with each genotype to find the allele frequencies. You cannot use the Hardy-Weinberg equations in these situations, but you *can* add up the alleles contributed by each genotype to the gene pool of the population. Here is an example.

    In a nonrandomly mating population of 80 individuals, 10 have the homozygous dominant genotype, 46 are heterozygous, and 24 have the homozygous recessive genotype for a particular trait. What are the allele frequencies for this trait in this population?

    Since there are 80 individuals in the population and each individual contributes two alleles for the trait to the population, there are a total of $80 \times 2 = 160$ alleles for the trait in this population. Each homozygous dominant individual contributes two copies of the dominant allele to the population for a total of $10 \times 2 = 20$ dominant alleles. Each heterozygous individual contributes one copy of the dominant allele for a total of $46 \times 1 = 46$ dominant alleles from the heterozygotes. The homozygous recessive individuals do not contribute any dominant alleles to the population. So the frequency of the dominant allele is $\frac{20 + 46}{160} = \frac{66}{160} = 0.413$. Each of the homozygous recessive individuals contributes two copies of the recessive allele to the population for a total of $24 \times 2 = 48$ recessive alleles. The heterozygous individuals each contribute one copy of the recessive allele to the population for a total of $46 \times 1 = 46$ recessive alleles. The frequency of the recessive allele is then $\frac{48 + 46}{160} = \frac{94}{160} = 0.588$.

Practice Questions

Multiple-Choice

1. In which populations is genetic drift more likely to occur?

    (A) large populations
    (B) small populations
    (C) populations with greater diversity
    (D) populations with a high degree of gene flow

2. A group of birds is flying south during their yearly migration when a hurricane with extremely strong winds occurs. Only 10% of the group survive the storm, reaching their winter nesting site and reproducing the next spring. This type of event is an example of

    (A) the bottleneck effect because only the fittest members of the population survived.
    (B) the bottleneck effect because the population size was rapidly reduced.
    (C) the founder effect because the group of birds from before the storm were not genetically diverse.
    (D) the founder effect because the birds colonized a new habitat and never migrated again back to their original habitat.

3. Which of the following best explains the conditions under which the founder effect occurs?

    (A) Random evolutionary change happens over multiple generations, causing a large population to separate into two smaller, genetically different populations.
    (B) A population's size is rapidly reduced, and the genetic diversity of the surviving population is not representative of the original, larger population.
    (C) A small group of individuals from a larger population colonizes a new habitat. The small group that colonizes the new habitat does not possess the genetic diversity of the original, larger population.
    (D) Environmental pressures select for the individuals with the adaptations that give them a greater likelihood of surviving and reproducing.

4. An isolated population in the United States are descendants of approximately 200 immigrants who arrived in the 1700s. Some individuals in this group of immigrants carried the allele for Ellis-van Creveld syndrome. Today, this allele occurs at a much higher frequency in the descendants of this group than in the general population in the United States. This difference in the frequency of the Ellis-van Creveld allele is an example of

    (A) the bottleneck effect.
    (B) the founder effect.
    (C) natural selection.
    (D) random mutations over multiple generations.

5. A scientist is studying the allele frequencies for a particular gene in a population of wild prairie dogs over a five-year period. The allele frequencies are shown in the table.

Year	Frequency of Allele $A1$	Frequency of Allele $A2$
1	0.00	1.00
2	0.00	1.00
3	0.10	0.90
4	0.13	0.87
5	0.15	0.85

Which of the following is the most likely cause of the observed change in allele frequencies?

    (A) artificial selection
    (B) natural selection
    (C) genetic drift
    (D) gene flow

6. The following table shows the frequency of three genotypes (*AA*, *Aa*, and *aa*) in a population.

Genotype	Frequency
AA	0.44
Aa	0.14
aa	0.42

The environment of the population changes so that individuals with the *Aa* genotype are more likely to survive and reproduce. Predict what would most likely happen to the frequencies of the *AA* and *aa* genotypes after 10 generations in this environment.

(A) The frequencies of both the *AA* and *aa* genotypes would increase.
(B) The frequencies of both the *AA* and *aa* genotypes would decrease.
(C) The frequency of the *AA* genotype would increase, and the frequency of the *aa* genotype would decrease.
(D) The frequency of the *AA* genotype would increase, and the frequency of the *aa* genotype would be 0.

7. Which of the following is a characteristic of a population that is in Hardy-Weinberg equilibrium?

(A) small population size
(B) gene flow
(C) sexual selection
(D) no mutations

8. Ebony body color is an autosomal recessive trait in fruit flies. In a large, randomly mating population of fruit flies that is in Hardy-Weinberg equilibrium, the frequency of ebony body color is 21%. What is the percentage of the population with the homozygous dominant genotype?

(A) 29%
(B) 45%
(C) 54%
(D) 79%

Questions 9 and 10
For a genetic disease that appears in homozygous recessive (*aa*) individuals, individuals who are heterozygous (*Aa*) for the disease are resistant to infections by certain parasites.

9. If 19% of the individuals in a population have the genetic disease, what is the frequency of individuals who are resistant to the parasites? Assume the population is in Hardy-Weinberg equilibrium.

(A) 0.436
(B) 0.492
(C) 0.564
(D) 0.810

10. Over half of the individuals living in a population where a particular parasite is prevalent are heterozygous for a recessive allele (*Aa*). Further studies reveal that individuals who are heterozygous for the recessive allele are resistant to the parasite, while individuals who are homozygous dominant or homozygous recessive are susceptible to the parasite. If the parasite was totally eliminated from this area, predict what would happen to the frequencies of the *AA*, *Aa*, and *aa* genotypes.

(A) The frequencies of the *AA* and *aa* genotypes would increase, and the frequency of the *Aa* genotype would decrease.
(B) The frequency of the *AA* genotype would decrease, and the frequencies of the *Aa* and *aa* genotypes would increase.
(C) The frequency of the *Aa* genotype would increase, and the frequencies of the *AA* and *aa* genotypes would decrease.
(D) The frequency of the *aa* genotype would increase, and the frequencies of the *AA* and *Aa* genotypes would decrease.

Short Free-Response

11. A scientist is studying the allele frequencies in a population over several generations to help determine if the population is undergoing evolution. The data are shown in the table.

Generation Number	Frequency of Allele A1	Frequency of Allele A2
1	0.81	0.19
2	0.74	0.26
3	0.62	0.38
4	0.49	0.51

Part A

(i) **Explain** why looking at allele frequencies over several generations could be used to determine whether or not a population is undergoing evolution.

Part B

(i) **Describe** the changes in alleles *A1* and *A2* from generation 1 to generation 4.

Part C

(i) **Evaluate** the scientist's hypothesis that this population is undergoing evolution based on the data available.

Part D

(i) **Explain** how the data would differ if individuals with the genotype *A1A2* were more likely to survive than individuals with the genotype *A1A1* or individuals with the genotype *A2A2*.

12. Wildfires in the western United States have drastically reduced the size of some animal populations and destroyed the habitats of other animal populations.

Part A

(i) **Describe** how wildfires can create the bottleneck effect in a population.

Part B

Ninety percent of the habitat of a spotted owl colony was destroyed by a wildfire. A small percentage of the surviving owl population migrated to a new habitat.

(i) **Describe** the type of genetic drift this illustrates.

Part C

The banana slug, *Ariolimax columbianus*, lives in the redwood forest in the mountains near Santa Cruz, California. A wildfire in these mountains in the summer of 2019 divided the slug's habitat with a large area of dry ash that the slugs cannot cross.

(i) **Predict** the effect this had on the genetic diversity of the banana slug population.

Part D

(i) **Justify** your prediction from Part C.

Long Free-Response

13. The California kangaroo rat, *Dipodomys ingens*, is considered an ecosystem engineer for its role in creating extensive burrowing systems that can change soil characteristics and provide habitats for other species. Fur color in *D. ingens* ranges from white to sandy to brown. Ecologists counted the number of *D. ingens* in an area before and five years after a large mudslide killed the majority of the kangaroo rats in the area. Data are shown in the table.

Fur Color	Number Before Mudslide	Number Five Years After Mudslide
White	75	0
Sandy	165	140
Brown	60	160

Part A

(i) **Describe** the probable cause of the change in frequency of fur color phenotypes in *D. ingens* after the mudslide event.

Part B

(i) **Identify** an appropriate control group for this study.

Part C

(i) **Explain** how the mudslide affected the frequencies of fur color phenotypes in *D. ingens*.

Part D

The burrows created by *D. ingens* provide habitats for many lizard species. In order to prevent future mudslides, retaining walls were constructed on the hillsides in the area, and the number of *D. ingens* in the area decreased by over 85%.

(i) **Predict** the effect of this on the lizard species diversity in the area.

(ii) **Justify** your prediction.

Answer Explanations

Multiple-Choice

1. **(B)** Genetic drift, the random loss of alleles, is more likely in small populations because in small populations, fewer individuals would have the opportunity to pass on a rare allele to the next generation than in a large population. Choice (A) is incorrect because it is the opposite of the correct answer. Greater diversity does not influence the chances of genetic drift, so choice (C) is incorrect. Choice (D) is incorrect because gene flow, or migration, does not necessarily cause genetic drift.

2. **(B)** The bottleneck effect occurs when the majority of a population dies and the surviving individuals do not possess all of the alleles from the original population. Choice (A) is incorrect because survival after bottleneck events may be random and is not dependent on an organism's fitness. The founder effect occurs when a small group of individuals leaves a larger population to colonize a new area. Since this scenario does not describe the founder effect, choices (C) and (D) are incorrect. Also, nothing in the question provides any information on the genetic diversity of the population before the hurricane, which further rules out choice (C). Choice (D) can be further ruled out because nothing in the question suggests that the birds never migrated back to their original habitat.

3. **(C)** The founder effect occurs when a small group of individuals from a large population leaves the larger group to colonize a new area and the small group does not possess the same allele frequencies as the larger group. Choice (A) is incorrect because the founder effect is not caused by random evolutionary change. The rapid reduction of a population's size, leading to a group of survivors that is not representative of the original, larger population, is the bottleneck effect. Thus, choice (B) is incorrect. Choice (D) is incorrect because it describes the process of natural selection, not the founder effect.

4. **(B)** This scenario is an example of the founder effect. Choice (A) is incorrect because the original population did not undergo a rapid decrease in population size. There is no evidence in the question that Ellis-van Creveld syndrome leads to decreased survival, so choice (C) is incorrect. Choice (D) is incorrect because it would be extremely unlikely that random mutations over several generations would lead to the appearance of the allele that causes Ellis-van Creveld syndrome.

5. **(D)** Initially, the population of wild prairie dogs did not possess the *A1* allele, so it is likely that it arrived in an individual prairie dog that migrated into the population sometime between years 2 and 3. Therefore, gene flow is the correct answer. Choice (A) is incorrect because these are wild prairie dogs, and there is no evidence of selective breeding of prairie dogs by humans. No information is given in the question about whether or not alleles *A1* or *A2* confer a survival advantage, so choice (B) is incorrect. No allele was lost in the population, so choice (C) is incorrect.

6. **(B)** If individuals with the *Aa* genotype are more likely to survive and reproduce (heterozygote advantage), the frequency of the *Aa* genotype would increase over time and the frequencies of the *AA* and *aa* genotypes would both decrease.

7. **(D)** The five conditions that are necessary for Hardy-Weinberg equilibrium are large population size, random mating, no selection of any kind, no gene flow, and no mutations. Choice (A) is incorrect because large populations, not small populations, are necessary for Hardy-Weinberg equilibrium. Migration (or gene flow) also violates one of the conditions necessary for Hardy-Weinberg equilibrium, so choice (B) is incorrect. Choice (C) is incorrect because in order for a population to be in Hardy-Weinberg equilibrium, no selection can occur.

8. **(A)** If 21% of the population has the recessive phenotype, $q^2 = 0.21$ and $q = 0.458$. Since $p + q = 1$, $p = 1 - 0.458 = 0.542$. The frequency of homozygous dominant individuals is $p^2 = (0.542)^2 = 0.29$, or 29%. Choice (B) is the frequency of the recessive allele and is incorrect. Choice (C)

is incorrect because 54% is the frequency of the dominant *allele*, not the homozygous dominant *genotype*. The frequency of the dominant *phenotype* is 79%, so choice (D) is also incorrect.

9. **(B)** If 19% of the individuals have the recessive genetic disease, $q^2 = 0.19$ and $q = 0.436$. Then since $p + q = 1$, $p = 0.564$. The heterozygous individuals are resistant to the parasites and have a frequency of $2pq = 2 \times 0.564 \times 0.436 = 0.492$. Choice (A) is the frequency of the recessive allele and is incorrect. Choice (C) is the frequency of the dominant allele and is incorrect. The frequency of individuals who do not have the recessive disease (some of whom may be heterozygous and others who may be homozygous dominant) is 81%, so choice (D) is incorrect.

10. **(A)** If the parasite was totally eliminated from this environment, there would no longer be any advantage to being heterozygous for the recessive allele. So over time, natural selection would increase the frequencies of the *AA* and *aa* genotypes and the frequency of the *Aa* genotype would decrease. Choice (B) is incorrect because the frequency of the *AA* genotype would increase (not decrease) and the frequency of the *Aa* genotype would decrease (not increase). If the parasite was eliminated, there would be no advantage to having the *Aa* genotype. So the frequency of the *Aa* genotype would no longer be favored and would be expected to decrease, not increase, and the frequencies of the *AA* and *aa* genotypes would increase. Thus, choice (C) is incorrect. While the frequency of the *aa* genotype would be expected to increase, the frequency of the *AA* genotype would also be expected to increase. So choice (D) is incorrect.

Short Free-Response

11. (A-i) If the allele frequencies in a population are changing, the population is evolving. So looking at allele frequencies in a population over time can help determine if the population is evolving.

(B-i) The frequency of the *A1* allele decreased from generation 1 to generation 4, and the frequency of the *A2* allele increased from generation 1 to generation 4.

(C-i) The scientist's hypothesis that this population is undergoing evolution is supported by the data because the allele frequencies are changing.

(D-i) If individuals with the *A1A2* genotype were more likely to survive than individuals with the *A1A1* or *A2A2* genotypes, over time, it would be expected that eventually the frequencies of the *A1* and *A2* alleles would approach 0.50.

12. (A-i) Wildfires could create a bottleneck effect by drastically reducing the population size so that all the individuals who carry rare alleles (that were present in the original population) might not survive. Those rare alleles would then be eliminated from the population.

(B-i) This scenario illustrates the founder effect since a very small percentage of the spotted owl population is colonizing a new habitat and that small group might not have the same allele frequencies as that of the original population.

(C-i) The genetic diversity of the banana slug population in this area likely decreased.

(D-i) Since the population of banana slugs was randomly separated into two smaller groups that could not interbreed, each of the subgroups would have less genetic diversity.

Long Free-Response

13. (A-i) The mudslide changed the environment of the area so that darker-colored kangaroo rats had a survival advantage after the mudslide and the lighter-colored kangaroo rats had a survival disadvantage after the mudslide.

    (B-i) An appropriate control group would be a similar habitat with a similar species in which a mudslide did not occur.

    (C-i) Five years after the mudslide, there were more brown-colored kangaroo rats in this area and no kangaroo rats with white fur. After the mudslide, brown-colored kangaroo rats had a survival advantage over the white-colored kangaroo rats. (The brown-colored kangaroo rats had fur color that was similar to the color of the surrounding habitat, so it was more difficult for predators to see the brown-colored kangaroo rats.) Therefore, the relative numbers of brown-colored kangaroo rats increased over time.

    (D-i and ii) If there are far fewer *D. ingens* in this area, there will be fewer burrows for the lizard species to occupy and the diversity of the lizard species in the area will decrease.

21

Phylogeny and Speciation

Learning Objectives

In this chapter, you will learn:
- → Phylogeny and Common Ancestry
- → Speciation
- → Modern-Day Examples of Continuing Evolution

Overview

The evolutionary history of life is driven by speciation. This chapter will review how scientists trace the evolutionary history of a species, the processes that lead to speciation, and modern-day examples of continuing evolution of species.

Phylogeny and Common Ancestry

Phylogeny is the history of the evolution of a species or group. Phylogeny shows lines of ancestry, common descent, and relationships among groups of organisms. **Phylogenetic trees** and **cladograms** are hypotheses about the history of evolution over time, with phylogenetic trees indicating the approximate time of evolutionary events. These phylogenetic trees and cladograms are created using morphological evidence from fossils and time estimates from **molecular clocks** (changes in DNA and protein sequences over time). Generally, evidence from molecular clocks is considered more accurate than morphological characteristics because molecular data are less influenced by convergent evolution or external geological events.

Morphological evidence from fossils shows traits that are gained, or lost, over time that can be used to construct phylogenetic trees. Shared characteristics are traits that are present in more than one lineage. **Shared derived characteristics** are found in a group of related organisms called a **clade** and set the clade apart from other organisms. Shared derived characteristics indicate homology among organisms in a clade and are evidence of their common ancestry.

Nodes on phylogenetic trees represent common ancestors. The more recent the common ancestor of two species, the closer their degree of relatedness. In a phylogenetic tree, the **outgroup** is the least closely related member of the tree. The **root** of a phylogenetic tree represents the common ancestor of all members of the tree.

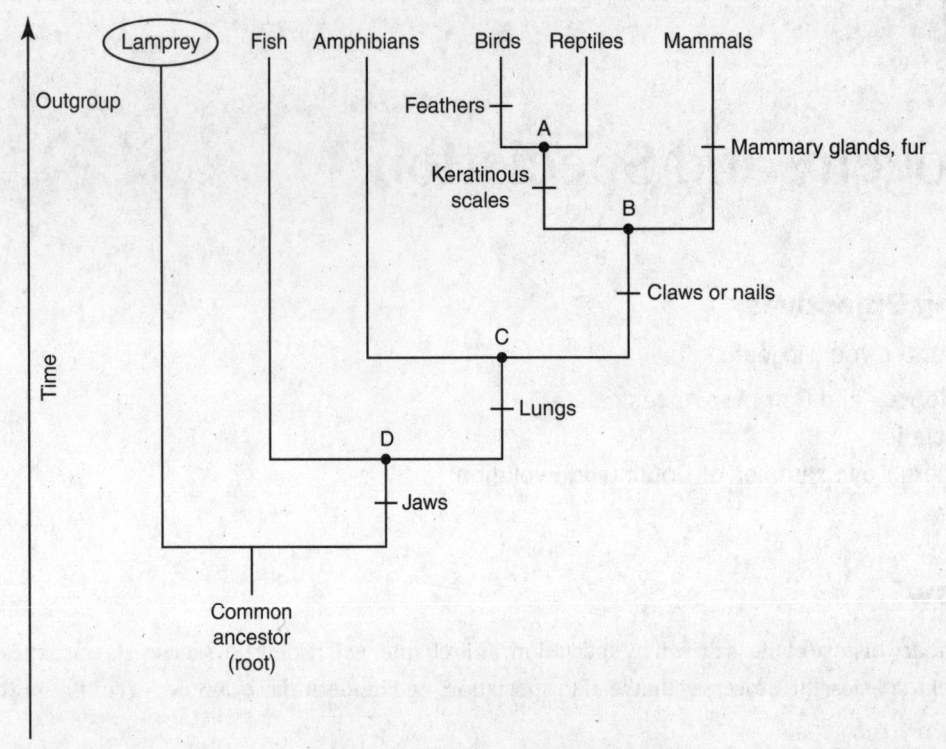

Figure 21.1 Phylogenetic Tree

In Figure 21.1, points A, B, C, and D represent nodes that indicate common ancestors. Point A represents the common ancestor of birds and reptiles, and point C represents the common ancestor of amphibians, birds, reptiles, and mammals. The phylogenetic tree indicates that birds and reptiles are more closely related than birds and amphibians because birds and reptiles share a more recent common ancestor (at point A) than do birds and amphibians (at point C). Keratinous scales are a shared derived characteristic that separate birds and reptiles into a clade. Keratinous scales are a characteristic shared by birds and reptiles that other organisms in this phylogenetic tree do not possess. Lampreys are the least closely related member of this phylogenetic tree and represent the outgroup in this example.

Phylogenetic trees and cladograms can also show speciation events (the evolution of new species) and extinction events (the death of all members of a species), as shown in Figure 21.2.

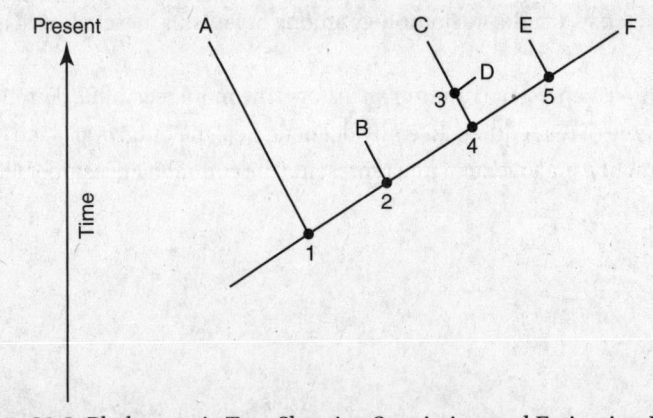

Figure 21.2 Phylogenetic Tree Showing Speciation and Extinction Events

In Figure 21.2, points 1, 2, 3, 4, and 5 represent speciation events that led to one ancestral lineage giving rise to two or more daughter lineages. Species B and D represent **extinct** species that did not survive to the present time, while A, C, E, and F represent **extant** species that have members that survived to the present time.

Organisms are linked by lines of descent, with a proposed common ancestor for all forms of life on Earth (**LUCA**—last universal common ancestor) estimated to have existed about 3.5 billion years ago. There are two major theories about how life on Earth originated:

1. Inorganic materials that were present in Earth's early atmosphere combined to make the building blocks of biological molecules. This theory is supported by evidence from the Miller-Urey experiment, in which a model of Earth's early atmosphere was constructed in a lab, and after a few weeks, amino acids and other components of biological molecules were found.
2. Another theory is that meteorites may have transported organic molecules (that are needed for life) to Earth. It is thought that early Earth was bombarded with meteorites. Evidence for this theory includes the Murchison meteorite (found in Australia in 1969), which contained sugars and over 70 different amino acids.

> **TIP**
> A common mistake made on the AP Biology exam is confusing the terms "extinct" and "extant." Extinct species have *no* surviving individuals in the present time, whereas extant species *DO* have individuals in the present time.

The common ancestor for all eukaryotic life is thought to have evolved about 2.7 billion years ago. Evidence for a common ancestor of all eukaryotes includes:

- Membrane-bound organelles in all eukaryotes
- Linear chromosomes in all eukaryotes
- All eukaryotes have genes that contain introns

None of these shared characteristics of all eukaryotes are present in prokaryotes, indicating that there was a common ancestor of all eukaryotes.

Speciation

The current definition of a species is the biological species concept—that a **species** is a group of organisms that are capable of interbreeding and producing viable and fertile offspring. It is important to note that the definition of a species is a human construct that is meant to help classify organisms. As more data are gathered about an organism, the definition as to which species it belongs to may change.

Speciation is the evolution of new species. Speciation occurs when two populations are reproductively isolated from each other. This reproductive isolation prevents interbreeding, and as the environments (where these isolated groups live) change, the evolution of new species may occur. Speciation can lead to **adaptive radiation**, the divergent evolution of organisms into separate species that occupy different ecological niches. Adaptive radiation is evident in Darwin's finches in the Galápagos Islands. These finch species are descendants of one species but have evolved over time to occupy the different available niches on the Galápagos Islands.

Rates of speciation can vary. If an environment is relatively stable, there will be less selective pressure on populations, and the rate of speciation will be slower. This slow and constant pace of speciation is called **gradualism**. If an environment rapidly changes, as it would after an asteroid strike, a volcanic eruption, or a rapid change in climate, rapid evolution may occur. A long period of stability in a species interrupted by periods of rapid evolution is called **punctuated equilibrium**.

Speciation can be allopatric or sympatric. In **allopatric speciation**, a larger population becomes geographically separated and the smaller subgroups diverge and become separate species over time. **Sympatric speciation** occurs in the same geographic area, but other factors lead to reproductive barriers between members of the groups. One mechanism of sympatric speciation is **polyploidy**, the replication of extra sets of chromosomes, which is a frequent method of sympatric speciation in plants. Plants that develop extra sets of chromosomes usually cannot interbreed with plants that maintained the original number of chromosomes and will thus become a separate species over time. **Sexual selection** in animals can also lead to sympatric speciation.

Reproductive barriers that can cause speciation can be prezygotic or postzygotic. **Prezygotic barriers** prevent the formation of a zygote, or a fertilized egg. **Postzygotic barriers** occur after the zygote is formed, and they prevent the zygote from developing into a viable and fertile adult organism.

Prezygotic barriers include:

- **Habitat isolation**—If organisms live in different habitats and do not come in contact with one another, they cannot mate and form zygotes and are thus reproductively isolated. An example of this would be the black-tailed deer found in the western United States and the white-tailed deer found in the eastern United States, which do not interbreed and are considered different species.
- **Temporal isolation**—Organisms can live in the same habitat, but if they are active at different times of the day or have breeding seasons during different times of the year, they will be temporally isolated and will not interbreed. Members of the cicada species *Magicicada tredecim* emerge to mate only once every 13 years, while members of the cicada species *Magicicada septendecim* emerge to mate once every 17 years. The mating cycles of these two species coincide only once every 221 years, which keeps them reproductively isolated.
- **Behavioral isolation**—Some species will interbreed only with others who perform compatible mating behaviors, such as mating calls or dances. Many bird species are reproductively isolated by their bird songs, which attract members of the same species but not others. The blue-footed booby birds of the Galápagos Islands will only mate with others that perform the correct mating dances.
- **Mechanical isolation**—If the sexual organs of the organisms are incompatible and prevent the transfer of gametes, the species will remain reproductively isolated. Many species of snails are mechanically isolated due to their reproductive organs being located in different parts of their shells. If their shell shapes are too different, the sex organs of the snails cannot successfully copulate.
- **Gametic isolation**—Even if two organisms are able to successfully copulate, if their gametes are incompatible, no zygote will be produced and the organisms will be reproductively isolated. Many examples of this are found in plants, in which the pollen of one plant species cannot successfully fertilize the ova of another plant species.

Postzygotic barriers include:

- **Reduced hybrid viability**—Two organisms that can form a zygote may still be reproductively isolated if that zygote does not survive to adulthood. A number of species of salamanders can interbreed and produce zygotes, but those zygotes do not survive to reproductive age, keeping the salamander populations reproductively isolated.
- **Reduced hybrid fertility**—Even if the zygote survives to adulthood, if the adult hybrid is infertile, the two species that created the hybrid will remain reproductively isolated. A male donkey can mate with a female horse to produce a mule. The hybrid mules are infertile, keeping horses and donkeys as separate species.
- **Hybrid breakdown**—In some plants, hybrids are viable and fertile, but with each subsequent generation, the hybrid becomes weaker and less robust and will cease to exist after a few generations. This hybrid breakdown keeps some plant species reproductively isolated.

Modern-Day Examples of Continuing Evolution

Life on Earth continues to evolve in Earth's changing environment. Populations evolve when the conditions in which they live change. Some examples of continuing evolution include:

- Antibiotic resistance in bacteria
- Tumor cells developing resistance to chemotherapy drugs
- Pesticide resistance in insects
- Evolution of previously unseen viruses and other pathogens
- Genomic changes in organisms over time

Practice Questions
Multiple-Choice

1. In the 1950s, Stanley Miller performed an experiment to investigate the possible origin of the molecules required for life on Earth. Water vapor, methane, hydrogen gas, and ammonia were placed in a flask, and electric charges were applied to the system to simulate atmospheric conditions that were thought to be prevalent at the time. After many weeks, amino acids were produced in the system. Which of the following hypotheses is best supported by the results of this experiment?

   (A) The molecules needed for life on Earth were brought to Earth by a meteorite.
   (B) The molecules needed for life could have formed from inorganic compounds in Earth's early atmosphere.
   (C) The first molecules needed for life (that were formed in Earth's early atmosphere) were RNA.
   (D) The first molecules needed for life (that were formed in Earth's early atmosphere) were carbohydrates.

2. Which of the following pieces of evidence best supports the hypothesis that birds are more closely related to reptiles than to other animals?

   (A) Fossils of birds and reptiles are first seen in rock layers from the same time period.
   (B) Birds and reptiles live in similar habitats and occupy the same trophic levels.
   (C) Both birds and reptiles are the only animals with amniotic eggs.
   (D) Molecular studies show that the DNA from birds has a greater degree of homology with the DNA from reptiles than with the DNA from other animals.

3. Generally, which type of data is considered most reliable when constructing phylogenies?

   (A) fossil evidence, because it shows when species originated
   (B) biogeography, because it shows organisms' habitats
   (C) morphological characteristics, because they show body structures
   (D) molecular evidence, because it is less prone to convergent evolution or changes caused by geological events

Questions 4 and 5

Five new bacterial species were discovered in the Mariana Trench in the Pacific Ocean. The glyceraldehyde 3-phosphate dehydrogenase (*GAPDH*) gene was sequenced in all five species, and the number of nucleotide differences in the *GAPDH* gene among the five species is shown in the table.

Number of Nucleotide Differences in the *GAPDH* Gene				
Species	I	II	III	IV	V
I	---	1	8	4	15
II		---	4	6	17
III			---	1	19
IV				---	20
V					---

4. Which of the following cladograms is best supported by the data in the table?

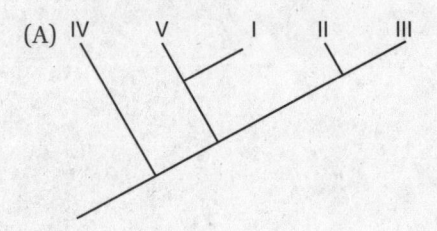

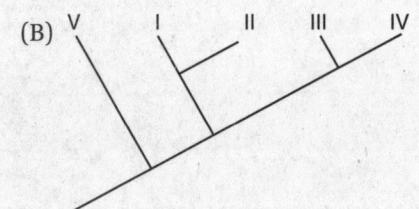

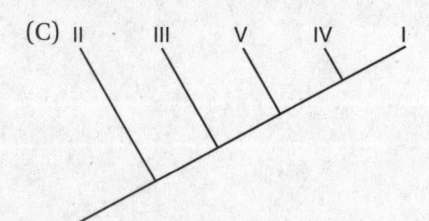

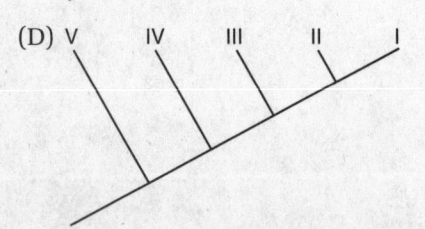

   (A) IV V I II III

   (B) V I II III IV

   (C) II III V IV I

   (D) V IV III II I

5. Based on the data in the table, which species is the outgroup?

    (A) II
    (B) III
    (C) IV
    (D) V

6. Which of the following best supports the existence of a common ancestor of all three domains of living organisms (Archaea, Bacteria, and Eukarya)?

    (A) All living organisms perform glycolysis in their cytoplasm.
    (B) All living organisms have membrane-bound organelles.
    (C) All living organisms have linear chromosomes.
    (D) All living organisms have genes that contain introns.

7. Some species of birds have unique bird songs that only attract members of the same species. This is an example of which type of reproductive isolation?

    (A) behavioral
    (B) gametic
    (C) habitat
    (D) mechanical

8. Orchids that belong to the genus *Dendrobium* will flower in response to certain weather stimuli. One species of *Dendrobium* flowers on the 8th day and closes on the 9th day after the weather stimuli, while another species of *Dendrobium* flowers on the 10th day and closes on the 11th day after the weather stimuli. This is an example of which type of reproductive isolation?

    (A) habitat
    (B) mechanical
    (C) temporal
    (D) hybrid breakdown

9. Which of the following is an example of mechanical reproductive isolation?

    (A) The sea urchin species *Strongylocentrotus purpuratus* and *Strongylocentrotus franciscanus* both live in the same marine habitat and release their gametes simultaneously into the surrounding water. However, the gametes cannot form a zygote.
    (B) In some snail species, the direction of the coil of the shell is controlled by a single gene. Snails with left-coiling shells cannot mate with snails with right-coiling shells because their copulating organs do not align.
    (C) The fruit fly species *Drosophila persimilis* breeds in the early morning, and the fruit fly species *Drosophila pseudoobscura* breeds in the late afternoon.
    (D) The bullfrog species *Rana draytonii* and *Rana catesbeiana* can mate and produce a zygote, but the zygote is not viable.

10. Tigers (*Panthera tigris*) and leopards (*Panthera pardus*) can mate and produce a zygote. The zygote will undergo cell division a few times, but this fails to result in the production of a viable embryo. This is an example of which type of reproductive isolation?

    (A) gametic
    (B) reduced hybrid viability
    (C) reduced hybrid fertility
    (D) hybrid breakdown

Short Free-Response

11. The following figure is a cladogram that represents the suspected ancestry of five species.

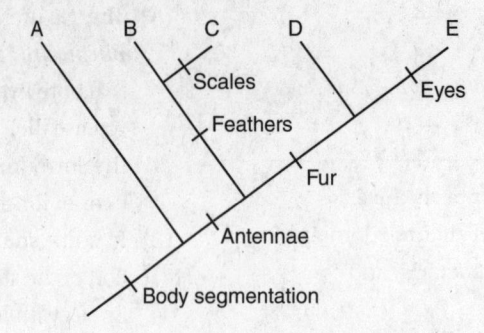

   Part A

   (i) **Describe** the characteristic(s) of the species that is the outgroup on this cladogram.

   Part B

   (i) **Describe** the similarities and differences in the characteristics present in species B and species D.

   Part C

   A new species (X) is discovered. Species X has body segmentation, antennae, and fur, but it does not have eyes or scales.

   (i) **Construct** a line that represents the ancestry of species X on the cladogram.

   Part D

   (i) **Explain** why species C is placed off of the same branch as species B and not the same branch as species D.

12. Two species of ground squirrels are separated by a river that they cannot cross. Genetic analyses of the two species of ground squirrels indicate that 99.3% of their DNA is homologous.

   Part A

   (i) **Describe** the type of reproductive isolation that separates the two species of ground squirrels.

   Part B

   (i) **Explain** how homology in DNA is used to determine ancestry.

   Part C

   Due to a severe drought, a decision is made to divert water away from the river to a reservoir that supplies water for a nearby city, and the river separating the two species dries up.

   (i) **Predict** the effect this will have on the two species of ground squirrels in the area.

   Part D

   (i) **Justify** your prediction from Part C.

Long Free-Response

13. The males of a particular species of bird attract mates with their bird songs. Male birds were observed, and the durations of their bird songs were recorded. The durations of the bird songs were classified into three groups: less than 120 seconds, 120–299 seconds, and 300–480 seconds. The number of females that approached the male birds within the 30-minute period, including during and immediately following the bird songs, was recorded. A graph of the data with 95% confidence intervals follows.

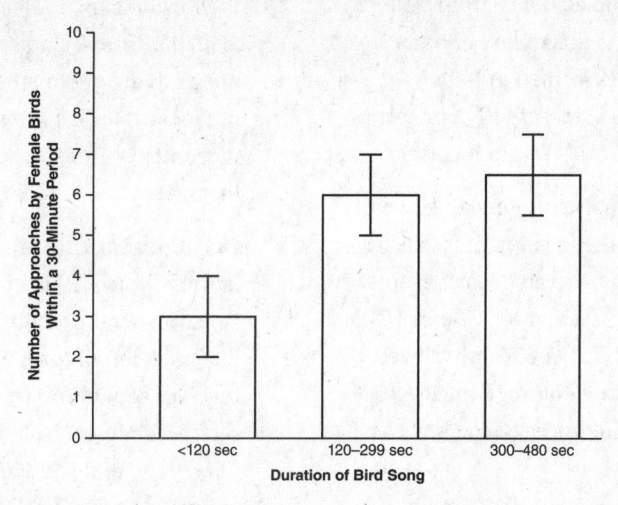

Part A

(i) Based on the data provided, **make a claim** about how the duration of the bird songs affects the number of females approaching the males.

Part B

(i) **Describe** the type of reproductive isolation that bird songs are an example of.

(ii) **Identify** one more example of this type of reproductive isolation.

Part C

(i) Analyze the data in the graph and **determine** whether there is a statistically significant difference between the number of females approaching males with bird songs in the 120–299 second range and the number of females approaching males with bird songs in the 300–480 second range.

Part D

(i) **Predict** the number of females that would approach a male with a bird song that is longer than eight minutes (480 seconds).

(ii) **Justify** your prediction.

Answer Explanations

Multiple-Choice

1. **(B)** Miller used molecules thought to have been present in Earth's early atmosphere, so the experiment supports this answer. Choice (A) is incorrect because there was no evidence from Miller's experiment that supports this theory about a meteorite. The molecules formed in Miller's experiment were amino acids, not RNA or carbohydrates, so choices (C) and (D) are incorrect.

2. **(D)** DNA evidence strongly supports a degree of ancestry of organisms. The location of fossils in rock layers can change due to geological events, so choice (A) is not the best answer. Choice (B) is incorrect because habitats and trophic levels do not necessarily indicate common ancestry. Mammals also have amniotic eggs, so choice (C) is incorrect.

3. **(D)** The DNA sequence of an organism is not affected by its environment or geological events, so it provides reliable evidence for ancestry of organisms. Fossils do not necessarily show when a species originated, so choice (A) is incorrect. Habitats and morphological characteristics can change during an organism's lifetime, so choices (B) and (C) are incorrect.

4. **(B)** Species V is the outgroup because it has the greatest number of amino acid differences from the other species. Species I and II have only one amino acid difference between them and are therefore closely related. Species III and IV have only one amino acid difference between them and are therefore closely related. Choice (A) is incorrect because species I is more closely related to species II (one amino acid difference) than species V (15 amino acid differences). Species II has fewer amino acid differences from the other species than does species V, so species II is not the outgroup. Thus, choice (C) is incorrect. Choice (D) is incorrect because species I has only four amino acid differences from species IV but eight amino acid differences from species III, so species IV should share a more recent common ancestor with species I.

5. **(D)** Species V is the outgroup because it has the greatest number of amino acid differences from the other four species.

6. **(A)** All living organisms perform glycolysis in their cytoplasm, indicating a common ancestor of all life-forms. Choice (B) is incorrect because not all living organisms have membrane-bound organelles. Prokaryotes do not have linear chromosomes or introns, so choices (C) and (D) are incorrect.

7. **(A)** Bird songs are an example of behavioral isolation. Choice (B) is incorrect because gametic isolation occurs when gametes are incompatible and cannot form a zygote. Habitat isolation involves separation by area, so choice (C) is incorrect. Mechanical isolation occurs when the genitalia of the male and female cannot successfully copulate, so choice (D) is incorrect.

8. **(C)** Temporal isolation is separation by time. Since the two species flower and are ready for pollination at different times, this is an example of temporal isolation. Habitat isolation is separation by area, so choice (A) is incorrect. Choice (B) is incorrect because mechanical isolation occurs when the male and female genitalia are incompatible. Hybrid breakdown occurs when two species can produce a viable and fertile offspring but each successive generation of the hybrid becomes weaker and weaker, so choice (D) is incorrect.

9. **(B)** Since their copulating organs do not align, these snails cannot exchange gametes, so this is an example of mechanical isolation. Choice (A) is an example of gametic isolation. Choice (C) is an example of temporal isolation. Choice (D) is an example of reduced hybrid viability.

10. **(B)** Since tigers and leopards can mate and produce a zygote, but doing so does not result in any viable offspring, this is an example of reduced hybrid viability. It is not gametic isolation since their gametes can form a zygote, so choice (A) is incorrect. There is no infertile adult hybrid in this scenario, so choice (C) is also incorrect. No viable hybrid was produced, so there is no chance of hybrid breakdown, making choice (D) incorrect.

Short Free-Response

11. (A-i) Species A is the outgroup, and according to the cladogram, the only characteristic it possesses is body segmentation.

    (B-i) Species B has body segmentation, antennae, and feathers. Species D has body segmentation, antennae, and fur.

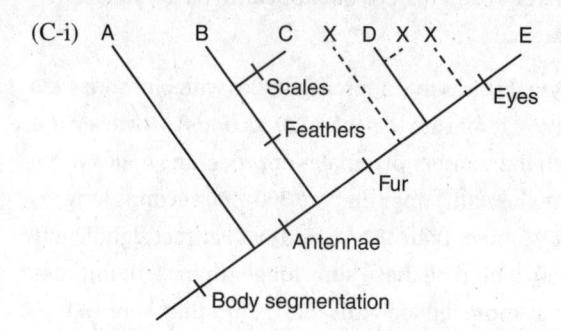

    (C-i)
    ```
    A     B     C  X  D X  X    E
                |            
                Scales        Eyes
              Feathers
                   Fur
                Antennae
             Body segmentation
    ```

    (All three possible positions for species X on the above cladogram would be considered acceptable.)

    (D-i) Species C has feathers, which is a characteristic that it has in common with species B. Species D has fur, which species C does not possess. Therefore, species C is placed off of the same branch as species B and not on the same branch as species D.

12. (A-i) This is a type of habitat isolation because the two species are separated by the river into different habitats.

    (B-i) The greater the percentage of homology of DNA between two species, the more recently they shared a common ancestor.

    (C-i) After the river that once separated the two species dries up, members of the two ground squirrel species will be able to interbreed with each other.

    (D-i) Since the degree of homology in the DNA between the two species is very high, they probably diverged relatively recently, so there is a good likelihood they will be able to interbreed and produce fertile offspring. With the elimination of the river that once separated them, they will be able to come into contact with one another and interbreed.

Long Free-Response

13. **(A-i)** Males with bird songs of shorter durations (less than 120 seconds) will be approached by fewer females than males with bird songs of 120 seconds or more in duration.

    (B-i and ii) Bird songs are a type of behavioral isolation. Another example of behavioral isolation is the mating dance.

    (C-i) The data are inconclusive because the 95% confidence intervals for the bird songs of 120–299 seconds and 300–480 seconds overlap. Therefore, there does not appear to be a statistically significant difference between the two groups.

    (D-i and ii) There is a large increase in the number of female birds that approach males with bird songs 120 seconds or longer (compared to those with songs that are less than 120 seconds). However, there does not appear to be a difference between the number of females approaching males with songs that are 120–299 seconds long and males with songs that are 300–480 seconds long. Therefore, it is unlikely that birds with songs longer than 480 seconds will attract significantly more females. Thus, a reasonable prediction would be that a song longer than eight minutes (480 seconds) would not attract significantly more female birds than songs that were 120–480 seconds long.

UNIT 8
Ecology

22

The Basics of Ecology

Learning Objectives

In this chapter, you will learn:
- → How Organisms Respond to Changes in the Environment
- → Energy Flow Through Ecosystems

Overview

The final unit in AP Biology brings together all that you have learned throughout the year in a discussion of ecology. In this unit, you need to know how organisms use energy and matter and how energy flows through ecosystems. The availability of energy (or lack thereof) in an ecosystem can determine the ecosystem's survival or demise. Ecosystems need to be able to adapt to disruptions in their environments. Communication among organisms is essential and allows them to respond to disruptions in their environments. These responses can lead to changes in populations and the evolution of populations. This chapter will first review how organisms respond to changes in their environment followed by a discussion of the flow of energy through ecosystems.

How Organisms Respond to Changes in the Environment

The survival of organisms often depends on their ability to respond to changes in their environment. Organisms can respond to environmental changes with behavioral or physiological mechanisms.

A change in the environment that triggers a response is called a **stimulus**. One example of a stimulus that can trigger a response is change in day length. Some birds respond to changes in day length by migrating, a **behavioral response**. Other animals may respond to changes in day length by slowing their metabolism to conserve energy, a **physiological response**.

Some stimuli are communicated among organisms. Organisms send signals to each other in response to changes in their environment. These signals can trigger changes in the behavior of other organisms. This communication among organisms can occur through different types of signals:

- **Audible signals**—Birds use audible signals to send warnings to other birds and to attract mates. Some primates use vocalizations to assert dominance or to warn of the presence of predators.
- **Chemical signals**—**Pheromones** are chemical signals released by some plants and animals to illicit a response in other organisms. Skunks release pungent chemicals to scare off potential predators. Female insects release chemicals (that males of the same species can detect) to initiate mating. Some plants release chemicals to warn neighbors when an herbivore is damaging them.
- **Electrical signals**—Sharks and rays send electrical signals through the water to locate prey species.
- **Tactile signals**—Touching between primates can be used to express affection or to indicate dominance. Some plants curl up the delicate parts of their body to shield them away from touch.
- **Visual signals**—Some species use warning coloration, a form of **aposematism**, to scare off potential predators. One example of this are the toxic amphibians in the rainforest that are often brightly colored.

Signaling among organisms can help them find mates, determine social hierarchies, and find needed resources. Natural selection will favor signals and responses that increase survival and the chances of successful reproduction. Over time, this selection can lead to changes in the population and evolution.

Cooperative behaviors can lead to increased fitness of individuals and populations. One example of this are murmurations in starlings. Murmurations occur when starlings fly in formations of hundreds, or sometimes even thousands, of birds. When flying in these large cooperative groups, starlings are more protected from predators, increasing their likelihood of survival. It is also thought that these large groups help the starlings keep warm.

Meerkats exhibit a wide range of cooperative behaviors. They huddle together for warmth and gather in groups to groom each other, picking parasites out of each other's fur. While some meerkats in the group are foraging for food, other meerkats will serve as lookouts and will send a loud signal when any predators approach. Meerkats have even been known to fight together in groups to ward off larger predators, like cobras.

Energy Flow Through Ecosystems

Organisms use energy to grow, reproduce, and maintain organization. Different species have different adaptations for maintaining energy levels and body temperatures.

Endotherms use thermal energy generated from their metabolism to maintain their body temperature. Mammals and birds are endotherms. **Ectotherms** do not have internal mechanisms for maintaining body temperature and obtain heat from their environment. They must change their behaviors to regulate their body temperature. For example, if a lizard's body temperature drops, it will move to a warm rock or into the sunlight to increase its body temperature. Fish, reptiles, and amphibians are ectotherms.

Metabolic rate is the total amount of energy an organism uses per unit time. Smaller organisms generally have a higher metabolic rate than larger organisms. As the size of an organism increases, its metabolic rate decreases. One reason for this is that smaller organisms have a greater surface area to volume ratio and therefore lose more heat to their environment. Smaller animals need higher metabolic rates to compensate for this loss of heat.

Access to energy is key to maintaining the health of an organism. Organisms are constantly expending energy to survive and obtaining energy from the food they eat (or the carbon-containing molecules they produce if the organism is photosynthetic). A net gain in energy can result in energy storage (such as in fat tissues of animals) or the growth of an organism. A net loss of energy can result in the loss of mass or even the death of the organism.

Changes in energy availability in an ecosystem, such as a reduction in sunlight or in the number of producers, can result in changes in population size. If energy becomes less available in an ecosystem (for example, if a large building reduces the amount of sunlight available), the producers' ability to perform photosynthesis will be reduced, and some producers will die. If the population size of producers in an ecosystem decreases, the entire ecosystem will be disrupted. An example of this is the rise in ocean temperatures, which has led to the death of producers in coral reefs (**coral bleaching**). This, in turn, has led to massive decreases in the number of species and in population sizes of these species in coral reef ecosystems.

Trophic levels represent steps in the food and energy transfers between organisms in an ecosystem. Organisms are classified into trophic levels based on their food and energy sources. Energy is captured by and then moves from producers to **herbivores** (primary consumers) to **carnivores** and **omnivores** (secondary, tertiary, and quaternary consumers).

Food chains show the transfer of energy between these trophic levels. It is important to note that in a food chain, the arrows show the direction of the transfer of energy. Figure 22.1 provides an example of a food chain.

Grass → Grasshopper → Bird → Snake → Hawk

Figure 22.1 A Food Chain

Most organisms do not rely on just one food source and are part of multiple food chains. **Food webs** show the interconnections between organisms in different food chains and provide a more complete representation of energy transfers in ecosystems than food chains. Figure 22.2 provides an example of a food web.

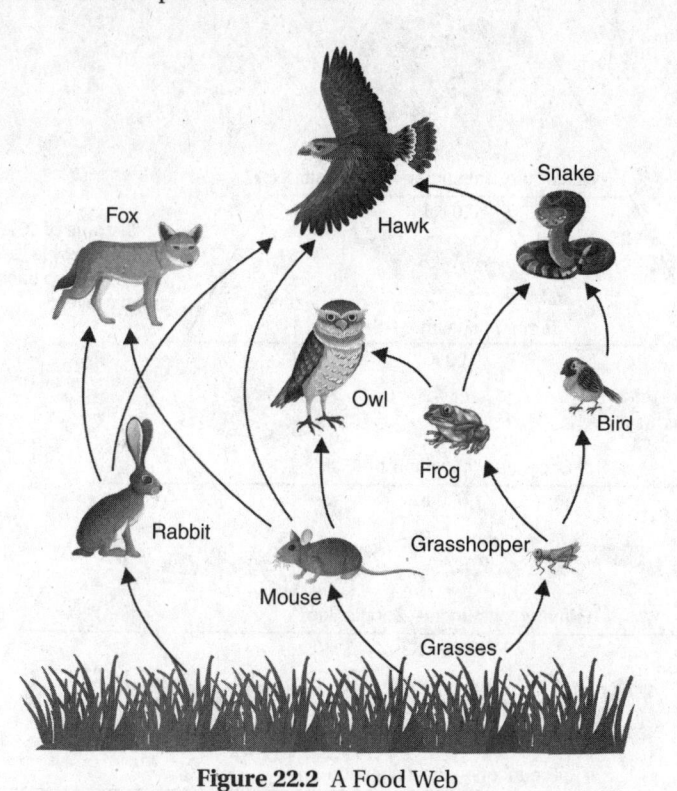

TIP

An easy way to remember how to draw arrows in a food web is to remember that an arrow points to the mouth of the predator that is eating the prey.

Figure 22.2 A Food Web

Autotrophs (producers) get energy from physical or chemical sources in their environment. **Photoautotrophs** get energy from sunlight. Plants are photoautotrophs. **Chemoautotrophs** obtain energy from small inorganic molecules in their environment. Most chemoautotrophs are bacteria found in extreme environments, such as deep-sea thermal vents or geothermal geysers. The rate at which these autotrophs use energy to convert inorganic matter into organic matter is **primary productivity**.

Heterotrophs get energy from carbon compounds made by other organisms. Heterotrophs can obtain energy from carbohydrates, lipids, or proteins by breaking down these macromolecules using hydrolysis reactions. Animals are heterotrophs.

Decomposers break down dead organic material, allowing the nutrients in dead organisms to be recycled through ecosystems. Many fungi and bacteria are decomposers. **Detritivores** are organisms that obtain energy by consuming the organic waste of dead plants and animals. Millipedes, centipedes, and earthworms are examples of detritivores. Both decomposers and detritivores are important in nutrient cycling in ecosystems.

Some organisms have unusual strategies for obtaining energy. A small number of organisms exhibit kleptoplasty. **Kleptoplasty** is when a heterotroph consumes an autotroph that it uses as a food source but removes the chloroplasts from the autotroph's cells and incorporates them into its own cells. The sea slug *Elysia crispata* consumes algae for food and incorporates the chloroplasts (from the algae it consumes) into its own cells. When the sea slug cannot find food to eat, it moves into the sunlight, and the chloroplasts it incorporated into its cells perform photosynthesis, providing food for the sea slug.

As energy moves between trophic levels, energy is lost. This energy is either lost as heat or it is consumed by the necessary metabolic processes that the organisms in that trophic level use. Because less energy is available as you move up trophic levels, higher trophic levels necessarily will have smaller population sizes, as shown in Figure 22.3.

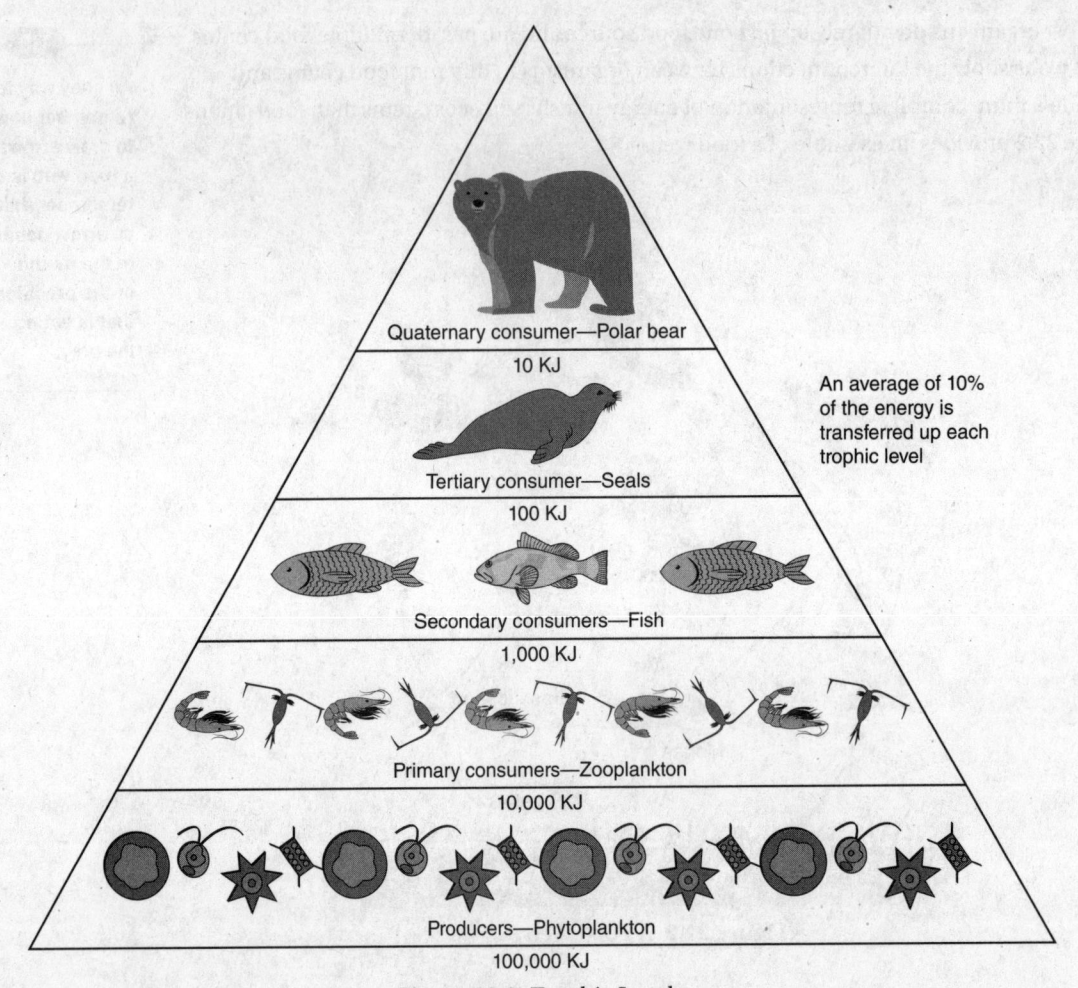

Figure 22.3 Trophic Levels

- Quaternary consumer—Polar bear — 10 KJ
- Tertiary consumer—Seals — 100 KJ
- Secondary consumers—Fish — 1,000 KJ
- Primary consumers—Zooplankton — 10,000 KJ
- Producers—Phytoplankton — 100,000 KJ

An average of 10% of the energy is transferred up each trophic level

One polar bear might eat 50 seals in one year. Each seal needs to consume about 15 times its body weight in fish in one year. As you move down the trophic levels in a food chain, the amount of biomass at the lower trophic levels necessarily increases, with the producers possessing the greatest biomass.

If the population size of the producers decreased, there may not be sufficient food or energy for the remaining trophic levels, and the food web may collapse. This is called **bottom-up** regulation of ecosystems. Photosynthetic phytoplankton are the bottom trophic level of many marine ecosystems. Zooplankton feed on the phytoplankton, and sea stars, fish, and even some whales feed on these zooplankton. Runoff from herbicides can pollute the waters, reducing the number of phytoplankton. This can lead to a reduction in the zooplankton population size and the animals that depend on the zooplankton for food, thereby reducing the biodiversity of the ecosystem. This is an example of bottom-up regulation.

Animals at higher trophic levels may help limit the population sizes at lower levels. If these top predators are removed from an ecosystem, the population sizes of other trophic levels may exceed the producers' ability to produce enough food to support them, and the food web may collapse. This is called **top-down** regulation. An example of top-down regulation can be seen in the rainforests of Venezuela. Manmade construction of dams created isolated islands of rainforests, some of which contained top predators like crocodiles, while others were left without top predators. After a few years, the biodiversity of the sections of rainforests without top predators was greatly reduced (with fewer saplings, trees, and other plants). Without crocodiles, plant-eating animal populations grew at a high rate, reducing the number of plants in these areas. This resulted in less habitats for the organisms that depend on these trees and plants and a reduced number of species in these areas. On the rainforest islands with top predators, crocodiles would eat some of the animals that feed on seeds and plants, limiting the population of these animals and preserving the number of plant species in these areas.

The availability of food and the energy it provides organisms affects these organisms' reproductive strategies. Different organisms use different reproductive strategies in response to energy availability. Organisms that live in unstable environments (that have less access to energy-containing compounds) will produce large numbers of offspring at a time. Because there is less access to energy-containing food in their environment, the survival rate of the offspring is lower than that of organisms living in a more stable environment. By producing more offspring, organisms increase the probability of at least one of their offspring passing on their alleles to the next generation of organisms. Organisms that live in more stable environments with greater access to energy-containing compounds will produce fewer offspring at a time. Since the likelihood of survival of their offspring is higher, not as many offspring need to be produced to ensure their alleles are passed on to the next generation. Some organisms can also alternate between sexual reproduction (which requires more energy) and asexual reproduction (which requires less energy) in response to energy availability. Reproductive strategies will be discussed in more detail in Chapter 23.

Practice Questions

Multiple-Choice

Questions 1 and 2

Refer to the following food web.

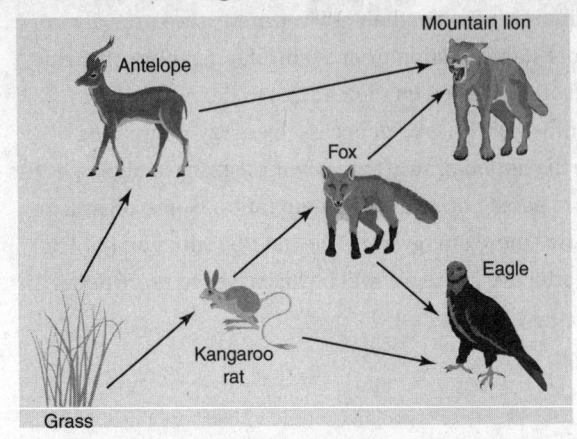

1. What is a role of the fox in this food web?

   (A) producer
   (B) primary consumer
   (C) secondary consumer
   (D) tertiary consumer

2. The application of rat poison in the area results in the elimination of the kangaroo rat population. Which of the following is the most likely effect of this event?

   (A) The biomass of the producers will increase.
   (B) The number of foxes in the ecosystem will likely increase.
   (C) The amount of food available to the antelope will decrease.
   (D) The mountain lions will eat fewer antelopes and more eagles.

3. Why is it unlikely that a food web would contain more than five trophic levels?

   (A) 90% of the nutrients are lost in each transfer to the next trophic level.
   (B) 90% of the energy is lost in each transfer to the next trophic level.
   (C) Decomposers can efficiently recycle nutrients for up to four trophic levels.
   (D) The biomass at the producer level cannot exceed the biomass at the primary consumer level.

4. Coyotes are a predator of skunks. Coyote urine has a strong scent and is marketed as a skunk repellent. Which type of signal is coyote urine?

   (A) audible
   (B) chemical
   (C) tactile
   (D) visual

5. Which of the following is an example of a physiological response to a stimulus?

   (A) A female bird moves toward a male bird that is singing an appropriate mating song.
   (B) A predator avoids a brightly colored frog in the rainforest.
   (C) A bear slows its metabolism in response to shortened day lengths and cooler temperatures.
   (D) A primate grooms its offspring.

6. Why are the metabolic rates of smaller organisms generally higher than those of larger organisms?

   (A) Smaller organisms have shorter life spans and therefore must accomplish more activities in less time. This requires a faster metabolic rate.
   (B) Smaller organisms have a higher surface area to volume ratio than larger organisms and lose more heat to the environment.
   (C) Smaller organisms are more likely to be ectothermic than larger organisms and therefore need a higher metabolic rate to compensate for this.
   (D) Smaller organisms consume less food per gram of body weight than larger organisms and therefore need a higher metabolic rate to compensate for this.

Questions 7 and 8

Refer to the following food web. The arrows show the direction of the flow of energy, and species are represented by Roman numerals.

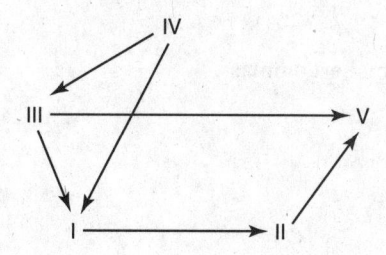

7. Which species in this food web is most likely a producer?

   (A) I
   (B) II
   (C) III
   (D) IV

8. Which species is both a primary consumer and a secondary consumer?

   (A) I
   (B) II
   (C) III
   (D) IV

Questions 9 and 10

The respiration rate (as measured by oxygen consumption per gram of body weight per minute) was measured in rats, grasshoppers, and newly discovered species animal X. The respiration rates were measured at 5°C and 30°C for all three species. Data are shown in the table.

Organism	Average Respiration Rate at 5°C (mL O_2/min/g)	Average Respiration Rate at 30°C (mL O_2/min/g)
Rat	0.158	0.076
Grasshopper	0.011	0.033
Species X	0.024	0.069

9. Which of the following conclusions about species X is best supported by the data?

   (A) Species X is ectothermic since its respiration rate is higher at a higher environmental temperature.
   (B) Species X is ectothermic since ectothermic animals always have an increased respiration rate at lower temperatures.
   (C) Species X is endothermic since its respiration rate at 30°C is closer to the respiration rate of the rat at 30°C than to the respiration rate of the grasshopper at 30°C.
   (D) Species X is endothermic since its respiration rate is higher at a higher environmental temperature.

10. This experiment is repeated with the same animals at a temperature of 37°C. Which of the following is the most likely result?

    (A) The rat will have a higher respiration rate than the grasshopper at 37°C because rats are ectothermic.
    (B) The grasshopper will have a higher respiration rate than species X at 37°C because grasshoppers are ectothermic and species X is endothermic.
    (C) Species X will have a higher respiration rate than the rat at 37°C because rats are ectothermic and species X is endothermic.
    (D) The rat will have a lower respiration rate than both the grasshopper and species X at 37°C because rats are endothermic and both grasshoppers and species X are ectothermic.

Short Free-Response

11. The following table shows the diet composition of members of an estuarine ecosystem (a shallow, coastal shelf where a freshwater river empties into a larger saltwater body). Higher percentages indicate a higher degree of reliance on a particular food source. An "X" indicates that the source is not a food source for that organism.

Food Source	Diet Composition (percent)					
	Snails	Oysters	Clams	Crabs	Fish	Birds
Algae	100	100	100	X	30	X
Snails	X	X	X	40	40	X
Oysters	X	X	X	30	20	10
Clams	X	X	X	30	10	10
Crabs	X	X	X	X	X	40
Fish	X	X	X	X	X	40

Part A

(i) **Describe** the role of algae in this ecosystem.

Part B

(i) **Represent** the relationships between these organisms by **constructing** a food web for this ecosystem.

Part C

(i) **Identify** the secondary consumers in this ecosystem.

Part D

(i) **Predict** the organism in this ecosystem that will have the smallest biomass.

(ii) **Explain** why you made this prediction.

12. In order to investigate the possible effects that chemicals in a certain insecticide have on plant growth, a researcher grew 100 plants from each of three different crop species. Fifty plants of each type were grown in a lab in the presence of the insecticide, and the other 50 plants of each type were grown in a lab without insecticide. All plants were grown to maturity and then dried, and the mean dry weight per plant and the standard error of the mean were calculated for each group. Data are shown in the table.

Crop	Mean Dry Weight per Plant Without Insecticide in Grams (± SEM*)	Mean Dry Weight per Plant in the Presence of Insecticide in Grams (± SEM)
A	12.40 ± 0.50	14.02 ± 0.56
B	8.63 ± 0.25	10.02 ± 0.29
C	15.64 ± 0.61	19.54 ± 1.32

*Standard Error of the Mean

Part A

(i) **Describe** the effect of the use of the insecticide on the mean dry weight of crop B.

Part B

(i) **Calculate** which crop (A, B, or C) had the greatest percent increase in mean dry weight in the presence of the insecticide. Show your work.

Part C

The researcher makes the claim that use of this insecticide increases the mean dry weight of all three crops.

(i) Using the data from the table, **evaluate** the researcher's claim.

Part D

(i) **Explain** why the use of an insecticide could increase the yield of some agricultural crops.

Long Free-Response

13. Plants thought to have evolved in temperate climates use C3 photosynthesis (also known as the light-independent cycle). Plants thought to have evolved in tropical climates use an additional step referred to as C4 photosynthesis. Scientists measured the photosynthetic rate of C3 plants (sugar beets) and C4 plants (sugarcane) in a sealed container under three different temperature conditions. Data are shown in the table.

Temperature (°C)	C3 plants CO_2 micromoles/ m^2/minute ($\pm$ SEM*)	C4 plants CO_2 micromoles/ m^2/minute ($\pm$ SEM)
20	37.0 $\pm$ 1.2	35.1 $\pm$ 1.1
30	35.2 $\pm$ 0.8	38.0 $\pm$ 1.0
40	26.8 $\pm$ 0.5	37.0 $\pm$ 1.2

*Standard Error of the Mean

Part A

(i) **Explain** why the rate of carbon dioxide consumption was used in this experiment.

Part B

(i) **Construct** a graph of the data, showing 95% confidence intervals.

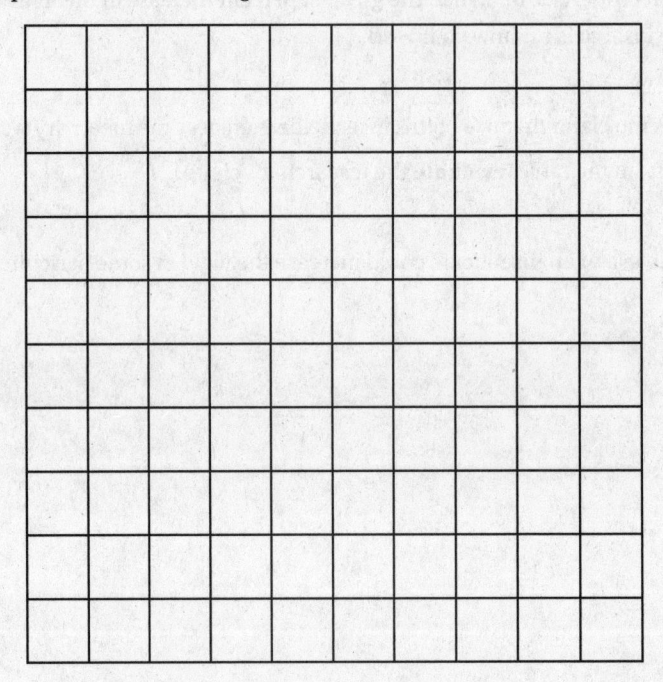

Part C

(i) Analyze the data and **determine** at which temperature(s) there was a statistically significant difference in carbon dioxide consumption in C3 and C4 plants.

Part D

Coleus is a popular C3 houseplant. A person has a *Coleus* plant inside her climate-controlled home, in which the temperature never exceeds 30° Celsius.

(i) If this plant were placed outside for seven days during a heat wave (in which the high temperature each day reached 39°C), **predict** the effect it would have on the plant.

(ii) **Justify** your prediction.

Answer Explanations

Multiple-Choice

1. **(C)** The fox is a secondary consumer because it gets its energy from a primary consumer (the kangaroo rat). Choice (A) is incorrect because the producer in this food web is grass; foxes are not photosynthetic and cannot be producers. Since the fox shown in this food web does not consume the producer, the fox cannot be a primary consumer, so choice (B) is incorrect. Tertiary consumers get their energy from secondary consumers. In the food web shown, the mountain lion and the eagle are tertiary consumers; the fox is not a tertiary consumer, so choice (D) is incorrect.

2. **(A)** Kangaroo rats eat grass. If the kangaroo rat population was eliminated, less grass would be consumed, and the biomass of the producers (the grass) would most likely increase. The foxes feed on the kangaroo rats, so the elimination of the kangaroo rats would result in less food for the foxes. In that case, the number of foxes would likely decrease, not increase, so choice (B) is incorrect. Choice (C) is incorrect because fewer kangaroo rats would mean more grass for the antelopes, not less. Mountain lions do not eat eagles in this food web, so choice (D) is incorrect.

3. **(B)** On average, 90% of the energy at a given trophic level is used in metabolic processes at that level or is lost as heat and is not transferred to the next trophic level. Choice (A) is incorrect because most nutrients are transferred to the next trophic level when food is consumed. Decomposers break down dead organic matter, regardless of the number of trophic levels, so choice (C) is incorrect. The biomass at the producer level must be larger than the biomass at the primary consumer level, so choice (D) is incorrect.

4. **(B)** Scent is a chemical signal. Choices (A), (C), and (D) are incorrect because audible, tactile, and visual signals depend on sound, touch, and sight, respectively.

5. **(C)** Adjusting metabolism in response to changing environmental conditions is a physiological response to a stimulus. Choices (A), (B), and (D) are examples of behaviors or behavioral responses.

6. **(B)** Smaller organisms do have a greater surface area to volume ratio than larger organisms and lose more heat to their environment per gram of body weight, so smaller organisms need a higher metabolism to help mitigate the heat loss. Not all small organisms have a shorter life span than larger organisms, so choice (A) is incorrect. Choice (C) is incorrect because not all small organisms are ectothermic. Also, ectothermic animals cannot internally control their metabolic rate; ectotherms are dependent upon environmental temperatures to influence their metabolic rate. Smaller organisms do not necessarily consume less food per gram of body weight; in fact, the higher metabolic rate of smaller organisms may require that these small organisms consume more food per gram of body weight, so choice (D) is incorrect.

7. **(D)** Producers transfer energy to other trophic levels and do not receive energy from any trophic level. IV is the only member of this food web that sends energy to other members but does not receive energy from other members of the food web.

8. **(A)** Primary consumers get energy from producers, and secondary consumers get energy from primary consumers (herbivores). In this food web, IV is a producer, I and III are both primary consumers, and I, II, and V are secondary consumers. Choice (A) is the only member of this food web that is both a primary and secondary consumer.

9. **(A)** Ectothermic animals have a higher metabolism at higher environmental temperatures. Since species X's consumption of oxygen is greater at 30°C than at 5°C, it is most likely ectothermic. Choice (B) correctly states that species X is ectothermic but gives incorrect reasoning; ectothermic animals have an increased respiration rate at *higher*, not lower, temperatures. Choice (C) is

incorrect because while species X's oxygen consumption at 30°C is closer to that of the rat than to that of the grasshopper, that does not determine whether or not species X is endothermic. Endothermic animals have higher metabolic rates at lower temperatures (so they can maintain a constant internal temperature). Therefore, choice (D) is also incorrect.

10. **(D)** At a higher temperature, the respiration rate of ectotherms (the grasshopper and species X, in this example) increases and the respiration rate of endotherms (the rat, in this example) decreases. Choice (A) is incorrect because rats are endothermic. Species X is most likely ectothermic, so choice (B) is incorrect. Choice (C) is incorrect because rats are endothermic, not ectothermic, and species X is most likely ectothermic.

Short Free-Response

11. (A-i) Algae are producers and are the only member of this ecosystem that can perform photosynthesis. Therefore, algae either directly or indirectly provide organic compounds to the other members of this ecosystem.

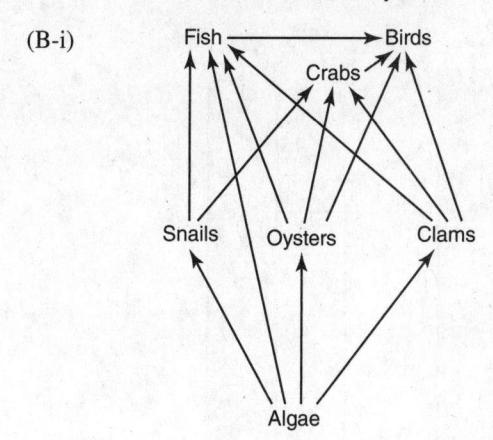

   (B-i)

   (C-i) Secondary consumers eat primary consumers. Primary consumers eat producers. The primary consumers are snails, oysters, clams, and fish. Therefore, the secondary consumers are crabs, fish, and birds.

   (D-i and ii) Birds should have the fewest numbers since they are a top predator in this ecosystem.

12. (A-i) The use of the insecticide significantly increased the mean dry weight of crop B. The upper limit of the 95% confidence interval for crop B without the insecticide is 9.13, and the lower limit of the 95% confidence interval for crop B in the presence of the insecticide is 9.44. So there is likely a significant difference between the two groups.

   (B-i) Percent increase in mean dry weight =

   $$\frac{\text{mean dry weight in the presence of insecticide} - \text{mean dry weight without insecticide}}{\text{mean dry weight without insecticide}} \times 100$$

   Crop A $= \frac{14.02 - 12.40}{12.40} \times 100 = 13\%$

   Crop B $= \frac{10.02 - 8.63}{8.63} \times 100 = 16\%$

   Crop C $= \frac{19.54 - 15.64}{15.64} \times 100 = 25\%$

   Crop C has the greatest percent increase in mean dry weight.

   (C-i) The researcher's claim is *not* supported by the data. This is because the 95% confidence intervals for plants from crop A grown with and without insecticide overlap: the upper limit of the 95% confidence interval for crop A without insecticide is 13.40 and the lower limit of the 95% confidence interval for crop A in the presence of insecticide is 12.90. Therefore, it cannot be concluded that there is a significant difference between those two groups for crop A.

   (D-i) Many insects consume plant material and are therefore primary consumers. The use of insecticides would decrease the number of primary consumers and could thereby increase the crop yield.

Long Free-Response

13. (A-i) Carbon dioxide is a reactant that is consumed during photosynthesis, so measuring the rate of carbon dioxide consumption could be a good way to compare the rate of photosynthesis between these two organisms.

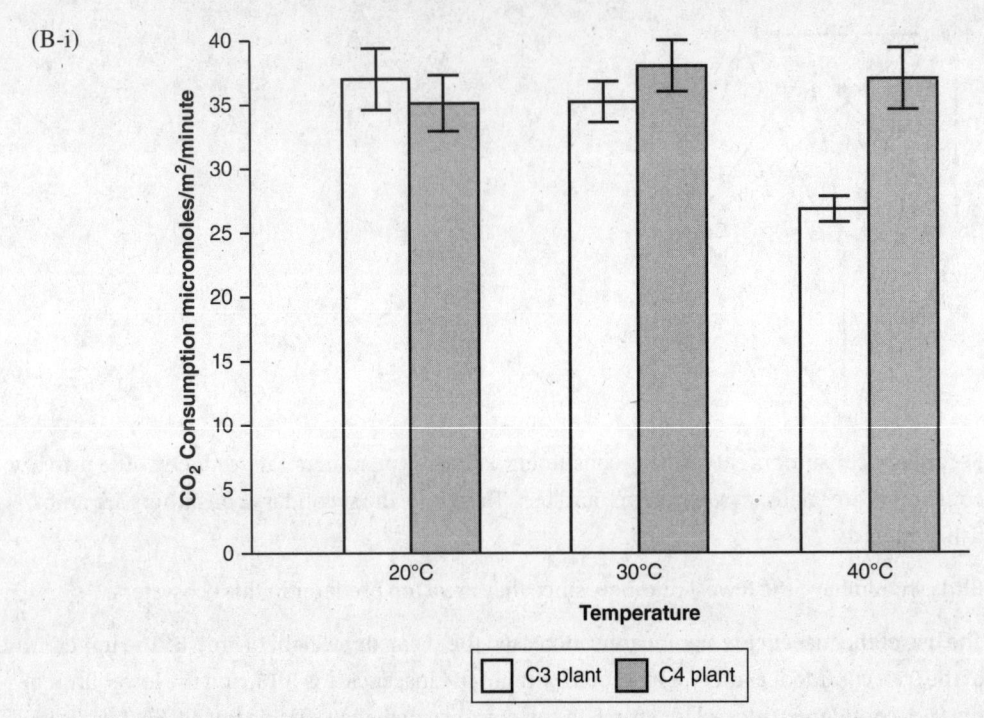

(B-i)

[Bar graph showing CO_2 Consumption micromoles/m²/minute vs Temperature for C3 plant and C4 plant at 20°C, 30°C, and 40°C. At 20°C: C3 ≈ 37, C4 ≈ 35. At 30°C: C3 ≈ 35, C4 ≈ 38. At 40°C: C3 ≈ 27, C4 ≈ 37. Error bars shown.]

(C-i) There is likely a statistically significant difference at a temperature of 40°C because the 95% confidence intervals for the C3 and C4 plants do not overlap at that temperature, but they do overlap at 20°C and 30°C.

(D-i and ii) The plant would likely consume less carbon dioxide and would perform less photosynthesis in a 39°C environment than in a 30°C environment. It would have fewer organic compounds with which to supply itself with energy. This is because the data show that a C3 plant will consume less carbon dioxide at 40°C (compared to at 30°C), so it will probably perform less photosynthesis when exposed to heat wave temperatures.

23

Population Ecology, Community Ecology, and Biodiversity

Learning Objectives

In this chapter, you will learn:
- → Population Ecology
- → *K*-Selected vs. *r*-Selected Populations
- → Community Ecology and Simpson's Diversity Index
- → Biodiversity

Overview

Populations are made of individual organisms that interact with one another and with their environment in complex ways. This chapter reviews the factors that affect the rate of growth of populations and the biodiversity of the communities in which they live.

Population Ecology

How organisms get the energy and matter they need to survive affects their population sizes and the rate of population growth. Population growth depends on a number of factors including:

- Population size (N)
- Birth rate (B)
- Death rate (D)

Population growth rates are calculated as the change in the population size over change in time: $\frac{dN}{dt}$. One way to calculate the rate of population growth is to simply compare the birth rate and the death rate in the population, as in the following equation:

$$\frac{dN}{dt} = B - D$$

If the birth rate exceeds the death rate in a population, the population size will increase. If the birth rate is less than the death rate in a population, the population size will decrease.

If there are no limiting factors on the growth of a population (there is abundant food and habitat, no predators are present, etc.), a population will experience **exponential growth** and can be described with the following equation:

$$\frac{dN}{dt} = r_{max}N$$

In this equation, N represents the population size, and r_{max} is the maximum per capita growth rate of the population. Note that the larger the population size, the higher the growth rate and the faster the population will grow. Exponential growth curves are usually J-shaped, as shown in Figure 23.1.

> **TIP**
>
> You do not need to memorize the equations presented in this chapter. These equations are included on the AP Biology Equations and Formulas sheet that you will have access to during the AP Biology exam. Spend your time applying your knowledge by answering as many practice problems as possible using these formulas.

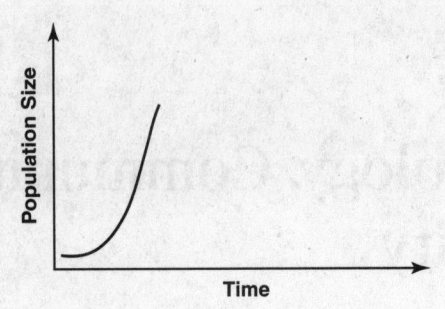

Figure 23.1 Exponential Growth Curve

Some populations will eventually exceed the resources available in their environment, and their growth will be limited by resource availability. Factors that limit the growth of these populations can be either **density-dependent factors** or **density-independent factors**. Some examples of density-dependent factors that limit population growth are disease, predation, and competition for food, habitat, or mates. Density-independent factors include temperature, precipitation, and natural disasters (such as forest fires or volcanic eruptions). These limitations result in **logistic growth** of a population. Logistic growth can be described by the following equation:

$$\frac{dN}{dt} = r_{max}N\left(\frac{K-N}{K}\right)$$

In this equation, N still represents the population size, and r_{max} again is the maximum per capita growth rate of the population. K represents the carrying capacity of the environment. **Carrying capacity** is defined as the maximum population that can be supported by the available resources in an environment. Notice that as the population size increases (N becomes closer to K), the rate of growth of the population will decrease. Smaller population sizes that are far below the carrying capacity of an environment will experience higher growth rates.

Logistic growth curves are S-shaped, as shown in Figure 23.2. Logistic growth curves start with a relatively flat **lag phase**, followed by a period of exponential growth (also called the **log phase**), which slows as the population reaches the carrying capacity of the environment. Logistic growth curves stabilize at or near the carrying capacity of the environment.

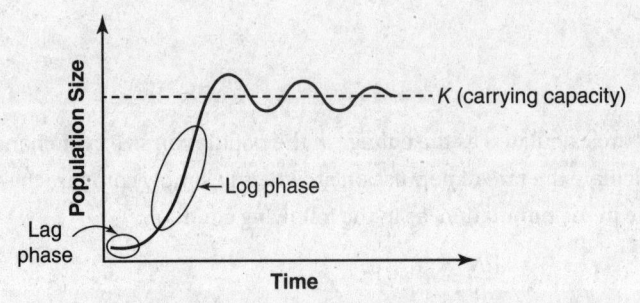

Figure 23.2 Logistic Growth Curve

K-Selected vs. *r*-Selected Populations

As discussed in Chapter 22, energy availability affects the reproductive strategies of populations. Populations that live in more stable environments and have more energy available tend to have *K*-selected reproductive strategies. ***K*-selected** populations possess relatively stable population sizes at or near the carrying capacity of their environment. *K*-selected populations usually reproduce more than once per lifetime, with few offspring per reproductive cycle. *K*-selected populations invest greater levels of parental care in their offspring, resulting in higher survival rates in their offspring. *K*-selected populations experience logistic growth and are sensitive to density-dependent factors. Most mammal and bird species are *K*-selected populations.

Populations that live in unstable environments and have less energy available have **r-selected** reproductive strategies. These populations reproduce at a younger age, often only once in their lifetime. Each reproductive cycle produces large numbers of offspring. However, r-selected populations invest little or no parental care in their offspring, leading to much lower survival rates in the offspring. r-selected populations experience "boom or bust" cycles, with periods of exponential growth leading to populations that far exceed the carrying capacity of an environment ("booms"), followed by rapid decreases in the population size ("busts"). The size of r-selected populations is usually not sensitive to population densities. Many fish and amphibian species have r-selected reproductive strategies.

Community Ecology and Simpson's Diversity Index

A **community** is a group of interacting populations living in the same habitat. Communities can be described by their species composition and species diversity. **Species composition** is the number of species that live in an area. **Species diversity** reflects the number of species in an area and the number of members of each of those species in the area. Species diversity gives a more accurate assessment of the variety of organisms found in an area. One way of representing species diversity is the equation for **Simpson's Diversity Index**:

$$\text{Diversity Index} = 1 - \Sigma \left(\frac{n}{N}\right)^2$$

where n = the total number of organisms of a *particular* species and N = the total number of organisms of *all* the species. The higher Simpson's Diversity Index, the more diverse the community.

Practice with Simpson's Diversity Index

The equation for Simpson's Diversity Index can be used to compare the diversity of two communities, as in the following example.

Community A is in a suburban backyard, and Community B is in a forest. The species composition and the number of organisms of each species in each community are shown in the following tables.

Community A	
Species	**Number of Organisms**
Grass	11,000
Camellia Shrubs	4
Elm Tree	1
Total Number of Organisms	11,005

Community B	
Species	**Number of Organisms**
Oak Trees	2
Earthworms	500
Ferns	15
Chipmunks	2
Ants	1,000
Mesquite Bush	5
Total Number of Organisms	1,524

Simpson's Diversity Index for Community A is:

$$\text{Diversity Index} = 1 - \left(\left(\frac{11,000}{11,005}\right)^2 + \left(\frac{4}{11,005}\right)^2 + \left(\frac{1}{11,005}\right)^2\right) = 9.08 \times 10^{-4}$$

Simpson's Diversity Index for Community B is:

$$\text{Diversity Index} =$$
$$1 - \left(\left(\frac{2}{1,524}\right)^2 + \left(\frac{500}{1,524}\right)^2 + \left(\frac{15}{1,524}\right)^2 + \left(\frac{2}{1,524}\right)^2 + \left(\frac{1,000}{1,524}\right)^2 + \left(\frac{5}{1,524}\right)^2\right) = 4.62 \times 10^{-1}$$

Community B has a diversity index more than 500 times larger than the diversity index of Community A.

Relationships Within Communities

Interactions and relationships among members of a community are important for the survival of organisms. The interactions among populations within a community can change over time. These changing interactions can influence how members of the community access the matter and energy they need to survive.

There are many different types of relationships among interacting members of a community:

- **Competition:** Organisms compete for resources, such as food, habitats, and mates. Competition can occur between two different species (interspecies competition) or between members of the same species (intra-species competition). Competition can lead to the demise of organisms in an ecosystem.
- **Predator/Prey:** Predator species eat prey species and depend on prey populations for food. Insufficient numbers of prey species will lead to declining numbers of predator species. As the number of prey increase, predator numbers will follow with an increase in numbers. Fluctuations in the numbers of predators generally follow fluctuations in the numbers of prey, as shown in Figure 23.3.

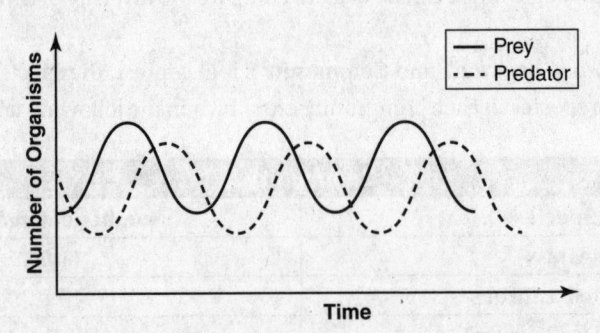

Figure 23.3 Prey-Predator Populations Over Time

- **Niche partitioning:** Competing species may coexist if they use the resources available in their habitat differently; this is known as niche partitioning. Competing species in a habitat may use dietary partitioning. An example of this is on the African savanna, where some animal species are grazers and eat grass, while other animal species in the same habitat are browsers and eat tree leaves. This allows them to coexist in the same habitat. Anole lizards exhibit habitat partitioning, with some anoles living in the forest canopy and others on the ground.
- **Trophic cascades:** A trophic cascade refers to the far-reaching effects of the reduction of one trophic level in a food web. For example, the decline of sea otter populations on the Pacific Coast led to increases in the sea urchin populations. Sea otters eat sea urchins. Without the presence of sea otters, the sea urchin populations exploded. Sea urchins eat kelp, and the increase in the numbers of sea urchins led to a decrease in kelp forests. Declines in kelp forests led to decreased habitats for other marine organisms in the area. Changes in just one level (the number of sea otters) had cascading effects throughout the food web.
- **Parasitism:** This is a symbiotic relationship where one species benefits from the relationship but the other species is harmed. Fleas that live on dogs or cats are parasites. The fleas benefit from consuming the blood of the host animal, and the dog or cat on which the flea lives is harmed.

- **Commensalism:** In commensalism, one species benefits and the other neither benefits nor is harmed. Cattle egrets are birds that follow cattle that are grazing. As the cattle graze, insects are disturbed and fly out of the grass; the cattle egrets eat these insects.
- **Mutualism:** Mutualistic relationships benefit both species. Oxpeckers are birds that eat the ticks and other parasites from the backs of rhinos and zebras. Oxpeckers benefit from this food source, and the rhinos and zebras benefit from the parasite removal.

These relationships among members of a community can be classified as positive (as in mutualism), negative (parasitism), or neutral (commensalism). However, some interactions can be classified in multiple categories. For example, mistletoe is a parasite on trees and negatively impacts the growth of the tree. Some bird species eat mistletoe berries, so mistletoe provides a benefit to those birds. Some mistletoe species also provide habitats for small animal species.

Biodiversity

Biodiversity refers to the variety of living organisms in an ecosystem. Ecosystems that have greater biodiversity are usually more resilient and adaptable to changes in their environment.

Biodiversity depends on both abiotic (nonliving) and biotic (living) factors. Abiotic factors, such as climate and water availability, will influence the types of species and the number of organisms of each species that are found in an ecosystem. Biotic factors, such as the number of producers, will limit how many consumers can survive in an ecosystem. The biodiversity of an ecosystem will influence the structure of the food chains and food webs found in that ecosystem.

Keystone species have a disproportionately large effect on an ecosystem compared to their numbers. When wolves were hunted to levels that neared extinction in Yellowstone National Park 100 years ago, deer populations in the park increased. This overpopulation of deer led to a decrease in the biodiversity in plant species. Other animal species that depended on these plant species declined in numbers. When wolves were reintroduced in the 1990s, wolves limited the deer populations. The populations of other animal species and plant species increased, increasing the biodiversity of the entire ecosystem. Wolves are a keystone species in the Yellowstone ecosystem. When a keystone species is removed from an ecosystem, the ecosystem may be in danger of collapsing.

Disruptions to ecosystems can change the structure of the ecosystem or even lead to its demise. **Invasive species** are species that are not native to a habitat. If an invasive species has no predators in a habitat, it can outcompete native species in the area, leading to the extinction of the native species. Cane toads are an invasive species that were introduced to Australia in the early 20th century in an effort to control the cane beetle, which was destroying sugarcane crops. Unfortunately, cane toads have no natural predator in Australia and grew in numbers, outcompeting the native species for food and habitats. In addition, cane toads are toxic when consumed by other animals. Larger animals that tried to eat cane toads died from the toxins the cane toads produced. Cane toads have led to the extinction of many native species in Australia.

Human impacts can also cause disruptions to ecosystems. Habitat destruction (as new cities are built) can lead to decreased biodiversity. When humans move into a previously uninhabited area, new diseases may be introduced into the ecosystem. Humans may also come into contact with previously unknown diseases when they move into new areas. Humans also generate pollution that can make water sources less habitable for other species. For example, overuse of chemical fertilizers can cause excess nitrogen and phosphorus to build up in nearby waterways. This excess of nitrogen and phosphorus can cause the overgrowth of algae and other organisms (**eutrophication**); this depletes the amount of oxygen in the waterway, making the water less habitable for other organisms and reducing the biodiversity of the waterway. The presence of pollutants can also result in **biomagnification**, the accumulation of toxic materials in organisms at higher trophic levels, which can have profound, deleterious effects.

Geological events, such as volcanic eruptions, can also disrupt ecosystems, leading to changes in biodiversity. Severe weather events, such as hurricanes, can decrease plant and animal diversity in an area. Prolonged droughts can change the biodiversity of an ecosystem.

These disruptions in ecosystems can lead to the evolution of populations. Environments will select for adaptations that provide an advantage in an environment. As the environment changes, the adaptations that are favored and selected for will also change. Adaptations are generated by mutations. Mutations are random events, but selection is not random. Rapid changes in the environment can accelerate the pace of evolution.

Practice Questions

Multiple-Choice

Questions 1 and 2

Refer to the figure, which shows the population of wild rabbits on an island as time progresses.

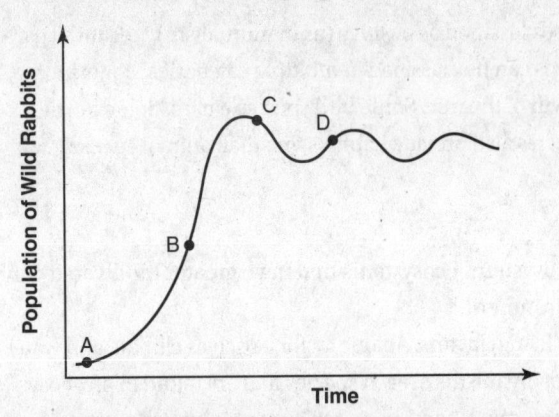

1. Which of the following best represents a time when the rabbit population was experiencing exponential growth?

    (A) A
    (B) B
    (C) C
    (D) D

2. Which of the following points best approximates the carrying capacity of the island for rabbits?

    (A) A
    (B) B
    (C) C
    (D) D

3. Wildlife management introduces 135 ducks to a reservoir in an effort to restore the duck population in the area. The growth rate of the population (r_{max}) is 0.074 ducks/year. Which of the following is the best estimate of the size of this duck population after one year?

    (A) 135
    (B) 141
    (C) 145
    (D) 152

4. In 1986, the quail population in Golden Gate Park in San Francisco was 140 birds. Hunting by feral cats reduced the number of quail to 32 birds by 2016. If the birth rate of quail over this period averaged 20.4 quail/year, what was the average death rate of quail during this period?

    (A) 2.4 quail/year
    (B) 3.6 quail/year
    (C) 24 quail/year
    (D) 36 quail/year

5. A population of 190 jackrabbits lives in McInnis Park in San Rafael, California. The carrying capacity of the park for jackrabbits is approximately 200 rabbits. If the r_{max} for the jackrabbit population is 1.5 surviving jackrabbits per year, what is the estimated increase in the number of jackrabbits in one year?

    (A) 10
    (B) 14
    (C) 285
    (D) 300

6. Which of the following describes a population that could experience exponential growth?

    (A) a population that was limited by density-dependent factors
    (B) a population whose size is beyond the carrying capacity of the environment
    (C) a population in which there are no limiting factors
    (D) a *K*-selected population

7. Acacia ants (*Pseudomyrmex ferruginea*) live on acacia trees (*Vachellia cornigera*). The ants obtain food and shelter from the acacia trees, and the ants drive away animals that would otherwise eat the leaves of the acacia trees. Which of the following best describes the relationship between *Pseudomyrmex ferruginea* and *Vachellia cornigera*?

    (A) niche partitioning
    (B) commensalism
    (C) mutualism
    (D) parasitism

8. Two species of lynx spiders, *Peucetia rubrolineata* and *Peucetia flava*, both inhabit the plant *Trichogoniopsis adenantha*, a type of sunflower found in Brazil. However, *Peucetia rubrolineata* inhabit plants in shaded areas, while *Peucetia flava* inhabit plants in open areas. This is an example of

   (A) competition.
   (B) mutualism.
   (C) niche partitioning.
   (D) predator/prey relationship.

9. *Cytisus scoparius*, also known as Scotch broom, was introduced into California from England in the mid-1800s as an easy-to-grow plant that helped stabilize soil on the hillsides. One Scotch broom plant can produce over 12,000 seeds a year. Scotch broom forms a dense shade canopy that prevents seedlings of other plants from growing. There are no natural predators for Scotch broom in California since its leaves are toxic to most animal species. Which of the following most accurately describes Scotch broom's role in the California ecosystem?

   (A) keystone species
   (B) invasive species
   (C) mutualistic species
   (D) native species

10. Tidal marshland A, which is adjacent to the San Francisco airport, has a Simpson's Diversity Index of 0.65. Tidal marshland B, which is two miles from the airport, has a Simpson's Diversity Index of 0.80. An oil spill occurs at a location at the halfway point between the two marshlands. If both marshlands are equally contaminated by the oil spill, which marshland is more likely to recover from the oil spill?

    (A) Tidal marshland A is more likely to recover, because its lower Simpson's Diversity Index means that few species will be affected.
    (B) Tidal marshland B is more likely to recover, because its higher Simpson's Diversity Index means it has a wider variety of species and can better absorb the effects of a disturbance.
    (C) Both ecosystems will be equally affected, because they are the same distance from the oil spill.
    (D) Neither ecosystem will be able to recover, because oil is toxic to all organisms.

Short Free-Response

11. A population of 45 geese are introduced to a lake. The birth rate of the population is 0.28 geese per year, and the death rate is 0.18 geese per year.

    Part A

    (i) **Calculate** the rate of population growth.

    Part B

    (i) **Determine** whether the population of geese is increasing or decreasing.

    Part C

    The carrying capacity for geese at this lake is about 80 geese.

    (i) **Make a prediction** about the population of geese relative to the lake's carrying capacity for geese in the next 10 years.

    Part D

    (i) **Justify** your prediction from Part C.

12. On an island north of the Arctic Circle, the Arctic fox preys upon puffins, and puffins consume Arctic grass.

    Part A

    (i) **Identify** the trophic levels to which the Arctic fox, puffin, and Arctic grass belong.

    Part B

    (i) **Explain** why the number of Arctic foxes on this island will most likely be less than the number of puffins in a healthy ecosystem.

    Part C

    (i) **Predict** the effect that the elimination of all Arctic foxes on this island would have on the puffins and Arctic grass.

    Part D

    (i) **Justify** your prediction from Part C.

Long Free-Response

13. Largemouth bass (*Micropterus salmoides*) eat dace fish (*Leuciscus leuciscus*). Dace fish eat daphnia. Daphnia eat algae in the lake. The number of largemouth bass and dace fish in a lake are recorded over a 10-year period. At year 5, great blue herons (*Ardea herodias*) are introduced to the lake. Great blue herons feed on young largemouth bass. The following graph shows the data regarding the relative population sizes of the largemouth bass and dace fish over the 10-year period.

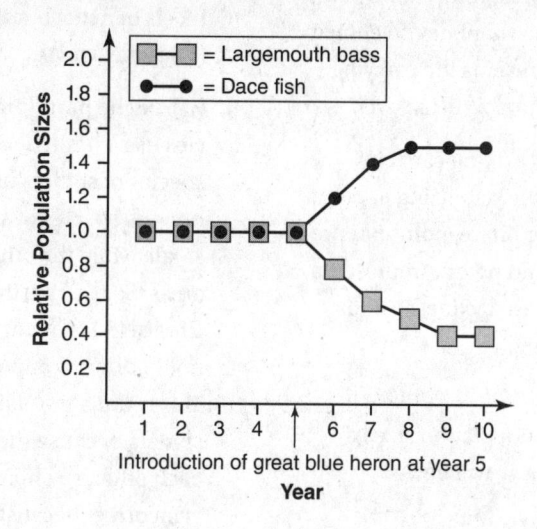

Part A

(i) **Describe** the trophic level of the dace fish in this ecosystem.

Part B

(i) **Identify** the independent variable in this experiment.

(ii) **Identify** the dependent variable in this experiment.

Part C

(i) Analyze the data to **determine** the effect of the introduction of great blue herons on the population sizes of the largemouth bass and the dace fish.

Part D

(i) **Predict** the effect of the introduction of great blue herons on the algae density of the lake.

(ii) **Justify** your prediction.

Answer Explanations

Multiple-Choice

1. **(B)** The most rapid population growth occurs during exponential growth, which is at point B. Choice (A) is incorrect because it represents growth during the lag phase. At point C, the population is starting to decline in numbers, so choice (C) is incorrect. Point D represents the carrying capacity, so choice (D) is also not the answer.

2. **(D)** Point D represents the carrying capacity because the population size is stabilizing at point D. Point A represents the lag phase, point B represents exponential growth, and point C represents a population that is starting to decline, so choices (A), (B), and (C) are incorrect.

3. **(C)** $\frac{dN}{dt} = r_{max}N = (0.074)(135) = 10$ ducks in one year. Since the starting population size is 135 ducks, after one year, there will be $135 + 10 = 145$ ducks.

4. **(C)** $\frac{dN}{dt} = B$ (birth rate) $- D$ (death rate). So the average death rate would be $D = B - \frac{dN}{dt}$. The death rate is
$$D = 20.4 \text{ quail/year} - \frac{(32-140) \text{ quail}}{30 \text{ years}} =$$
$20.4 - (-3.6) = 20.4 + 3.6 = 24$ quail/year.

5. **(B)** Since there is a carrying capacity in this environment, use this equation:
$$\frac{dN}{dt} = r_{max}N\left(\frac{K-N}{K}\right) = (1.5)(190)\left(\frac{200-190}{200}\right)$$
$= 14.25$. Choice (B) is the closest estimate to this number and is therefore the correct answer.

6. **(C)** Populations grow exponentially if there are no limiting resources or predators that would limit the population size. If the population was limited by density-dependent factors, its growth would be logistic, so choice (A) is incorrect. Choice (B) is incorrect because if the size of the population was beyond the carrying capacity of the environment, its numbers would be declining, not growing. K-selected populations experience logistic growth, so choice (D) is also incorrect.

7. **(C)** Since both the ants and the acacia trees benefit from the relationship, the symbiosis between them is mutualistic. Niche partitioning is using different parts of a resource, so choice (A) is incorrect. Choice (B) is incorrect because commensalism benefits one member of the symbiosis but the other member neither benefits nor is harmed. In parasitism, one member of the symbiosis benefits but the other is harmed, so choice (D) is incorrect.

8. **(C)** Niche partitioning occurs when different species use a limited resource differently. The two species of spiders use the same species of plant, but one species of spiders only uses plants in the shade while the other only uses plants in open areas. So they partition the resource differently. Choice (A) is incorrect because the spiders are not in direct competition for the same resource. This is not a mutualistic relationship between the spiders because they do not provide a benefit to each other, so choice (B) is incorrect. Choice (D) is incorrect because the spiders do not prey on each other.

9. **(B)** Scotch broom is not native to California. It outcompetes other plants and has no natural predators in California, so it is an invasive species. Choice (A) is incorrect because if a keystone species is removed, the entire ecosystem may collapse. Removing Scotch broom would most likely help the ecosystem, not cause it to collapse. Scotch broom does not benefit other organisms in this ecosystem; therefore, it is not a mutualistic species and choice (C) is incorrect. Choice (D) is incorrect because Scotch broom is not native to California.

10. **(B)** Ecosystems with more biodiversity are more likely to be able to recover from ecological disturbances. Choice (A) is incorrect because a lower Simpson's Diversity Index means less biodiversity and that ecosystem would be less likely to recover from a disturbance. The difference in the two marshlands' biodiversity will result in different abilities to recover from the disturbance, so choice (C) is incorrect. Choice (D) is incorrect because while oil is toxic to all organisms, some organisms may be able to recover from exposure to oil.

Short Free-Response

11. (A-i) The rate of population growth is $\frac{dN}{dt}$ = birth rate − death rate = 0.28 − 0.18 = 0.10 geese per year.

    (B-i) This population of geese is increasing because the birth rate is greater than the death rate.

    (C-i) It is expected that the geese population will exceed the carrying capacity of the lake within the 10-year period.

    (D-i) The growth rate of the geese population is 0.10, or 10%. If that growth rate continues over the next 10 years, the population of geese will exceed the lake's carrying capacity before the end of the 10-year period.

12. (A-i) The Arctic fox is a secondary consumer. The puffin is a primary consumer. The Arctic grass is a producer.

    (B-i) On average, only about 10% of the energy is available to the next trophic level. So the number of puffins must exceed the number of Arctic foxes or else the Arctic foxes would not be able to meet their energy needs.

    (C-i) The elimination of all Arctic foxes would increase the number of puffins and decrease the density of the Arctic grass population.

    (D-i) Since Arctic foxes prey on puffins, the elimination of the Arctic foxes would increase the number of puffins. Since puffins consume the Arctic grass, this increase in puffins would decrease the density of the Arctic grass on the island.

Long Free-Response

13. (A-i) The dace fish are secondary consumers because they eat daphnia, which are primary consumers (that eat algae, which are the producers). Dace fish are also the prey of the largemouth bass.

    (B-i) The independent variable is the introduction of the great blue herons.

    (B-ii) The dependent variable is the relative population sizes of the largemouth bass and the dace fish.

    (C-i) The introduction of the great blue herons reduced the relative population size of the largemouth bass by about 40% and increased the relative population size of the dace fish by about 40%. This is because the great blue herons fed on the largemouth bass, reducing their numbers. As a result of this decrease in largemouth bass, the number of dace fish increased since there were no longer as many predators (largemouth bass) to contend with.

    (D-i and ii) The algae density in the lake would increase. This is because algae are consumed by daphnia and daphnia are consumed by dace fish. If the numbers of dace fish increase (following the introduction of the great blue herons), more daphnia will be consumed and there will be fewer daphnia to consume the algae. Thus, the algae density in the lake will increase.

Lab Review

24
Labs

Learning Objectives

In this chapter, you will learn:

→ 13 Common Lab Experiments that Cover All Units of AP Biology

Overview

The AP Biology course is designed to help you develop skills in six core science practices:

1. Explaining biological concepts
2. Analyzing visual representations
3. Asking questions and designing experiments to test those questions
4. Representing and accurately describing data
5. Using statistical tests to analyze data
6. Using evidence to develop arguments that support or reject claims and hypotheses

These core science practices are woven throughout the course content and throughout the course labs. Designing an experiment and analyzing the experimental data are key skills assessed on the AP Biology exam.

On Section II of the exam, each of the long free-response questions (#1 and #2) will likely involve analyses of lab experiments. Question 1 may assess your ability to interpret and evaluate experimental results. You might be required to describe and explain a biological process or model and identify design procedures from the experiment. You might also need to analyze data and make and justify predictions. Question 2 may ask you to construct a graph of experimental results using confidence intervals or error bars. You might again need to show that you can analyze the data and predict what could happen based on the data provided. One of the four short free-response questions might also revolve around a lab investigation, requiring you to describe the biological concept that is being illustrated by the experiment, identify the controls or variables, predict the results of the investigation, and justify your predictions.

The more opportunities you have to practice these skills, the better prepared you will be for the AP Biology exam. At least eight inquiry-based lab experiences are required in AP Biology, but no specific labs are required for the AP Biology course. This chapter describes the 13 common labs that you should be familiar with.

> These labs are not requirements for the course, so do not worry if your teacher did not go over these specific labs. Questions about the specifics of these procedures are not typically tested on the AP Biology exam. The *science practice skills you develop* when you complete lab experiments are what will help you succeed on the AP Biology exam.

The skills assessed by these labs are skills you should practice before you take the AP Biology exam. At the end of the AP Biology course, you should be able to do the following:

- Identify the key components of an experimental design (controls, independent variable(s), dependent variable(s), etc.)
- Describe how to collect data
- Graph data appropriately with 95% confidence intervals
- Use data to develop conclusions
- Apply conclusions to larger biological concepts
- Predict how changes to an experiment may affect the results
- Justify your predictions

Lab 1: Artificial Selection

One of the key driving forces in evolution is selection. Natural selection is the process by which organisms that are better adapted to their environment survive and reproduce at a greater rate. In artificial selection, humans choose organisms with desirable traits, and only these organisms are allowed to survive and reproduce.

While many different organisms can be used to study artificial selection, Fast Plants (*Brassica rapa*) are an ideal model organism for studying evolution in the classroom because they are easy to grow and have a short generation time (about 40 days). Also, since Fast Plants do not self-pollinate, the experimenter can control which plants pollinate each other to produce offspring, which allows for artificial selection. In this experiment, the expression of a quantifiable trait in a parental generation and in the F1 generation is compared after artificial selection.

Any quantifiable trait can be used in this experiment. One of the most common traits used is the number of trichomes on plant leaves. Trichomes are short, hairlike appendages on the margins, or edges, of the leaves. Trichomes deter predation of the plant by herbivores. The number of trichomes per leaf can vary greatly in a population of Fast Plants, as shown in Figure 24.1.

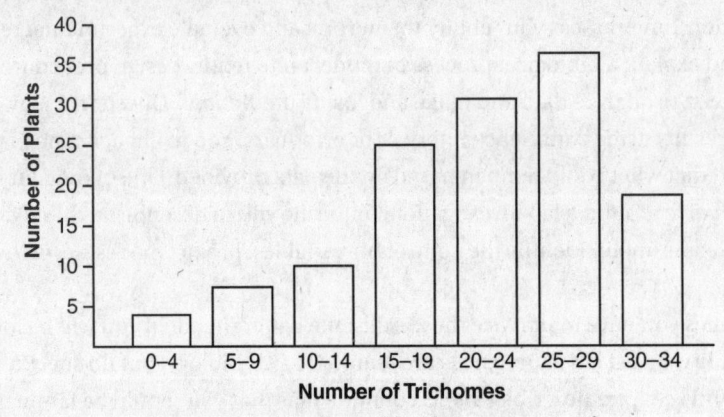

Figure 24.1 Number of Trichomes per Leaf

Imagine that a class decides to artificially select for plants with the most trichomes. Fast Plants are grown from stock seeds (parental generation), which reflect normal variation in the population. On day 11 of the plants' life cycle, the students count the number of trichomes per leaf. The students record this data and use a spreadsheet to calculate the mean and the standard error of the mean of the data. When the plants flower, students cross-pollinate only the plants that ranked in the top 10% of the number of trichomes per leaf measurement that was taken on day 11. These plants then develop seeds, and the seeds are harvested. These seeds are planted (to create the F1 generation) and grown to day 11 of their life cycle, at which time the students count the number of trichomes per leaf. The students record this data in a spreadsheet, which calculates the mean and the standard error of the mean of the data set. The number of trichomes per leaf on day 11 from both the parental generation and the F1 generation are plotted on a bar graph, which includes 95% confidence intervals, as shown in Figure 24.2.

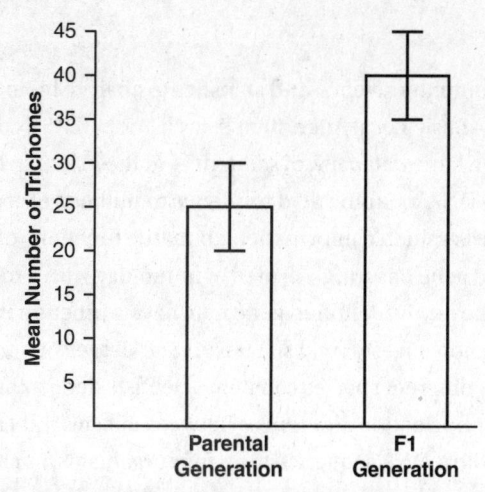

Figure 24.2 Mean Number of Trichomes in Parental and F1 Generations

If the 95% confidence intervals of the parental generation and the F1 generation do not overlap, it is likely there is a statistically significant difference in the number of trichomes per leaf between the parental and F1 generations. If the 95% confidence intervals do overlap, the data are inconclusive, and it is not possible to say whether or not there is a statistically significant difference between the two groups.

Lab 2: Hardy-Weinberg

If the allele frequencies in a population are changing, the population is likely evolving. Evolution can be studied in a population by using mathematical models or computer simulations to observe changing allele frequencies in populations.

Hardy-Weinberg equilibrium and the Hardy-Weinberg equations describe the characteristics of a population that is not evolving and has stable allele frequencies. A population is in Hardy-Weinberg equilibrium if the following five conditions are met:

1. Large population size
2. Random mating
3. No gene flow
4. No selection of any kind
5. No mutations

If a population is in Hardy-Weinberg equilibrium, the following equations apply:

$$p + q = 1$$
$$p^2 + 2pq + q^2 = 1$$

where p represents the frequency of the dominant allele in the gene pool of the population and q represents the frequency of the recessive allele in the population. p^2 represents the frequency of the homozygous dominant genotype, $2pq$ represents the frequency of the heterozygous genotype, and q^2 represents the frequency of the homozygous recessive genotype.

One way to study Hardy-Weinberg equilibrium is to build a spreadsheet to model how a gene pool would change in a hypothetical population from one generation to the next. Another way to study this would be to use one of the widely available computer simulations online. Adjusting the parameters of a spreadsheet or online simulation allows for the exploration of the effects of selection, mutation, migration, genetic drift, or heterozygote advantage on allele frequencies in the gene pool of a population.

Lab 3: BLAST

Bioinformatics uses the power of computer science and statistics to analyze biological data. One of many bioinformatics tools available is **BLAST**—Basic Local Alignment Search Tool. BLAST compares molecular data on DNA and amino acid sequences to a large databank of sequences at the National Institutes of Health (NIH). The BLAST algorithm can compare any DNA or amino acid sequence to millions of sequences in the databank in just a few seconds. This can give scientists valuable information about the functions of similar genes. For example, if a DNA sequence of a newly identified gene has a high degree of homology with a mouse gene that is involved in cell signaling, this could indicate that the newly identified gene may have a function in cell signaling.

Species with a more recent common ancestor will share a greater degree of homology in their genes. Using BLAST to compare genes shared by different species can help scientists determine the degree of relatedness and construct phylogenetic trees. Fewer nucleotide differences between species indicate a more recent common ancestor. For example, compare a short DNA sequence from four organisms, as shown in the following table.

Position			3					9						13				18		
Species I	A	G	C	G	C	T	T	A	A	C	C	G	A	A	T	T	A	T	C	G	G
Species II	A	G	C	G	C	T	T	A	G	C	C	G	A	A	T	T	A	C	C	G	G
Species III	A	G	C	G	C	T	T	A	A	C	C	G	T	A	T	T	A	T	C	G	G
Species IV	A	G	G	G	C	T	T	A	G	C	C	G	T	A	T	T	A	T	C	G	G

Comparing these sequences shows that species I and II have two nucleotide differences, species I and III have one nucleotide difference, and species I and IV have three nucleotide differences. A possible phylogenetic tree based on this data might look like Figure 24.3.

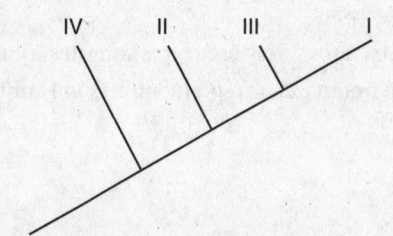

Figure 24.3 Phylogenetic Tree for Species I, II, III, and IV

In this lab, you might first be asked to develop a hypothesis about the placement of a fossil on a phylogenetic tree based on its morphology. Then four DNA sequences (like the ones in the table above) that are isolated from a fossil are uploaded to BLAST to determine which species have DNA with the greatest degree of homology to the DNA sequences isolated from the fossil. This additional molecular data is then used to evaluate the original hypothesis about the placement of the fossil on the phylogenetic tree, and if supported by this molecular data, revisions to the hypothesis are made.

Lab 4: Diffusion and Osmosis

Just as a ball on a hill will move from an area of higher potential energy to an area of lower potential energy, water will move from an area of higher water potential to an area of lower water potential. **Water potential** can be defined as the potential energy of water in a solution, or the ability of water to do work. The more water there is in a solution, the higher its water potential will be. The less water there is in a solution, the lower its water potential will be. Water potential measures the tendency of water to move from one place to another. Water potential is a result of the effects of both pressure potential (Ψ_p) and solute potential (Ψ_s), as shown in the following equation:

$$\Psi = \Psi_p + \Psi_s$$

Most living systems are open to the atmosphere and are in pressure equilibrium with their environment. In these cases, the pressure potential is zero, and the total water potential is due solely to the solute potential.

As solute is added to a solution, the relative amount of water in the solution decreases. As the amount of solute in a solution increases, the solute potential, and consequently the total water potential, decreases. Further details on how solute potential is calculated are discussed in Chapter 6.

This investigation can be divided into multiple parts, as in the following scenario.

Part 1: Artificial cells are made from agar or gelatin cubes impregnated with an indicator dye that can be used to track the movement of a solution into the artificial cell. Cubes of three different sizes, with three different surface area to volume ratios, are used, as described in the table.

Length (L) of One Side of the Block (cm)	Surface Area in cm^2 (= $6 \times L^2$)	Volume in cm^3 (= L^3)	Surface Area to Volume Ratio (SA/V)
1	6	1	6
2	24	8	3
3	54	27	2

The cubes are placed into a solution, and the dye in the cubes is used to track the movement of the solution into the cells. This part of the lab demonstrates that smaller cells, with their higher surface area to volume ratios, are more efficient at taking in materials from their environment.

Part 2: In this part of the investigation, dialysis tubing is used as a model for the semipermeable cell membrane. A solution of glucose and starch is placed into dialysis tubing and is tied off at both ends to create a model cell. This model is placed into a beaker of water to which iodine has been added. The dialysis tubing that is used has pores through which relatively small molecules can pass but much larger molecules cannot.

When iodine comes into contact with starch, the iodine-starch complex turns deep blue or blue-black in color, allowing one to monitor the movement of iodine and starch. The movement of glucose can be monitored in a number of ways, including the use of the Benedict's assay for glucose or simple glucose test strips.

The experimental results, as shown in Figure 24.4, indicate that while water, glucose, and iodine can pass through the pores in the semipermeable dialysis tubing, the starch molecules cannot.

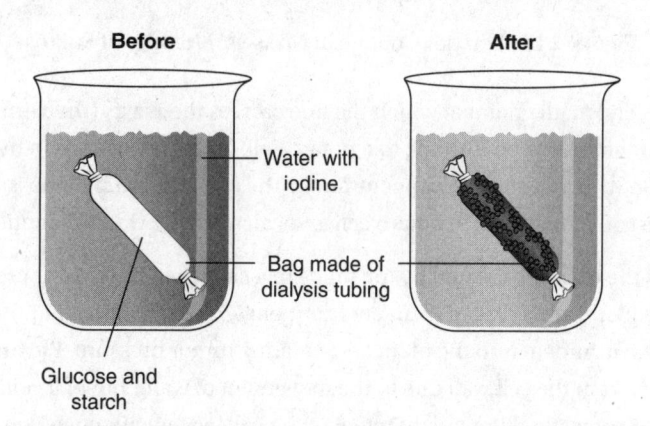

Figure 24.4 Diffusion Across a Semipermeable Membrane

Part 3: In this part of the investigation, dialysis tubing is again used to create a model of a cell with a semipermeable membrane. Each of six pieces of dialysis tubing are filled with one of six different unknown solutions with various sucrose concentrations (0.0, 0.2, 0.4, 0.6, 0.8, and 1.0 molar). The sucrose molecule is approximately twice as large as a glucose molecule and is too large to pass through the pores in the dialysis tubing. Each model cell is massed and then placed in a beaker of distilled water for at least 30 minutes. Then they are massed again, and the percent change in mass for each bag is calculated. Water will move into the dialysis bags. Bags that contain a solution with the highest sucrose concentration (lowest water potential) will gain more mass and have a greater percent change in mass during the allotted time than bags that contain a solution with a lower sucrose concentration (and a relatively higher water potential). By comparing the percent changes in mass of each bag, the concentration of sucrose in each bag can be identified.

Part 4: In this part of the investigation, the movement of water in plant cells is measured. Cubes of potato (or any other root vegetable) of identical sizes are made and then massed. Each cube is placed in a solution with a different concentration of sucrose (0.0, 0.2, 0.4, 0.6, 0.8, and 1.0 molar). Plant cells have a cell wall in addition to a cell membrane, so these cubes are placed in the solutions at least overnight to allow for movement of water between the plant cells and their environment. Then each cube is removed from the solution and massed, and the percent change in mass is calculated. Cubes with a higher water potential than the surrounding solution are expected to lose water to their environment and therefore have a negative percent change in mass. Cubes with a lower water potential than the surrounding solution are expected to gain water from their environment and therefore have a positive percent change in mass. A graph is made plotting the molarity of each solution on the *x*-axis and the percent change in mass on the *y*-axis, as shown in Figure 24.5.

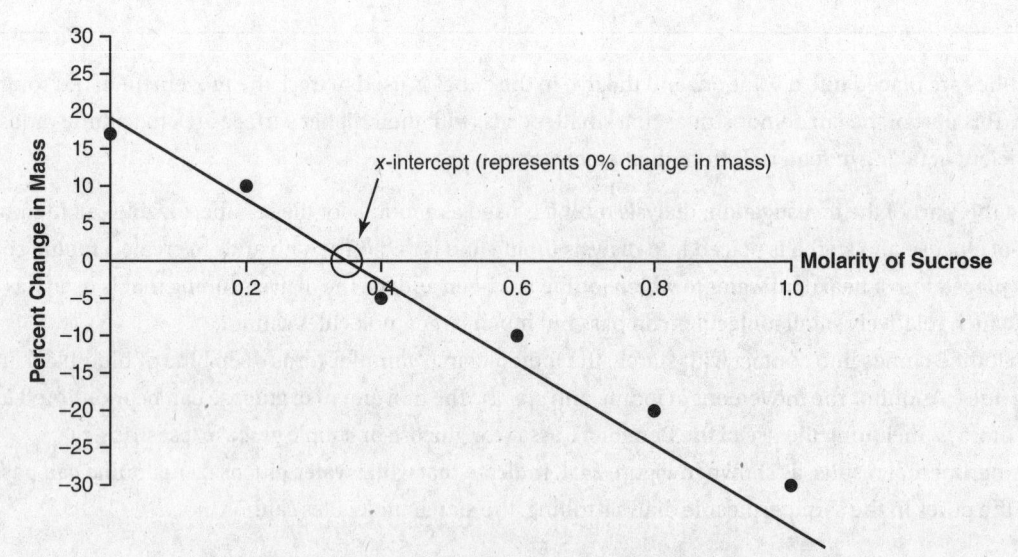

Figure 24.5 Percent Change in Mass vs. Molarity of Sucrose

When a line of best fit is drawn, the point at which the line crosses the *x*-axis (the *x*-intercept) represents the concentration of a solution that would be isotonic to a potato cell, with a 0% change in mass expected if a potato cell was placed in a solution with that sucrose concentration. The concentration of the isotonic solution can be used to estimate the solute concentration of a potato cell and calculate the water potential of a potato cell.

Part 5: Turgor pressure is the pressure created by the fluids in a cell when the fluid is pressing against the cell membrane or cell wall. If a plant cell is placed in an environment with a higher water potential than that of the cell, water will move from the environment into the plant cell, creating turgor pressure. **Plasmolysis** is the shrinking of the cytoplasm of a cell away from the cell wall due to the movement of water out of a cell. If a plant cell is placed in an environment with a lower water potential than that of the cell, water will move from the higher water potential in the cell to the lower water potential in its surroundings, and plasmolysis will occur. If a plant is placed in an environment with a lower water potential (for example, if you forget to water your plant), plasmolysis will occur and the plant will wilt.

In this part of the investigation, plant cells are placed in an environment with a lower water potential than that of the plant cell and in an environment with a higher water potential than that of the plant cell. Observations of changes in the cytoplasm are made under a light microscope.

Lab 5: Photosynthesis

Life on Earth depends on photosynthesis. Photosynthetic organisms capture light energy and store it in organic compounds, which are consumed by other organisms. The overall equation for photosynthesis is the following:

$$6CO_2 + 6H_2O \xrightarrow{\text{light energy}} C_6H_{12}O_6 + 6O_2$$

The rate of photosynthesis can be measured by the rate of consumption of the reactants (CO_2 or H_2O) or by the rate of production of the products ($C_6H_{12}O_6$ or O_2).

In this lab investigation, the floating leaf disk technique is used to measure the production of oxygen during photosynthesis. A hole punch or sturdy straw is used to create 10 equal-sized disks from a leaf. Spinach works well for this because it has a high concentration of chlorophyll and is a relatively soft leaf with no waxy cuticle that could impede the movement of materials into or out of the leaf. The disks are placed in a syringe that contains a 1% sodium bicarbonate solution. The bicarbonate enriches the carbon dioxide content of the solution.

The syringe is capped, and the plunger of the syringe is pulled out about 1 or 2 cm to create a vacuum in the syringe. The interior of a leaf contains air spaces between the chloroplast-containing mesophyll cells. When a vacuum is formed in the syringe, the air in the spaces between the mesophyll cells is removed and replaced with the sodium bicarbonate solution. When enough of the air has been replaced with the sodium bicarbonate solution, the density of the leaf disk increases and it sinks.

The sunken leaf disks are transferred to a beaker that contains sodium bicarbonate solution. When exposed to a light source, the leaf disks perform photosynthesis: oxygen is produced and replaces the sodium bicarbonate solution in the spaces between the mesophyll cells. When enough oxygen has been produced, displacing the sodium bicarbonate solution inside the leaf, the leaf disks float. At one-minute intervals, the number of floating leaf disks is recorded. The time needed until 50% of the leaf disks float can be used as an indirect assay of the rate of photosynthesis. A sample data table and a graph of the data (Figure 24.6) follow.

Time (minutes)	Number of Floating Leaf Disks
0	0
1	0
2	0
3	0
4	0
5	2
6	2
7	4
8	5
9	6
10	7

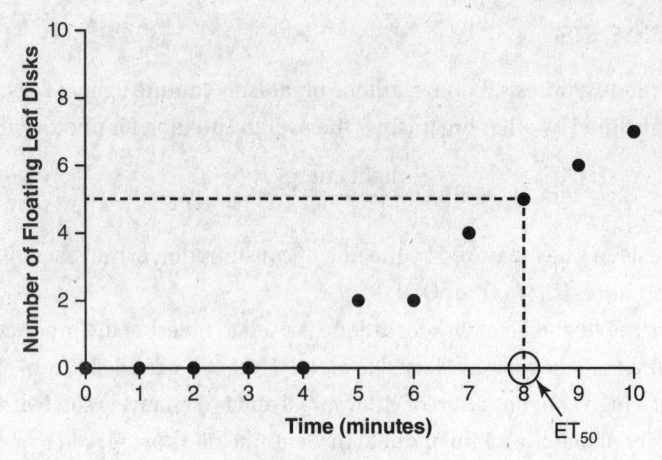

Figure 24.6 Number of Floating Leaf Disks vs. Time (Minutes)

The conditions of the experiment can be varied to investigate the effects of different independent variables on the time required for the leaf disks to float. The time required for 50% of the leaf disks to float (ET_{50}) under each condition can be compared to learn how different conditions affect the rate of photosynthesis.

Lab 6: Cellular Respiration

The energy stored in the chemical bonds of glucose during the process of photosynthesis is released by the process of cellular respiration. The overall equation for cellular respiration is the following:

$$C_6H_{12}O_6 + 6O_2 \rightarrow 6CO_2 + 6H_2O + ATP$$

The rate of cellular respiration can be measured by the rate of consumption of the reactants (glucose and oxygen) or the rate of production of the products (carbon dioxide and water).

A **respirometer** is a sealed chamber with an attached measuring pipette that is used to measure the change in volume due to the respiration of the organism. A small respiring organism (germinating seeds, fruit flies, and crickets are commonly used) is placed in a respirometer. Glass or plastic beads of roughly equal volumes to the respiring organisms are placed in a second respirometer as a control.

At room temperature, glucose and water are in the solid and liquid states, respectively, so their contributions to any change in volume in the respiring system is minimal. As oxygen gas is consumed during respiration, carbon dioxide is produced, balancing out any potential change in the volume of the gases present. However, potassium hydroxide (KOH) solution reacts with carbon dioxide gas to form a solid potassium carbonate salt, as shown in the following equation:

$$2KOH_{(aq)} + CO_{2(g)} \rightarrow K_2CO_{3(s)} + H_2O_{(l)}$$

If a few drops of potassium hydroxide solution are added to the respirometer, any carbon dioxide produced will be precipitated into the potassium carbonate salt. The result is that any change in volume in the gases in the respirometer is the result of the consumption of oxygen gas. Thus, this allows the rate of cellular respiration to be assayed.

The respirometers can be placed in environments with different temperatures to observe the effect of temperature on the rate of respiration. Sample data is shown in the table and in Figure 24.7.

Oxygen Consumption (mL)

Time (minutes)	Control at 4°C	Non-Germinating Seeds at 4°C	Germinating Seeds at 4°C	Control at 25°C	Non-Germinating Seeds at 25°C	Germinating Seeds at 25°C
0	0	0	0	0	0	0
10	0	0	0.05	0	0	0.25
20	0	0	0.10	0	0.05	0.55
30	0	0	0.15	0	0.10	0.90

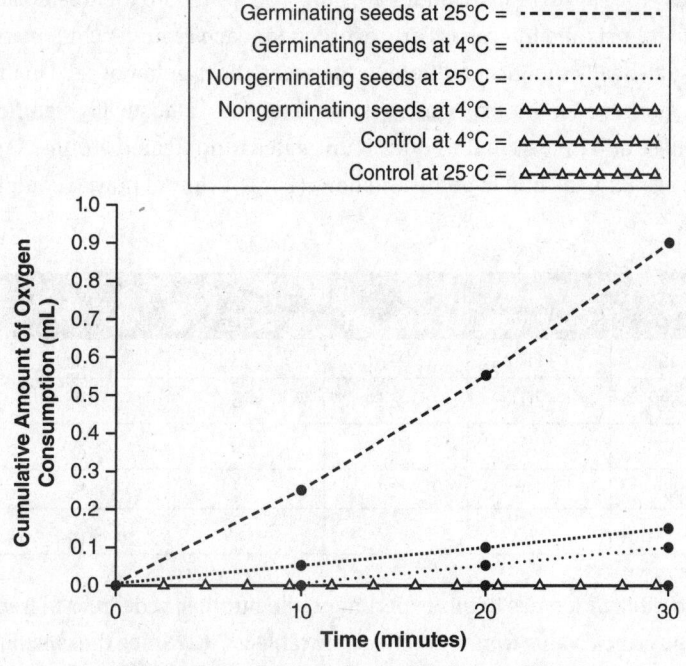

Figure 24.7 Oxygen Consumption vs. Time

Lab 7: Mitosis and Meiosis

Mitosis is the process of cell division in somatic (body) cells that produces genetically identical daughter cells. Mitosis is important in asexual reproduction, growth, and tissue repair. Meiosis is the process used in sexually reproducing organisms to produce haploid gametes. Meiosis ensures that the number of chromosomes remains consistent from generation to generation while also generating genetic diversity through genetic recombination or crossing-over. The roles of mitosis and meiosis in a living organism's life cycle are represented in Figure 24.8.

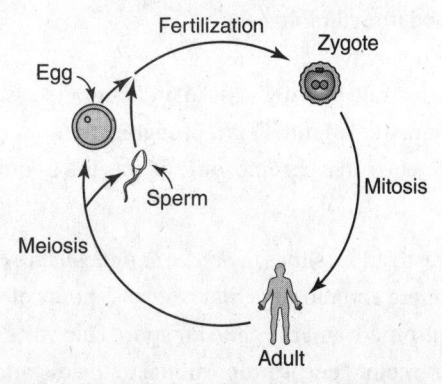

Figure 24.8 Roles of Mitosis and Meiosis

As in Lab 4, this investigation can be divided into multiple parts, as in the following scenario.

Part 1: In this part of the investigation, the cell cycle and mitosis can be modeled using beads, pipe cleaners, clay, or a variety of other tools. After modeling the cell cycle and mitosis, you should have a better understanding of:

- The part of the cell cycle during which the genetic information of a cell is duplicated
- Why chromosomes condense before they are separated
- How the sister chromatids in a chromosome are separated and the result if a pair of sister chromatids fail to separate

Part 2: In this part of the lab, cells are treated with a chemical and are compared to untreated cells to find out whether the chemical alters the rate of mitosis. Slides are prepared from both the treated and untreated groups, and the number of cells in mitosis and interphase are recorded for each group. A chi-square test is used to determine whether there is a statistically significant difference between the number of cells in mitosis and interphase in each group. The null hypothesis for this experiment is that there is no statistically significant difference in the number of cells in interphase and mitosis in the control (untreated) and treated groups. Sample data and the chi-square calculation to evaluate the null hypothesis follow. (For a review of how to complete a chi-square calculation, review Chapter 2).

Group	Observed	Expected	$o - e$	$(o - e)^2$	$\frac{(o - e)^2}{e}$
Control Interphase	110	107	3	9	0.084
Control Mitosis	20	23	−3	9	0.391
Treated Interphase	120	123	−3	9	0.073
Treated Mitosis	30	27	3	9	0.333
				Chi-Square =	0.881

Since there are four possible outcomes in this experiment, the number of degrees of freedom (df) is $4 - 1 = 3$. Using a *p*-value of 0.05, the critical value from the chi-square table is 7.81. Since the calculated chi-square value of 0.881 is less than the critical value of 7.81, fail to reject the null hypothesis.

Part 3: In this part of the lab, karyotypes from normal and cancerous cells are compared. A **karyotype** is a picture of the chromosomes in a dividing cell. Karyotypes can show abnormalities in the number of chromosomes in a cell caused by errors in cell division. Recall that control of cell division is monitored at checkpoints. These checkpoints monitor for proper ratios of cyclins and cyclin-dependent kinases and for accurate DNA replication to prevent errors in cell division. Factors that regulate cell division are discussed further in Chapter 11.

Part 4: In this part of the investigation, the process of meiosis is modeled using beads, pipe cleaners, clay, or a variety of other tools. After modeling the process of meiosis, you should have a better understanding of:

- How haploid gametes are formed from diploid cells
- How meiosis generates genetic diversity
- The events in meiosis that demonstrate Mendel's law of independent assortment
- The events in meiosis that demonstrate Mendel's law of segregation
- How the distance between two genes on a chromosome affects the frequency of genetic recombination between the two genes

Part 5: In the final part of this lab investigation, slides of *Sordaria fimicola* are observed. The frequency of recombination events (between the centromere and the gene that controls spore color in *Sordaria*) is used to calculate the genetic distance between the centromere and the gene for spore color. The percentage of recombination events between the centromere and the spore color gene is proportional to the distance between them.

Lab 8: Bacterial Transformation

Biotechnology has transformed medicine by making medications (such as insulin) more readily available and creating personalized therapies to treat some cancers and genetic disorders. Many of these medical advances involve inserting new DNA into cells. One technique for inserting DNA into cells is **transformation**, the uptake and expression of foreign DNA by a cell. In this lab, a gene carried on a **plasmid**, a small circular piece of DNA, is inserted into bacterial cells. Heat shock is the procedure used to insert the plasmid into the bacterial cells.

Plasmids in nature often contain genes for antibiotic resistance. These genes for antibiotic resistance can be used as selectable markers to detect which bacteria have absorbed the plasmid. For example, say you perform a bacterial transformation procedure using a plasmid that contains a gene for resistance to the antibiotic ampicillin. If you grow the bacteria used on agar plates that contain the antibiotic ampicillin, only bacteria that have absorbed the plasmid will grow on the agar plates.

Heat shock was one of the first techniques used to insert foreign DNA into cells and is a key step in this investigation. In this lab, bacterial cells are placed in a test tube with a transformation solution and the plasmid DNA. This tube is then incubated on ice, exposed to a brief heat shock in a water bath, and then returned to the ice. The cell membrane of a bacterial cell is composed of a phospholipid bilayer. When the cell experiences a heat shock, temporary microscopic "cracks" are formed in the cell membrane through which the plasmid DNA may enter. Not all bacteria will take in the plasmid, but manipulating the heat shock under different conditions can lead to higher or lower numbers of plasmids entering the cells.

If the bacteria are exposed to a heat shock for too long, the cells may lyse, and no transformed bacteria will result. If the amount of time of the heat shock is too short, the plasmids may not be able to enter the bacteria cells. So it is important to have appropriate controls for this investigation so that one can ascertain which bacteria were successfully transformed and to confirm that twhe heat shock procedure did not kill the bacteria. The following table and Figure 24.9 show the components of a typical bacterial transformation experiment (that used a plasmid that contains a gene for ampicillin resistance) and the expected results.

Test Tube	Plasmid Present?	Agar Plate Components	Growth Detected?
A (control)	No	LB agar only	Yes, a "lawn" of bacteria, where the number of colonies are too numerous to count (TNTC)
B (control)	No	LB agar with ampicillin	No colonies
C	Yes	LB agar only	Yes, a "lawn" of bacteria, where the number of colonies are too numerous to count (TNTC)
D	Yes	LB agar with ampicillin	Yes, colonies detected, but not as many as on plate C

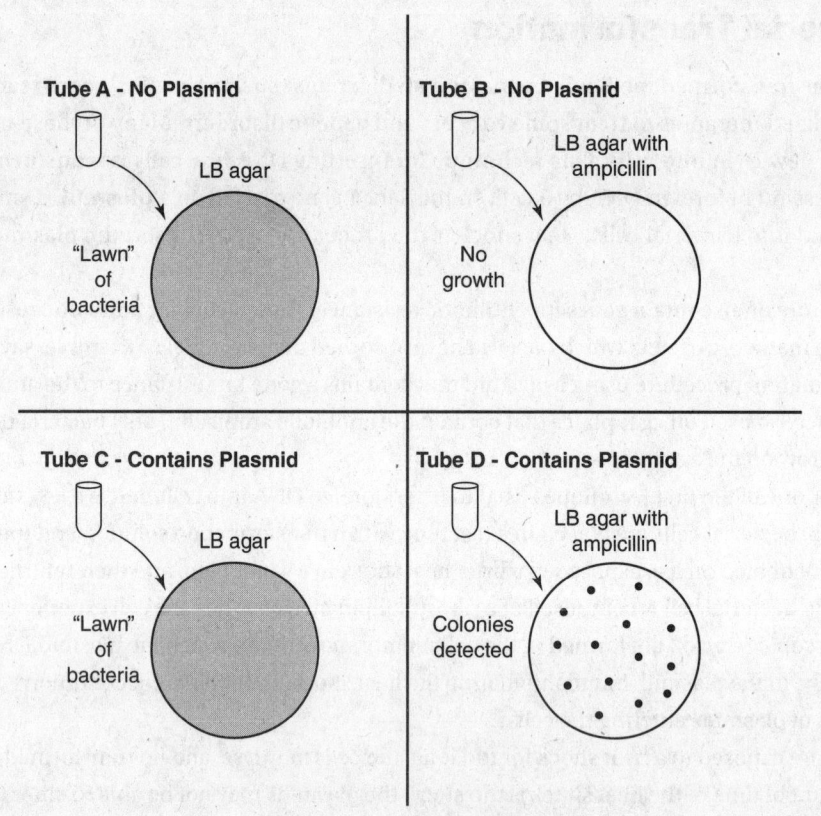

Figure 24.9 Expected Results

- Test tube A does not contain the plasmid. Its contents are grown on LB agar, upon which both bacteria with or without the plasmid will grow, so it is expected that there will be robust growth of bacteria on this plate.
- Test tube B does not contain the plasmid. Its contents are grown on LB agar to which ampicillin has been added. Ampicillin will kill any bacteria that have not absorbed the plasmid and therefore are not resistant to ampicillin. It is expected that no bacteria will be seen on this plate. If a colony of bacteria is seen on this plate, it could indicate that there was a very rare spontaneous mutation that conferred ampicillin resistance to those bacteria, but that is a highly improbable, though not impossible, event.
- Test tube C does contain the plasmid. Its contents are grown on LB agar, upon which both bacteria with or without the plasmid will grow, so it is expected that there will be robust growth of bacteria on this plate.
- Test tube D does contain the plasmid. Its contents are grown on LB agar to which ampicillin has been added. Ampicillin will kill any bacteria that have not absorbed the plasmid and therefore are not resistant to ampicillin. It is expected that any colonies seen on this plate will contain the plasmid with the resistance gene and will be considered "transformed."

Lab 9: Restriction Enzyme Analysis of DNA

The genetic code is shared by all living organisms, and it is a powerful piece of evidence for the common ancestry of all forms of life on Earth. The genetic code is also redundant, meaning multiple codons can code for the same amino acids during translation. Because of the redundancy of the genetic code, individuals may have the same proteins with the same amino acid sequences but with different DNA sequences. **Restriction enzymes** cut DNA at specific DNA sequences. As the restriction enzyme cuts DNA, differences in DNA sequences result in differently sized fragments of DNA from different individuals. These DNA fragments of different sizes can be separated by gel electrophoresis.

In this lab, a scenario might involve a description of a crime scene and suspects. DNA from evidence at the crime scene is isolated and cut with restriction enzymes. DNA is isolated from the suspects and cut with the same restriction enzymes. The resulting DNA fragments are placed on an agarose gel and are separated by gel electrophoresis. The sugar-phosphate backbone of DNA has a slightly negative charge and is attracted to the positive charge of the cathode in the gel electrophoresis apparatus. Smaller fragments of DNA will travel farther in the agarose gel than longer fragments. The gel is stained with a dye that allows visualization of the DNA fragments, as shown in Figure 24.10.

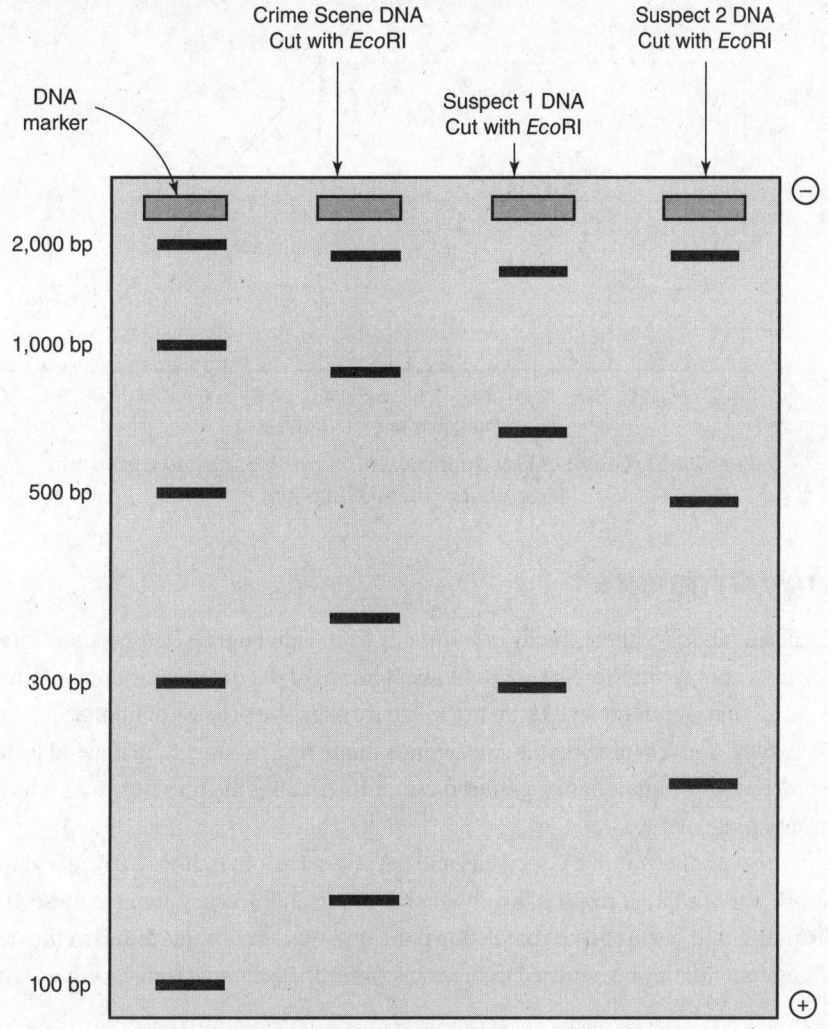

Figure 24.10 DNA Fragments on Electrophoresis Gel

If the pattern of DNA fragments from the suspect does not match the pattern from the DNA found in evidence at the crime scene, it is unlikely the individual was at the crime scene. However, if the pattern of DNA fragments does match, it does not prove guilt; it just shows that there is a possibility that the suspect was at the crime scene.

If a DNA marker that contains fragments of known sizes is run along with the DNA samples, the size of the DNA fragments in the other samples can be estimated. The distance migrated by the DNA marker fragments of known sizes are plotted on the x-axis of a graph, and the sizes of these known DNA marker fragments are plotted on the y-axis. A line of best fit is drawn between these points to create a standard curve (as shown by the solid line in Figure 24.11). Using the standard curve shown in Figure 24.11, if a fragment of unknown size migrated 4.0 cm in the gel, then interpolated to the line of best fit, the size of the fragment can be estimated from the standard curve to be about 420 base pairs (shown by the dotted line in Figure 24.11).

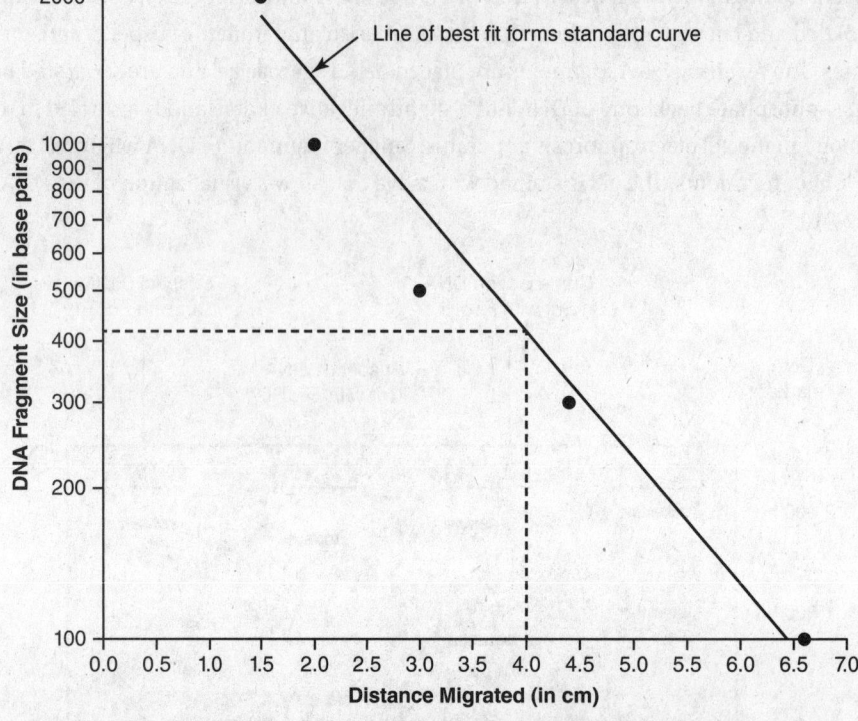

Figure 24.11 Graph of Electrophoresis Data with Standard Curve and Interpolation of an Unknown

Lab 10: Energy Dynamics

All life on Earth obtains its energy either directly or indirectly from light energy. Photosynthetic organisms (producers) capture light energy and store it in the chemical bonds of the organic products of photosynthesis. The total amount of organic compounds created by an organism through the process of photosynthesis is called the gross primary productivity. Some of the organic compounds made by a producer are needed by the organism to power its own life processes. The amount of organic compounds left after the producer meets its own energy needs is called its net primary productivity.

Primary consumers eat photosynthetic organisms and get energy through the breakdown of the organic compounds by the process of cellular respiration. Other consumers in the ecosystem eat these primary consumers. Eventually, organisms die, and decomposers break down the organic compounds found in the dead organisms so that the nutrients they contain may be returned to the ecosystem. These energy transfers in ecosystems are called **energy dynamics**.

In this lab, Fast Plant (*Brassica rapa*) seeds are grown in a sealed ecosystem. As the Fast Plants grow, eggs from the cabbage white butterfly (*Pieris rapae*) are introduced and allowed to hatch. The resulting larvae consume the leaves of the Fast Plants (which are a member of the cabbage family). The larvae grow to maturity. Then the contents of the ecosystem are removed and dried, and the dry weights of each component are recorded. Conversion factors for each component are used to calculate the energy stored in each trophic level. For example, Fast Plants contain about 4.35 kilocalories of energy per gram of dry weight, and butterfly larvae contain about 5.5 kilocalories of energy per gram of dry weight. The droppings (frass) of the butterfly larvae contain about 4.76 kilocalories of energy per gram of dry weight. Using these conversions, the amount of energy available from each component can be calculated, as shown in the table.

Organism	Dry Weight (grams)	Energy per Gram of Dry Weight (kilocalories per gram)	Energy Contained (kilocalories)
Fast Plants	2.20	4.35	9.57
Larvae	0.27	5.5	1.49
Frass	0.50	4.76	2.38

Different conditions can be used (access to light, number of seeds, number of eggs, etc.) to see how variations in different factors affect the final results.

Lab 11: Transpiration

Transpiration is the loss of water through the stomata in the leaves of a plant. **Stomata** are openings in the surface of a plant leaf through which gases can be exchanged and water may evaporate. Plants need to open their stomata to take in the carbon dioxide needed for photosynthesis, but each time the stomata are open, water is lost from the plant. Plants need to constantly balance their needs for carbon dioxide and water; this is referred to as the **transpiration-photosynthesis compromise**.

Transpiration also helps pull water up from the roots of a plant to its leaves. When water evaporates from the stomata, the water potential of the leaves decrease. Water flows from areas of higher water potential to areas of lower water potential. As water evaporates through the stomata in the leaves, the cohesion of water molecules (due to water's ability to form hydrogen bonds) will pull water up from the stems and roots of the plant, which have a higher water potential than the leaves of the plant. This process is called **transpirational pull**.

Different environmental factors can affect the rate of water loss through transpiration. If the environment that surrounds a plant has a lower water potential than that of the plant, the plant will lose water to its environment. The lower the water potential in the environment around the plant, the greater the rate of water loss from transpiration. Several variables can be used to measure which has the greatest effect on the rate of water loss due to transpiration:

- Increased humidity
- Increased light intensity
- Increased wind

In this lab, a plant can be attached to a potometer, which measures water loss in the plant. The amount of total water loss over a defined time period is measured and can be graphed, as shown in Figure 24.12.

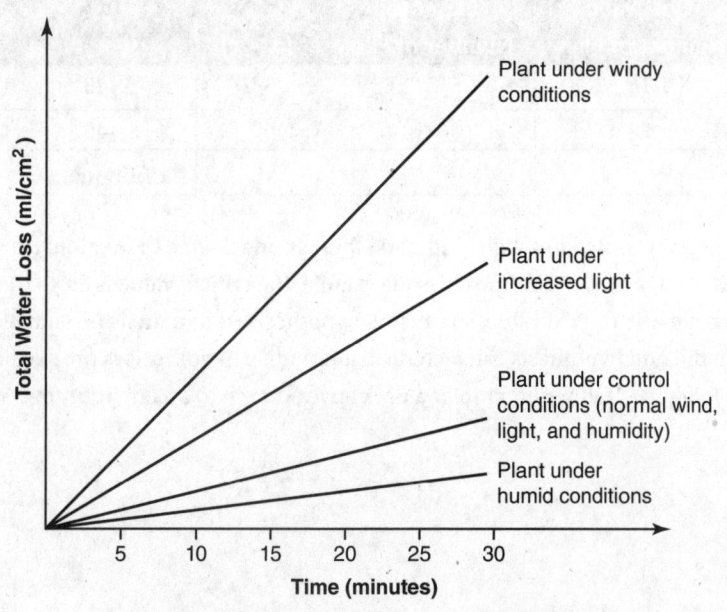

Figure 24.12 Total Water Loss vs. Time

Lab 12: Fruit Fly Behavior

Fruit flies (*Drosophila melanogaster*) have been used as a model organism in scientific studies for over 100 years. Their small size and low maintenance requirements make them very useful in labs around the world. In this lab, *Drosophila* preferences for a wide variety of experimental conditions can be tested.

A choice chamber can be constructed from easily obtainable items, such as plastic Petri dishes, plastic bottles, or graduated cylinders. An example of a choice chamber is shown in Figure 24.13.

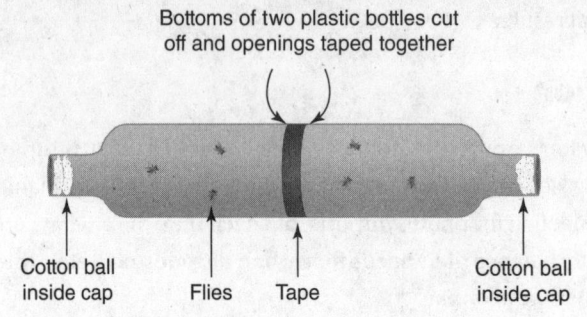

Figure 24.13 Choice Chamber

Drosophila are placed in the middle of the choice chamber, and different environmental conditions are created at each end of the chamber. Several conditions can be created with common classroom materials:

- Light and dark—cover one half of the choice chamber with foil or black paper
- Food preference—soak the cotton balls in different fruit juices and place them at opposite ends of the choice chamber
- Temperature—place one side of the choice chamber on an ice pack

As with all lab investigations, it is important to use appropriate controls and repeat the experiment multiple times in order to obtain the most reliable data.

After a period of time (usually at least 10 minutes), the number of *Drosophila* on each side of the choice chamber are recorded. The null hypothesis for the experiment is that there would be no statistically significant difference in the preferences of the *Drosophila*, with roughly equal numbers on each side of the choice chamber. A chi-square test is performed to find out if the null hypothesis is rejected or if you must fail to reject the null hypothesis. A sample data set with the chi-square calculation is shown in the following table.

Condition	Observed Number of Flies	Expected Number of Flies	$(o - e)$	$(o - e)^2$	$\frac{(o - e)^2}{e}$
Light	14	21	−7	49	2.333
Dark	28	21	7	49	2.333
				Chi-Square =	4.666

Since there are two possible outcomes, light and dark, there is one degree of freedom (df = number of possible outcomes − 1). Using a *p*-value of 0.05 and the chi-square table, the critical value is 3.84. The calculated chi-square value of 4.666 is greater than the critical value, so the null hypothesis (that there is no statistically significant difference) is rejected. When the null hypothesis is rejected, an alternative hypothesis is proposed. In this case, an alternative hypothesis might be "*Drosophila* prefer a dark environment to an environment with light."

Lab 13: Enzyme Activity

Enzymes speed up the rate of chemical reactions by lowering the activation energy of the reaction. The shape of an enzyme is specific to its substrate and crucial to its function. Enzymes bind to a substrate at the active site and form an enzyme-substrate complex that lowers the activation energy of the reaction. Environmental factors can change the shape of an enzyme and consequently change its function.

Peroxidase is an enzyme that breaks down hydrogen peroxide, a by-product of cellular respiration. Peroxidase can be obtained from many sources, including potatoes, liver, and turnips. In this lab, peroxidase from turnips is used to catalyze the breakdown of hydrogen peroxide. Peroxidase catalyzes the breakdown of hydrogen peroxide according to the following reaction:

$$2H_2O_2 \xrightarrow{\text{peroxidase}} 2H_2O + O_2$$

The rate of this reaction can be measured by the rate of the disappearance of the reactant or the rate of production of the products. The chemical guaiacol has a high affinity for oxygen. When guaiacol binds to oxygen, it forms the chemical tetraguaiacol, which is brown in color. The more oxygen produced in the reaction, the darker brown the solution formed will be. By adding guaiacol to the reaction, a colorimetric assay of the relative amounts of oxygen produced can be performed. A spectrophotometer set to measure absorbance at 470 nm, the wavelength most absorbed by tetraguaiacol, can be used to quantify the color of each tube and therefore the relative amount of oxygen produced. If a spectrophotometer is not available, a simple color chart can be used to quantify the relative amount of oxygen produced.

Different environmental conditions can be changed to measure their effects on the production of oxygen by this enzyme-catalyzed reaction. Some common independent variables in this experiment include exposing the enzyme to different pH levels or temperatures. A sample data set from a temperature experiment is shown in the table and in Figure 24.14.

Absorbance Measured at 470 nm							
Temperature (°C)	5	15	25	35	45	55	65	75
Absorbance	0.001	0.151	0.236	0.301	0.345	0.218	0.013	0.000

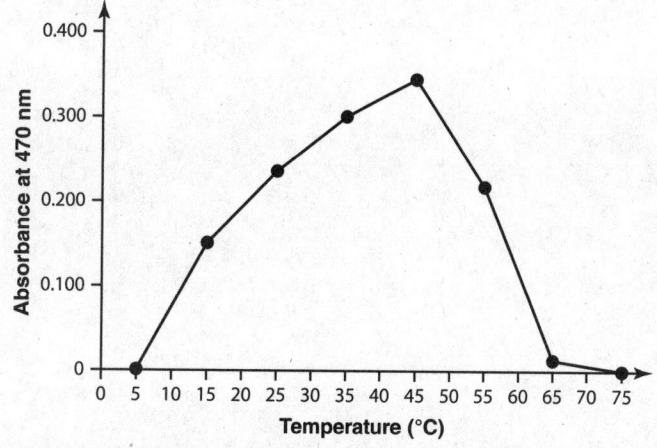

Figure 24.14 Absorbance at 470 nm vs. Temperature in Degrees Celsius

Practice Questions

Multiple-Choice

1. A cube of sugar beet is placed in a beaker of distilled water that is open to the atmosphere, as shown in the figure.

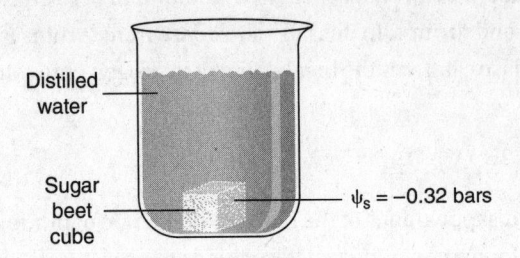

   [Diagram: beaker containing distilled water with sugar beet cube at bottom, labeled $\psi_s = -0.32$ bars]

   Which of the following statements correctly describes the figure?

   (A) The water potential of the distilled water is 0 bars, and the total water potential of the cube of sugar beet is also 0 bars.
   (B) The water potential of the distilled water is 0 bars, and the total water potential of the cube of sugar beet is −0.32 bars.
   (C) The water potential of the distilled water is +0.32 bars, and the total water potential of the cube of sugar beet is −0.32 bars.
   (D) The water potential of the distilled water is −0.32 bars, and the total water potential of the cube of sugar beet is +0.32 bars.

2. A student wants to study the effects of different colors of light on the rate of photosynthesis. The student uses the floating leaf disk method and places 10 leaf disks from a green plant that contains chlorophyll in each of three beakers. The first beaker is placed under white light, the second beaker is placed under green light, and the third beaker is placed under red light. All light sources have equal light intensity and are kept at the same temperature. The number of floating leaf disks at the end of each minute is recorded for 10 minutes and graphed. The estimated time for 50% of the leaf disks to float (ET_{50}) is calculated. Which color of light is expected to produce the highest ET_{50} value?

   (A) white light, because it contains all colors of the spectrum of visible light and chlorophyll will absorb all the wavelengths of light in the visible light spectrum
   (B) green light, because chlorophyll absorbs the most energy from wavelengths of green light
   (C) red light, because chlorophyll reflects red light and does not absorb energy from wavelengths of red light
   (D) green light, because chlorophyll reflects green light and does not absorb energy from wavelengths of green light

3. The table shows cumulative milliliters of oxygen consumed by *Drosophila* placed in respirometers at different temperatures.

Time (minutes)	Cumulative Milliliters of O_2 Consumed		
	5°C	15°C	25°C
0	0	0	0
5	0.02	0.20	0.31
10	0.08	0.40	0.55
15	0.17	0.58	0.75
20	0.20	0.71	0.83
25	0.25	0.83	0.98

What is the rate of oxygen consumption of the *Drosophila* at 25°C from 10 minutes to 20 minutes?

(A) $0.028 \frac{mL}{minute}$

(B) $0.0392 \frac{mL}{minute}$

(C) $0.83 \frac{mL}{minute}$

(D) $0.98 \frac{mL}{minute}$

4. Which of the following is a correct statement about mitosis and meiosis?

(A) Mitosis produces genetically identical haploid body cells, and meiosis produces genetically different diploid gametes.
(B) Mitosis produces genetically different haploid body cells, and meiosis produces genetically identical diploid gametes.
(C) Mitosis produces genetically identical diploid body cells, and meiosis produces genetically different haploid gametes.
(D) Mitosis produces genetically different diploid gametes, and meiosis produces genetically identical haploid body cells.

Questions 5 and 6

Two tubes that contain *E. coli* are transformed by heat shock. Tube 1 contains *E. coli* and a plasmid that contains the gene for antibiotic resistance to streptomycin. Tube 2 contains *E. coli* but does not contain any additional plasmid. After the transformation procedure, the contents of the tubes are plated on agar plates, as described in the table.

Plate	Description
A	Agar in plate contains LB; 100 microliters from Tube 1 are placed on this plate.
B	Agar in plate contains LB and streptomycin; 100 microliters from Tube 1 are placed on this plate.
C	Agar in plate contains LB; 100 microliters from Tube 2 are placed on this plate.
D	Agar in plate contains LB with streptomycin; 100 microliters from Tube 2 are placed on this plate.

5. On which plate would you expect to see the least amount of bacterial growth after incubation for 24 hours?

(A) Plate A, because the agar in the plate does not contain any streptomycin
(B) Plate B, because only a small percentage of the *E. coli* will absorb the plasmid
(C) Plate C, because Tube 2 did not contain the plasmid
(D) Plate D, because Tube 2 did not contain the plasmid, and streptomycin will kill the *E. coli*

6. What is the most likely explanation for the presence of bacterial colonies on plate D?

(A) *E. coli* in Tube 2 absorbed the plasmid and became resistant to streptomycin.
(B) *E. coli* in Tube 2 learned how to catabolize the antibiotic.
(C) Some of the *E. coli* in Tube 2 had a rare spontaneous mutation that conferred resistance to streptomycin.
(D) Some of the *E. coli* in Tube 2 consumed streptomycin as a nutrient source.

7. A map of the sites where restriction enzymes cut a particular segment of DNA is shown below.

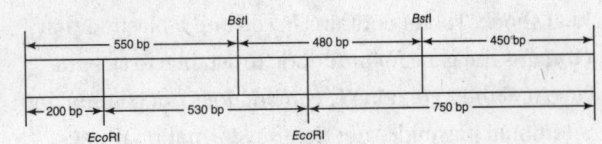

```
|←—— 550 bp ——→|Bstl|←—— 480 bp ——→|Bstl|←—— 450 bp ——→|
|←200 bp→|←—— 530 bp ——→|←———— 750 bp ————→|
         EcoRI              EcoRI
```

If the 1,480 bp DNA fragment shown is cut with both the restriction enzymes *Eco*RI and *Bst*I, what would be the size of the DNA fragments produced?

(A) 450 bp, 480 bp, and 550 bp
(B) 200 bp, 530 bp, and 750 bp
(C) 180 bp, 200 bp, 300 bp, 350 bp, and 450 bp
(D) 200 bp, 450 bp, 480 bp, 530 bp, 550 bp, and 750 bp

Questions 8 and 9

A population is in Hardy-Weinberg equilibrium.

8. The frequency of a recessive phenotype in this population is 20%. What will be the frequency of the recessive allele after three generations?

(A) 0.200
(B) 0.447
(C) 0.494
(D) 0.800

9. What is the expected frequency of the heterozygotes after three generations?

(A) 0.200
(B) 0.447
(C) 0.494
(D) 0.800

10. A student wants to study the effects of salt (NaCl) concentrations on the activity of the enzyme peroxidase. The student sets up four tubes, each with the same starting concentration of the substrate (hydrogen peroxide), a pH of 7, and a temperature of 21°C. What would be an appropriate control in this experiment?

(A) a tube with the same starting concentration of the substrate, a pH of 7, a temperature of 21°C, but a different salt (KCl)
(B) a tube with the same starting concentration of the substrate, a pH of 7, a temperature of 21°C, but with no salt (no NaCl)
(C) a tube with the same starting concentration of the substrate, a pH of 7, and a temperature of 37°C
(D) a tube with water that replaces the enzyme solution

Short Free-Response

11. A student creates a sealed terrarium that contains Fast Plants (*Brassica rapa*) and cabbage white butterfly (*Pieris rapae*) eggs in order to study the energy dynamics of a model ecosystem. On December 10th, the terrarium is placed in the classroom under a light source attached to a timer so that the terrarium will be exposed to 12 hours of light each day. On December 17th, winter break starts, and the terrarium is left unattended under the light source. Unfortunately, when school reopens two weeks later, it is discovered that there was a power outage at the school that started on December 19th, and no light was provided to the terrarium from December 19th through January 3rd.

    Part A
    (i) **Describe** the function of the light source in this experiment.

    Part B
    (i) **Explain** how plants capture light energy.

    Part C
    (i) **Predict** what the terrarium might look like on January 3rd.

    Part D
    (i) **Justify** your prediction from Part C.

12. The DNA sequence in part of the protein hemoglobin from four species are shown in the table.

Species	DNA Sequence						
I	AAA	ACC	GTT	GGG	CCC	CAG	CAG
II	AAA	ACC	GTT	GGC	CCC	CAG	CAG
III	AAT	ACC	GTT	GGG	CCC	CAG	CGG
IV	AAA	ACC	GTG	GGC	CCG	CAG	CGG

    Part A
    (i) **Explain** why two organisms with the same amino acid sequence in a protein can have different DNA sequences for the gene coding for that protein.

    Part B
    (i) **Explain** how DNA sequences can be used to create phylogenetic trees.

    Part C
    (i) Using the data in the table, **construct** a cladogram that represents the most likely phylogenetic relationships among these organisms.

    Part D
    (i) **Identify** two other types of data that could be used to create phylogenetic trees.

Long Free-Response

13. In an experiment, students grow 100 *Brassica rapa* plants for 10 days. The mean plant height on day 10 and the standard error of the mean are calculated and shown in the table (as Generation 1). The 10 tallest plants are allowed to continue to grow, while the remaining 90 plants are eliminated from the population. When these 10 tallest plants flower, they are cross-pollinated, seed pods form, and their seeds are harvested. These seeds are grown to day 10, and the heights of the resulting plants are recorded. The mean plant height on day 10 and the standard error of the mean are calculated and shown in the table (as Generation 2).

Generation	Mean Plant Height on Day 10 (cm)	Standard Error of the Mean
1	15.8	1.1
2	20.3	0.9

Part A

(i) **Explain** the type of selection that was performed in this experiment.

Part B

(i) Using the axes provided, **construct** a graph of the data, showing the means and 95% confidence intervals.

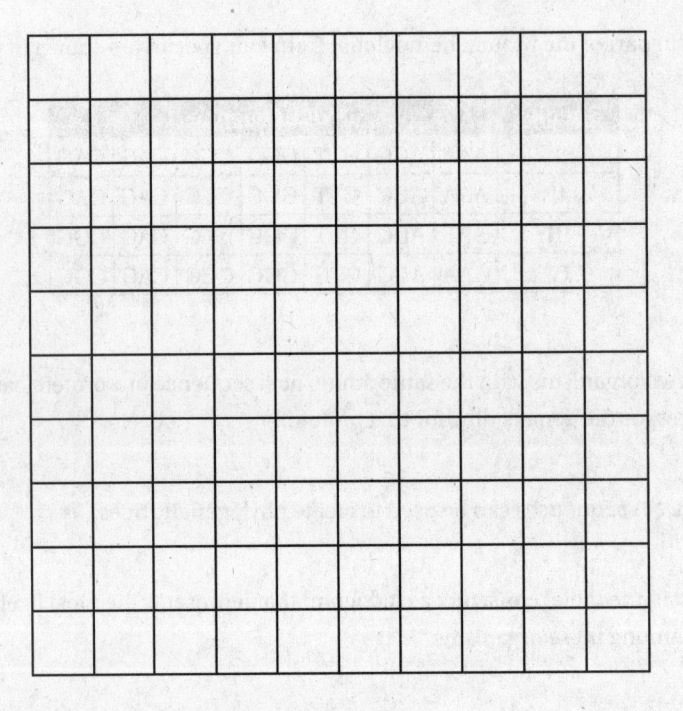

Part C

(i) Using the graph you constructed in Part B, analyze the data, and **determine** whether there is a statistically significant difference in the mean plant height on day 10 between Generation 1 and Generation 2.

Part D

On day 10 of Generation 2, the 10 shortest plants are allowed to continue to grow to flowering. These 10 plants are cross-pollinated, and their seeds are collected and planted to create Generation 3.

(i) **Predict** the mean plant height for Generation 3 on day 10.

(ii) **Justify** your prediction.

Answer Explanations

Multiple-Choice

1. **(B)** Total water potential is the sum of the pressure potential and the solute potential. Since the beaker is open to the atmosphere and in equilibrium with the atmospheric pressure, the pressure potential on the sugar beet cube is 0 bars, making its total water potential equal to its solute potential, −0.32 bars. Choice (A) is incorrect because although the water potential of distilled water is 0 bars, the total water potential of the sugar beet cube is not also 0 bars. The water potential of distilled water that is open to the atmosphere is 0 bars, which rules out choices (C) and (D).

2. **(D)** ET_{50} values are high when the rate of photosynthesis is low. Chlorophyll appears green because it reflects, not absorbs, wavelengths of green light, so green light will provide the least amount of energy for photosynthesis. White light does contain all colors of the visible spectrum and would provide the most light energy for photosynthesis, so it would have the lowest, not the highest, ET_{50} value. Thus, choice (A) is incorrect. Choice (B) is incorrect because chlorophyll reflects, not absorbs, green light. Chlorophyll does absorb energy from red light, so choice (C) is incorrect.

3. **(A)** The rate of oxygen consumption at 25°C from 10 minutes to 20 minutes $= \frac{0.83 \text{ mL} - 0.55 \text{ mL}}{20 \text{ min} - 10 \text{ min}} = 0.028 \frac{\text{mL}}{\text{min}}$. Choice (B) is incorrect because it is the overall rate of oxygen consumption over the 25-minute period. Choice (C) is the cumulative oxygen consumption at 20 minutes, so it is incorrect. Choice (D) is incorrect because it is the cumulative oxygen consumption at 25 minutes.

4. **(C)** The following table summarizes some of the major differences between mitosis and meiosis:

Mitosis	Meiosis
Produces genetically identical cells	Produces genetically different cells
Produces diploid cells	Produces haploid cells
Forms somatic (body) cells	Forms gametes

Choices (A) and (B) are incorrect because mitosis produces diploid (not haploid) cells and meiosis produces haploid (not diploid) cells. Choice (D) is incorrect because mitosis produces genetically identical (not different) cells and meiosis produces genetically different (not identical) cells.

5. **(D)** Streptomycin will kill any bacteria that are not resistant to the antibiotic. Since Tube 2 did not contain the plasmid with the streptomycin resistance gene and plate D contains agar with streptomycin, it is very likely that no *E. coli* will grow on plate D. Choices (A) and (C) are incorrect because both of those plates do NOT contain streptomycin, so many, many *E. coli* colonies will be found on those plates. Choice (B) is incorrect because although this plate does contain agar with streptomycin, Tube 1 received the plasmid that confers resistance to streptomycin, so some *E. coli* will grow on that plate. However, not every *E. coli* cell will absorb the plasmid, so plate B will have fewer bacterial colonies than plates A or C.

6. **(C)** The *E. coli* in Tube 2 did not receive the plasmid, so no growth is expected on plate D; however, there is a very small probability that a spontaneous mutation occurred in some of the *E. coli* cells that gave them resistance to streptomycin. Choice (A) is incorrect because no plasmid was added to Tube 2. There is no evidence that *E. coli* catabolize the antibiotic and individual bacteria do not learn how to catabolize an antibiotic, so choice (B) is incorrect. Streptomycin is not used as a nutrient source for *E. coli*, so choice (D) is incorrect.

7. **(C)** The following figure shows the fragments that would be produced if the DNA was digested with both *Eco*RI and *Bst*I:

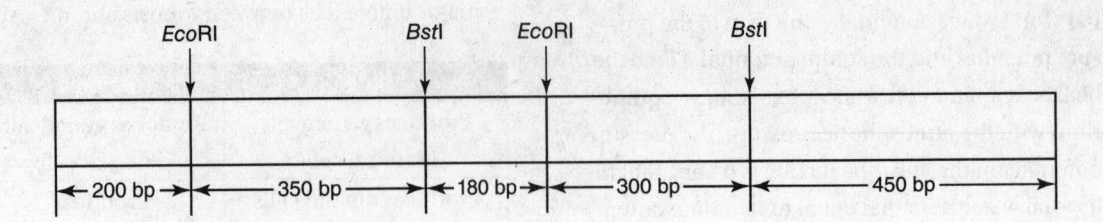

```
         EcoRI      BstI    EcoRI        BstI
           ↓         ↓        ↓           ↓
    |─────────|─────────|─────|─────────|─────────────|
    ←─200 bp─→←─350 bp─→←180 bp→←─300 bp─→←───450 bp───→
```

Choice (A) is incorrect because it lists the fragments that would be produced if the DNA was digested with only *Bst*I. Choice (B) is incorrect because it lists the fragments that would be produced if the DNA was digested with only *Eco*RI. Choice (D) is incorrect because it does not take into account that when both restriction enzymes are used, several of the larger original fragments will be cut and produce smaller fragments.

8. **(B)** The frequency of the recessive phenotype $= q^2$. If $q^2 = 0.20$, then the frequency of the recessive allele is $q = 0.447$. If the population is in Hardy-Weinberg equilibrium, the allele frequencies will not change, so q will still be 0.447 after three generations.

9. **(C)** $p + q = 1$, so if $q = 0.447$, then $p = 0.553$. The frequency of the heterozygotes is $2pq = 2(0.553)(0.447) = 0.494$. If the population is in Hardy-Weinberg equilibrium, the allele frequencies will not change, so the frequency of the heterozygotes will still be 0.494 after three generations.

10. **(B)** An appropriate control for an experiment investigating the effects of different salt concentrations would keep the same starting concentration of the substrate, the same pH and temperature conditions, but would not include salt. Choice (A) is incorrect because using a different salt (with possibly different properties) would confound the results of the experiment. Changing the temperature could affect the results, so choice (C) would not be an appropriate control. Replacing the enzyme solution with water would be an appropriate control for studying the effects of enzyme concentration but would not be an appropriate control for studying the effects of salt concentrations, so choice (D) is incorrect.

Short Free-Response

11. (A-i) The light source simulates sunlight and functions as a source of light energy for photosynthesis.

    (B-i) Plants capture light energy when photons excite electrons in pigments, such as chlorophyll.

    (C-i) There would be less biomass and probably some dead plants and larvae on January 3rd.

    (D-i) Without light energy, the plants could not perform photosynthesis. Without photosynthesis, there would be no stored organic molecules for the larvae to eat.

12. (A-i) The genetic code is redundant: more than one codon can code for an amino acid. So two organisms could have the exact same amino acid sequence for a protein but have slightly different DNA sequences in the gene coding for the protein.

    (B-i) The more homology (common sequences) between the DNA sequences of organisms, the more recently they shared a common ancestor. This information could be used to develop phylogenetic trees.

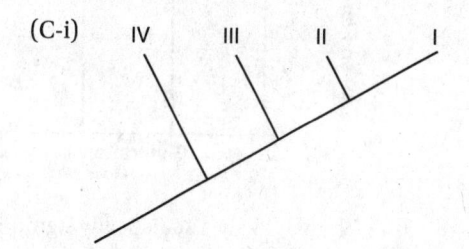

    (C-i)

    (D-i) Morphology (body characteristics) and fossil evidence could also be used to generate phylogenetic trees.

Long Free-Response

13. (A-i) Artificial selection was used in this experiment because the students selected for desired traits and chose which plants would survive and reproduce.

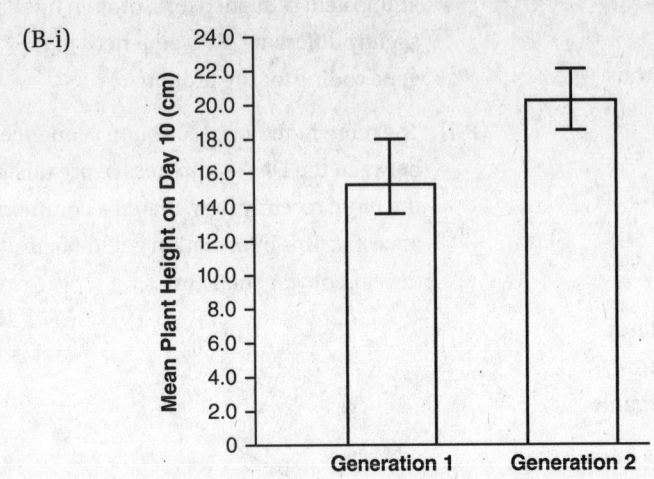

    (B-i)

    [Bar graph showing Mean Plant Height on Day 10 (cm) for Generation 1 (~15.4 cm) and Generation 2 (~20.3 cm) with error bars.]

    (C-i) There is likely a statistically significant difference between the two groups because the 95% confidence intervals (CI) of Generation 1 and Generation 2 do not overlap (the upper limit of the 95% CI for Generation 1 is $15.8 + 2(1.1) = 18$ and the lower limit of the 95% CI for Generation 2 is $20.3 - 2(0.9) = 18.5$).

    (D-i and ii) The mean plant height for Generation 3 would likely be less than the mean plant height for Generation 2 because during this instance of artificial selection, only the plants with the combination of gene(s) for short height were allowed to breed.

Practice Tests

ANSWER SHEET
Practice Test 1

1. Ⓐ Ⓑ Ⓒ Ⓓ
2. Ⓐ Ⓑ Ⓒ Ⓓ
3. Ⓐ Ⓑ Ⓒ Ⓓ
4. Ⓐ Ⓑ Ⓒ Ⓓ
5. Ⓐ Ⓑ Ⓒ Ⓓ
6. Ⓐ Ⓑ Ⓒ Ⓓ
7. Ⓐ Ⓑ Ⓒ Ⓓ
8. Ⓐ Ⓑ Ⓒ Ⓓ
9. Ⓐ Ⓑ Ⓒ Ⓓ
10. Ⓐ Ⓑ Ⓒ Ⓓ
11. Ⓐ Ⓑ Ⓒ Ⓓ
12. Ⓐ Ⓑ Ⓒ Ⓓ
13. Ⓐ Ⓑ Ⓒ Ⓓ
14. Ⓐ Ⓑ Ⓒ Ⓓ
15. Ⓐ Ⓑ Ⓒ Ⓓ
16. Ⓐ Ⓑ Ⓒ Ⓓ
17. Ⓐ Ⓑ Ⓒ Ⓓ
18. Ⓐ Ⓑ Ⓒ Ⓓ
19. Ⓐ Ⓑ Ⓒ Ⓓ
20. Ⓐ Ⓑ Ⓒ Ⓓ
21. Ⓐ Ⓑ Ⓒ Ⓓ
22. Ⓐ Ⓑ Ⓒ Ⓓ
23. Ⓐ Ⓑ Ⓒ Ⓓ
24. Ⓐ Ⓑ Ⓒ Ⓓ
25. Ⓐ Ⓑ Ⓒ Ⓓ
26. Ⓐ Ⓑ Ⓒ Ⓓ
27. Ⓐ Ⓑ Ⓒ Ⓓ
28. Ⓐ Ⓑ Ⓒ Ⓓ
29. Ⓐ Ⓑ Ⓒ Ⓓ
30. Ⓐ Ⓑ Ⓒ Ⓓ
31. Ⓐ Ⓑ Ⓒ Ⓓ
32. Ⓐ Ⓑ Ⓒ Ⓓ
33. Ⓐ Ⓑ Ⓒ Ⓓ
34. Ⓐ Ⓑ Ⓒ Ⓓ
35. Ⓐ Ⓑ Ⓒ Ⓓ
36. Ⓐ Ⓑ Ⓒ Ⓓ
37. Ⓐ Ⓑ Ⓒ Ⓓ
38. Ⓐ Ⓑ Ⓒ Ⓓ
39. Ⓐ Ⓑ Ⓒ Ⓓ
40. Ⓐ Ⓑ Ⓒ Ⓓ
41. Ⓐ Ⓑ Ⓒ Ⓓ
42. Ⓐ Ⓑ Ⓒ Ⓓ
43. Ⓐ Ⓑ Ⓒ Ⓓ
44. Ⓐ Ⓑ Ⓒ Ⓓ
45. Ⓐ Ⓑ Ⓒ Ⓓ
46. Ⓐ Ⓑ Ⓒ Ⓓ
47. Ⓐ Ⓑ Ⓒ Ⓓ
48. Ⓐ Ⓑ Ⓒ Ⓓ
49. Ⓐ Ⓑ Ⓒ Ⓓ
50. Ⓐ Ⓑ Ⓒ Ⓓ
51. Ⓐ Ⓑ Ⓒ Ⓓ
52. Ⓐ Ⓑ Ⓒ Ⓓ
53. Ⓐ Ⓑ Ⓒ Ⓓ
54. Ⓐ Ⓑ Ⓒ Ⓓ
55. Ⓐ Ⓑ Ⓒ Ⓓ
56. Ⓐ Ⓑ Ⓒ Ⓓ
57. Ⓐ Ⓑ Ⓒ Ⓓ
58. Ⓐ Ⓑ Ⓒ Ⓓ
59. Ⓐ Ⓑ Ⓒ Ⓓ
60. Ⓐ Ⓑ Ⓒ Ⓓ

PRACTICE TEST 1

Practice Test 1

Section I: Multiple-Choice

TIME: 90 MINUTES

DIRECTIONS: For each question or incomplete statement, select the choice that best answers the question or completes the statement.

1. The following figure shows the interaction of water molecules with sodium chloride.

   Which of the following statements best explains the arrangement of the molecules in this figure?

   (A) Water is nonpolar, and it can easily dissolve ionic compounds by surrounding both negative ions and positive ions with a nonpolar ring of water molecules.
   (B) Water is polar, and it can easily dissolve ionic compounds by surrounding negative ions with partially positive hydrogen atoms and by surrounding positive ions with partially negative oxygen atoms.
   (C) Water and sodium chloride are both nonpolar molecules, and this allows sodium chloride to dissolve in water.
   (D) Hydrogen bonds form between water molecules and the ions in sodium chloride.

2. The following figure shows the interaction between water molecules and a micelle (a spherical formation of lipids dissolved in an aqueous environment).

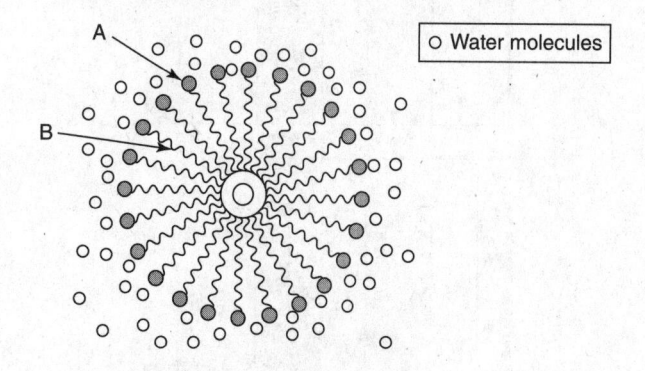

   Which of the following correctly describes the parts of the micelle labeled A and B?

   (A) Part A represents the hydrophilic phosphate head of a phospholipid, and part B represents the hydrophobic lipid tail of a phospholipid.
   (B) Part A represents the hydrophobic phosphate head of a phospholipid, and part B represents the hydrophilic lipid tail of a phospholipid.
   (C) Part A represents the glycosidic head of a glycosphingolipid, and part B represents the hydrophilic tail of a phospholipid.
   (D) Part A represents the inorganic phosphate compound, and part B represents the fatty sterol cholesterol.

Questions 3-5

A student conducts an experiment to find out how many drops of different liquids can be placed on the surface of a penny before the liquids flow over the sides of the penny. Data are shown in the following table.

	Water	Rubbing Alcohol	Olive Oil
Mean Number of Drops of Liquid Before Overflow ($\pm 1\ SE_{\bar{x}}$)	22 ± 1.2	14 ± 2.2	6 ± 2.1

3. Which of the following graphs most accurately represents the data, including 95% confidence intervals?

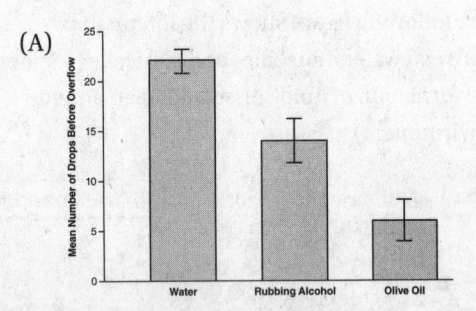

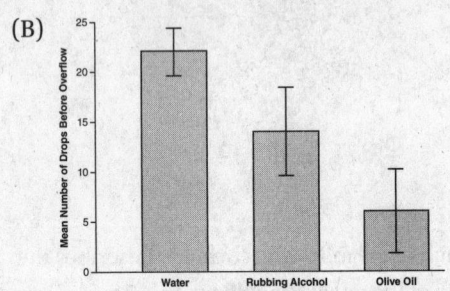

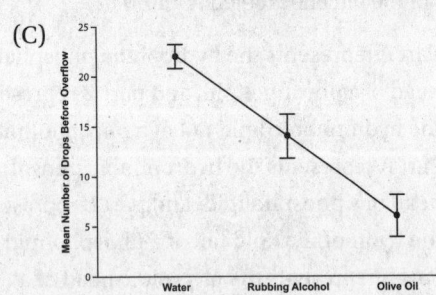

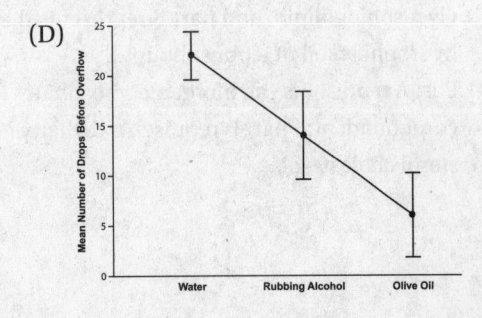

(A) [bar graph with small error bars]

(B) [bar graph with larger error bars]

(C) [line graph with small error bars]

(D) [line graph with larger error bars]

4. Which two liquids are least likely to have a statistically significant difference between their means?

   (A) water and rubbing alcohol (because their 95% confidence intervals do overlap)
   (B) water and olive oil (because their 95% confidence intervals do not overlap)
   (C) water and rubbing alcohol (because their 95% confidence intervals do not overlap)
   (D) rubbing alcohol and olive oil (because their 95% confidence intervals do overlap)

5. The student repeats the experiment, but before doing so, first coats the surface of the penny with a thin layer of surfactant (a chemical compound that disrupts the formation of hydrogen bonds). Predict the effect the surfactant will have on the number of drops of water that can be placed on the surface of the penny before overflow occurs, and justify your prediction.

   (A) The number of drops of water will decrease due to fewer hydrogen bonds forming between the water molecules.
   (B) The number of drops of water will decrease due to fewer polar ends forming within the water molecules.
   (C) The number of drops of water will increase due to increased polar covalent bonds within the water molecules.
   (D) The number of drops of water will increase due to increased surface tension.

6. Which of the following biological molecules is the carrier of genetic information?

   (A) carbohydrates
   (B) lipids
   (C) nucleic acids
   (D) proteins

7. Which of the following correctly describes what would be needed to break down a polypeptide chain into its 15 component amino acids?

   (A) dehydration reactions that would produce 15 water molecules
   (B) dehydration reactions that would consume 14 water molecules
   (C) hydrolysis reactions that would produce 15 water molecules
   (D) hydrolysis reactions that would consume 14 water molecules

8. Which of the following is a correct statement about protein structure?

   (A) Primary structure is formed by hydrophobic interactions.
   (B) Secondary structure is formed by hydrogen bonds.
   (C) Tertiary structure is formed by peptide bonds.
   (D) Quaternary structures are found in all proteins.

Questions 9–11

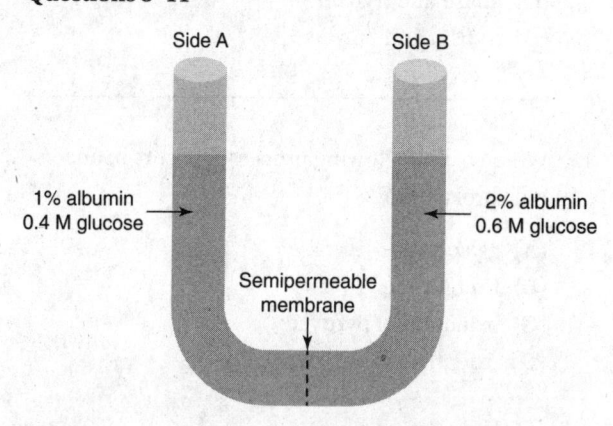

Side A: 1% albumin, 0.4 M glucose
Side B: 2% albumin, 0.6 M glucose
Semipermeable membrane

The two sides of the U-tube apparatus are separated by a semipermeable membrane. The aqueous solution on side A of the U-tube contains 1% albumin and 0.4 M glucose. The aqueous solution on side B contains 2% albumin and 0.6 M glucose.

9. Which of the following statements is correct?

   (A) Side A initially has a higher water potential than side B because side A has a lower solute concentration than side B.
   (B) Side B initially has a higher water potential than side A because side B has a higher solute concentration than side A.
   (C) Side A and side B initially have the same water potential because they are both aqueous solutions.
   (D) Side A and side B initially have the same water potential because they both contain the same types of solutes.

10. Glucose may pass through the semipermeable membrane in the U-tube, but albumin may not. If the solutions on each side of the U-tube apparatus are allowed to equilibrate for 30 minutes, which of the following is the most likely result?

    (A) The concentration of glucose on side B will increase.
    (B) The concentration of albumin on side A will increase.
    (C) The concentration of glucose on side A will increase.
    (D) The concentration of albumin on side B will increase.

11. Which of the following correctly predicts and explains the movement of water in the U-tube after 30 minutes?

    (A) Water will move from side A to side B because the water potential on side A will be higher than the water potential on side B.
    (B) Water will move from side A to side B because the water potential on side B will be higher than the water potential on side A.
    (C) Water will move from side B to side A because the water potential on side A will be higher than the water potential on side B.
    (D) Water will move from side B to side A because the water potential on side B will be higher than the water potential on side A.

Questions 12–14

Refer to the following table.

Enzyme	Location of Action in Digestive Tract	pH of Location	Substrate It Breaks Down
Amylase	Mouth	6.8	Starch
Pepsin	Stomach	2.5	Protein
Lipase	Small intestine	6.9	Fats
Trypsin	Small intestine	6.9	Protein

12. Which of the following graphs most likely reflects the activity of pepsin at different pHs?

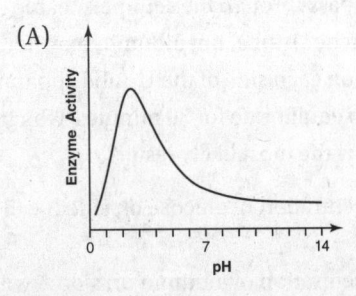

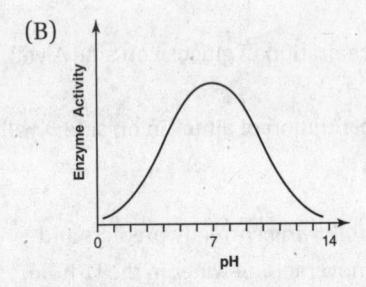

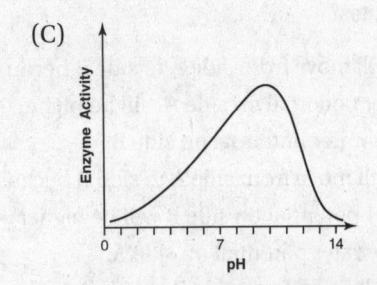

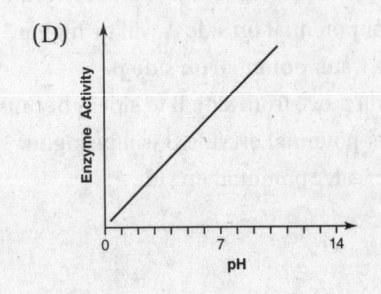

(A) [graph: peak near pH 2]

(B) [graph: peak near pH 7]

(C) [graph: peak near pH 8-9]

(D) [graph: linear increase]

13. A person eats enough antacids to raise the pH of both his mouth and his stomach acid. The activities of which enzymes would most likely be affected by this?

(A) amylase and pepsin
(B) pepsin and lipase
(C) lipase and trypsin
(D) trypsin and amylase

14. The pyloric sphincter controls the flow of the contents of the stomach (chyme) into the small intestine. The pyloric sphincter prevents the acidic chyme from the stomach from entering the small intestine until the pancreas is ready to secrete bicarbonate into the small intestine, which neutralizes the acids in the chyme. If the pyloric sphincter was damaged and could not control the flow of chyme into the small intestine, which enzymes would most likely have reduced activity during digestion?

(A) amylase and pepsin
(B) pepsin and lipase
(C) lipase and trypsin
(D) trypsin and amylase

15. Which of the following processes occurs in anaerobic prokaryotes?

(A) glycolysis
(B) Krebs cycle
(C) oxidation of pyruvate
(D) oxidative phosphorylation

Questions 16–18

In a classic experiment to determine which wavelengths of light produce the greatest amounts of photosynthesis in photosynthetic algae, light was passed through a prism to shine different colors of light onto the photosynthetic algae *Spirogyra* (that was placed on a microscope slide), as shown in the following figure.

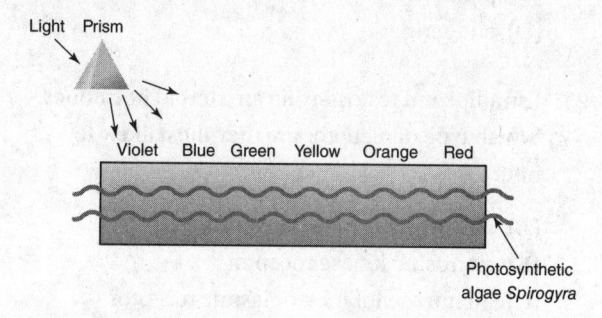

Aerobic bacteria (that need oxygen to survive) were layered on top of the algae. The wavelengths (in nm) of different colors of visible light are noted in the table.

Color of Visible Light	Wavelength (nm)
Violet	380–450
Blue	450–495
Green	495–570
Yellow	570–590
Orange	590–620
Red	620–750

16. The greatest numbers of aerobic bacteria were found on the slide where the wavelength of light was between 420–470 nm and 660–700 nm. What conclusion could be drawn from this data?

    (A) Algae produce the same amount of oxygen when exposed to any color of light.
    (B) Algae produce oxygen only when in the presence of aerobic bacteria.
    (C) Algae produce the most oxygen when exposed to green light.
    (D) Algae produce the most oxygen when exposed to blue-violet or red light.

17. Further research revealed that the light-absorbing pigment in the algae *Spirogyra* (that was used in this experiment) was chlorophyll, which absorbs the most light energy in the wavelength ranges of 420–470 nm and 660–700 nm. Some photosynthetic red algae use the pigment phycoerythrin, which absorbs the most light energy in the wavelength range of 495–570 nm. If this experiment was repeated with photosynthetic red algae, near which color of light on the slide would you expect to see the greatest number of bacteria?

    (A) violet
    (B) green
    (C) orange
    (D) red

18. What is the role of light energy in photosynthesis?

    (A) to activate the expression of genes needed for photosynthesis
    (B) to trigger a cell signaling pathway that produces carbon dioxide
    (C) to excite the electrons in the photosystems
    (D) to fix carbon dioxide during the Krebs (citric acid) cycle

Questions 19 and 20

Equal numbers of germinating mung beans were placed in sealed chambers of equal volumes with probes that measured the levels of oxygen and carbon dioxide gas over time. One chamber was kept at a temperature of 4° Celsius, and the other chamber was kept at a temperature of 25° Celsius for the duration of the experiment.

19. What is the independent variable in this experiment?

    (A) the number of germinating mung beans
    (B) the temperature of the chambers
    (C) the amount of carbon dioxide produced
    (D) the amount of oxygen consumed

20. This graph shows the carbon dioxide levels in both chambers during a 30-minute period.

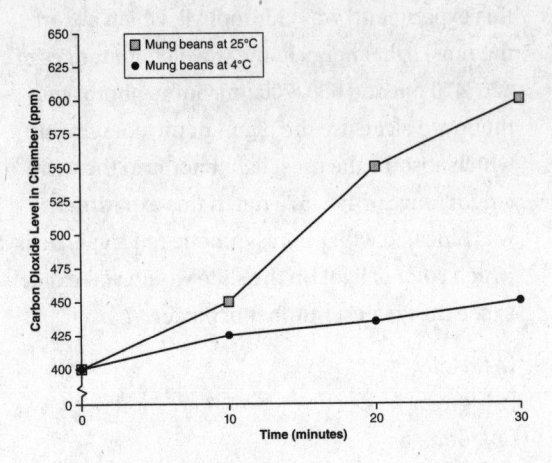

Which of the following correctly states the results of this experiment?

(A) The mung beans at 4° Celsius increased the level of carbon dioxide in the chamber at a rate of 6.67 ppm CO_2 per minute and had a higher respiration rate than the mung beans at 25° Celsius.
(B) The mung beans at 4° Celsius increased the level of carbon dioxide in the chamber at a rate of 1.67 ppm CO_2 per minute and had a higher respiration rate than the mung beans at 25° Celsius.
(C) The mung beans at 25° Celsius increased the level of carbon dioxide in the chamber at a rate of 6.67 ppm CO_2 per minute and had a higher respiration rate than the mung beans at 4° Celsius.
(D) The mung beans at 25° Celsius increased the level of carbon dioxide in the chamber at a rate of 1.67 ppm CO_2 per minute and had a higher respiration rate than the mung beans at 4° Celsius.

21. A cell secretes a growth factor that slowly diffuses to nearby cells, stimulating the growth of those cells. This is an example of _____ signaling.

(A) autocrine
(B) juxtacrine
(C) paracrine
(D) endocrine

22. Quorum sensing is used by bacteria to sense the density of bacteria in the area. Bacteria release a signaling molecule that travels short distances to receptors on nearby bacteria. This would best be described as which type of cell signaling?

(A) synaptic
(B) cytokine
(C) paracrine
(D) endocrine

23. Estradiol and testosterone are steroid hormones. Which type of receptor are they most likely to bind to?

(A) a G-protein-linked receptor
(B) a tyrosine kinase receptor
(C) an intracellular cytoplasmic receptor
(D) a membrane ion channel receptor

24. cAMP and Ca^{2+} ions are examples of _____.

(A) cell membrane receptors
(B) ligands
(C) intracellular cytoplasmic receptors
(D) secondary messengers

25. Gap junctions in animal cells and plasmodesmata in plant cells are examples of which type of signaling?

(A) autocrine
(B) juxtacrine
(C) paracrine
(D) endocrine

26. How does a cell that has just completed the S stage of the cell cycle compare to the same cell at the start of the S stage?

(A) It has twice the amount of DNA and twice the number of chromosomes.
(B) It has twice the amount of DNA and the same number of chromosomes.
(C) It has the same amount of DNA and twice the number of chromosomes.
(D) It has the same amount of DNA and the same number of chromosomes.

27. During which stage of the cell cycle are you most likely to find cells that have fully differentiated and are not actively dividing?

    (A) G0
    (B) G1
    (C) S
    (D) G2

28. _____ are present at near-constant levels during the cell cycle and are dependent on rising levels of _____ to become active.

    (A) Cyclins; mitosis-promoting factor
    (B) Cyclins; cyclin-dependent kinases
    (C) Cyclin-dependent kinases; cyclins
    (D) Cyclin-dependent kinases; phosphatases

29. The three major events of the division of the genetic material during cell replication are (1) replication of the genetic material, (2) alignment of chromosomes, and (3) separation of chromosomes. These events happen during which stages (respectively)?

    (A) S, G1, G2
    (B) S, metaphase, anaphase
    (C) G2, metaphase, telophase
    (D) prophase, metaphase, anaphase

30. Which of the following would affect the greatest number of steps in the signal transduction process?

    (A) an antibody irreversibly binding to a cell membrane receptor
    (B) deletion of the gene for the enzyme adenylyl cyclase
    (C) inactivation of a cytoplasmic protein kinase
    (D) introduction of an inhibitor of a protein phosphatase

Questions 31 and 32

The ability to taste phenylthiocarbamide (PTC) is a dominant trait. Nontasters of PTC are recessive. In a population that is *not* in Hardy-Weinberg equilibrium, there are 520 homozygous dominant tasters, 400 heterozygous tasters, and 80 nontasters of PTC.

31. What is the frequency of the nontaster allele?

    (A) 0.28
    (B) 0.44
    (C) 0.56
    (D) 0.72

32. What is the frequency of individuals who are carriers of the nontasting allele?

    (A) 0.08
    (B) 0.40
    (C) 0.50
    (D) 0.52

Questions 33 and 34

Refer to the following table.

Cell	Type of Chromosome	Presence of Mitochondria	Presence of Chloroplasts
A	Circular	No	No
B	Linear	No	No
C	Linear	Yes	No
D	Linear	Yes	Yes

33. Which of the cells in this table shares characteristics of both prokaryotic and eukaryotic cells?

    (A) A
    (B) B
    (C) C
    (D) D

34. Which of the organisms in the table is most closely related to modern plants?

    (A) A
    (B) B
    (C) C
    (D) D

35. Two species of deer live on different mountain ranges separated by 500 miles. Which type of reproductive isolation does this scenario represent?

    (A) habitat isolation
    (B) temporal isolation
    (C) behavioral isolation
    (D) mechanical isolation

36. Sea urchins release their sperm and eggs into the water. However, the sperm from red sea urchins cannot fertilize the eggs from purple sea urchins. Which type of reproductive isolation is this?

    (A) mechanical isolation
    (B) gametic isolation
    (C) reduced hybrid fertility
    (D) hybrid breakdown

37. A template strand of DNA has the following sequence:

    3'-AAT TCC GGA TCG-5'

    What is the complementary strand created during DNA replication?

    (A) 3'-TTA AGG CCT AGC-5'
    (B) 5'-TTA AGG CCT AGC-3'
    (C) 3'-UUA AGG CCU AGC-5'
    (D) 5'-UUA AGG CCU AGC-3'

38. Which of the following statements best explains the mechanism for DNA replication?

    (A) DNA replication is reductive, because half of the total DNA present is copied.
    (B) DNA replication is semiconservative, because each DNA strand serves as a template for a new strand during replication.
    (C) DNA replication is dispersive, because the two resulting DNA molecules are random mixtures of parent and daughter DNA.
    (D) DNA replication is conservative, because one resulting molecule is identical to the original and the other consists of two new strands.

39. During DNA replication, DNA "unwinds" to form two template strands: the leading strand and the lagging strand. Which of the following statements about these strands is *true*?

    (A) On the lagging strand, short fragments are formed when the new strand of DNA is synthesized.
    (B) The leading strand of DNA is synthesized discontinuously.
    (C) DNA polymerase can only synthesize DNA on the leading strand.
    (D) The lagging strand can only be synthesized once the leading strand has been completed.

40. Which of the following is the enzyme used during transcription that directly generates the transcript?

    (A) helicase
    (B) ligase
    (C) DNA polymerase
    (D) RNA polymerase

41. Which of the following statements about introns is *true*?

    (A) Introns are translated by ribosomes.
    (B) Introns have no function in the chromosomes.
    (C) Introns are found in eukaryotes.
    (D) Introns are found only in prokaryotes.

42. Streptomycin, a commonly used antibiotic, interferes with bacterial ribosomes. Which process would most likely be negatively affected by streptomycin?

    (A) DNA replication
    (B) posttranscriptional modification of mRNA
    (C) transcription
    (D) translation

43. A scientist is studying a eukaryotic gene that is 15,000 base pairs long. The scientist isolates the mRNA (produced by this gene) from the cytosol and finds that it is only 12,000 base pairs long. Why is the isolated mRNA not the same length as the DNA that codes for it?

   (A) The scientist made a mistake and isolated the wrong mRNA.
   (B) Introns are removed from eukaryotic mRNAs, so the mRNA is shorter than the DNA.
   (C) RNA is less stable than DNA and is more likely to degrade in the cell.
   (D) The poly-A tail on the mRNA results in an mRNA that is shorter than the gene.

44. In the cross $AABbCc \times AaBbCC$, what is the probability of producing an offspring with the genotype $AaBbCc$?

   (A) 0
   (B) $\frac{1}{8}$
   (C) $\frac{1}{4}$
   (D) $\frac{1}{2}$

45. Based on this pedigree, what is the most likely mode of inheritance of the trait?

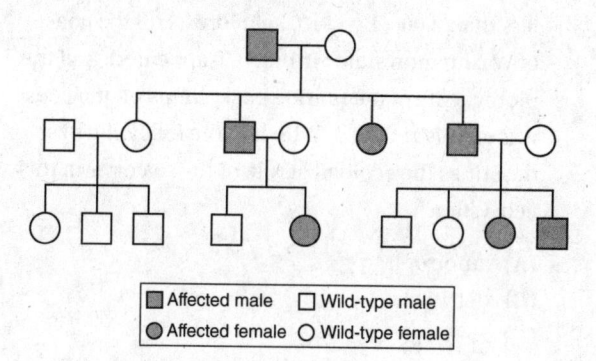

   ■ Affected male   □ Wild-type male
   ● Affected female ○ Wild-type female

   (A) autosomal dominant
   (B) sex-linked dominant
   (C) sex-linked recessive
   (D) mitochondrial inheritance

Questions 46–48

Color blindness in humans is a sex-linked recessive trait. A female who is not color blind, but whose father was color blind, has children with a male who is not color blind.

46. What is the likelihood that their son will be color blind?

   (A) 0%
   (B) 25%
   (C) 50%
   (D) 100%

47. What is the likelihood that their daughter will be color blind?

   (A) 0%
   (B) 25%
   (C) 50%
   (D) 100%

48. Which of the following is a likely pedigree for this family? Note that circles indicate females, and squares indicate males. Shaded circles and squares indicate individuals with the trait.

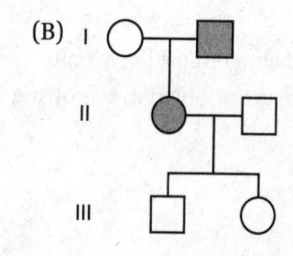

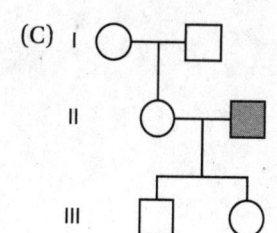

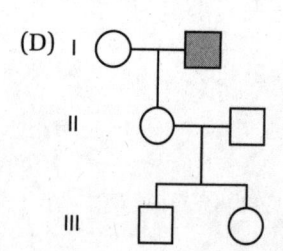

49. Which of the following is a density-independent factor that can limit population size?

    (A) competition
    (B) disease
    (C) predation
    (D) weather

Questions 50 and 51

Refer to the following graph.

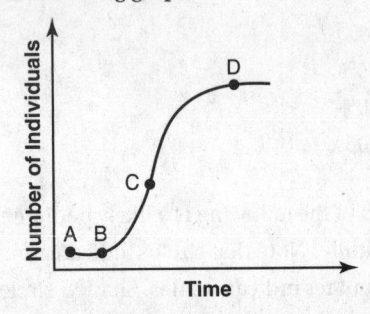

50. Which point on the graph represents the carrying capacity of the environment for this population?

    (A) A
    (B) B
    (C) C
    (D) D

51. Which point on the diagram most likely represents a point in which the population is growing exponentially?

    (A) A
    (B) B
    (C) C
    (D) D

52. Sea anemones, which live as sessile forms attached to rock surfaces, have stinging nematocysts that release paralyzing toxins (upon contact) into their prey, such as fish. Tentacles move the paralyzed prey into the gastric cavity for ingestion.

    Clown fish are covered with a thin layer of mucus that prevents them from being paralyzed by sea anemone toxins. Clown fish live among the tentacles, which provide protection. Predators that attempt to eat clown fish are stung by the sea anemone nematocysts and are then consumed by the sea anemone.

    Which of the following correctly describes the relationship between sea anemones and clown fish?

    (A) commensalism
    (B) competition
    (C) mutualism
    (D) parasitism

53. In the marine ecosystems on the Pacific Coast of Washington state, sea otters eat sea urchins. Sea urchins eat kelp. Kelp forests provide food and habitat for a wide variety of marine organisms. Due to overhunting from 1750 to 1910, sea otter populations were reduced from over 100,000 to less than 2,000. By 1910, kelp forests off the coast of Washington state virtually disappeared, and the biodiversity of the marine ecosystems on the coast was greatly reduced. Which of the following best describes the ecological role of the sea otter in this ecosystem?

    (A) autotroph
    (B) detritivore
    (C) keystone species
    (D) prey species

54. Which of the following is a correct statement about energy in ecosystems?

    (A) A net gain in energy can result in the loss of mass and the death of the organism.
    (B) Energy is recycled through ecosystems.
    (C) Changes in energy availability can disrupt ecosystems.
    (D) Heterotrophs capture the energy in sunlight.

55. Which of the following statements about population growth is *false*?

    (A) Reproduction without limiting factors results in the exponential growth of a population.
    (B) Density-dependent limits to populations results in logistic growth of a population.
    (C) The carrying capacity of an environment depends on resource availability.
    (D) K-selected populations can far exceed their environment's carrying capacity.

56. The synthesis of the amino acid tryptophan in bacteria is controlled by the repressible *trp* operon. The following figure represents components of the *trp* operon.

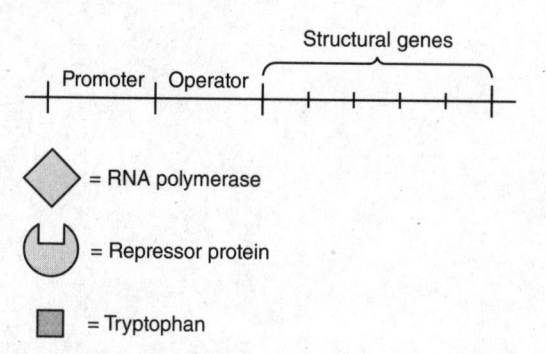

    Which of the following best represents the *trp* operon when the amino acid tryptophan is present in the bacteria's environment?

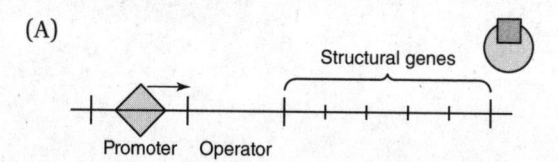

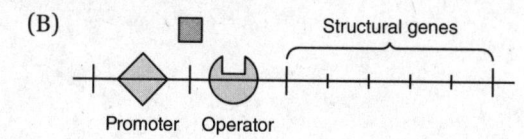

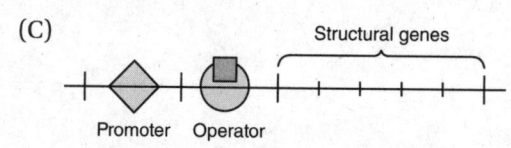

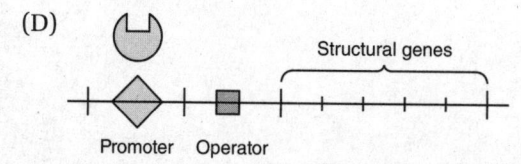

    (A)
    (B)
    (C)
    (D)

57. The digestion of the sugar lactose in bacteria is controlled by the inducible *lac* operon. This figure represents the components of the *lac* operon.

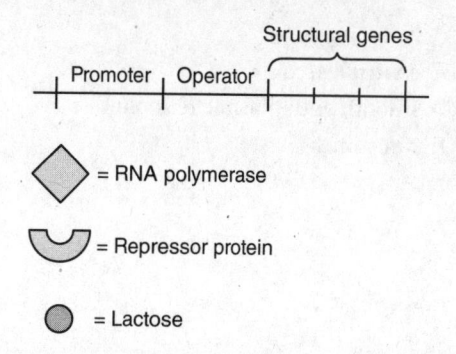

    Which of the following best represents the *lac* operon when the sugar lactose is present in the bacteria's environment?

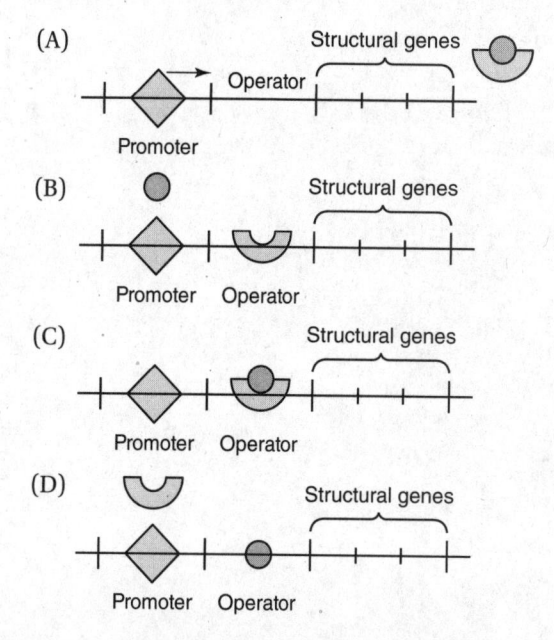

    (A)
    (B)
    (C)
    (D)

58. Transmembrane proteins span across the entire cell membrane. Where would the hydrophobic amino acids in a transmembrane protein most likely be found?

    (A) in contact with the polar phosphates of the phospholipids
    (B) on the outer surface of the membrane in contact with the cell's aqueous surroundings
    (C) in contact with the nonpolar fatty acid chains of the phospholipids
    (D) on the inner surface of the membrane in contact with the cell's cytosol

59. In the alpha cells in the pancreas, which produce the peptide hormone insulin, there would likely be a high concentration of which of the following?

    (A) lysosomes
    (B) centrioles
    (C) smooth endoplasmic reticulum
    (D) ribosomes

60. A geneticist discovers a previously unknown genetic disorder. Pedigree analysis shows that both males and females can inherit this disorder from their mother but not from their father. Affected females can pass on the allele for this disorder to the next generation, but affected males will not pass on this disorder to the next generation. What is the most likely mode of inheritance of this disorder?

    (A) sex-linked recessive
    (B) mitochondrial inheritance
    (C) autosomal recessive
    (D) autosomal dominant

STOP If there is still time remaining, you may review your answers.

Section II: Free-Response

TIME: 90 MINUTES

DIRECTIONS: Answer each of the following six free-response questions using complete sentences. Allow approximately 20–25 minutes each for the long free-response questions (Questions 1 and 2) and approximately 5–10 minutes each for the short free-response questions (Questions 3, 4, 5, and 6).

1. A scientist completes an analysis of the amino acid sequences of cytochrome *c* in six different species. The scientist compares how many amino acid differences there are between the cytochrome *c* protein in humans and that of each of the six different species in the table. The number of amino acid differences is shown in the table.

Species	Number of Amino Acid Differences from Human Cytochrome *c*
Chicken	13
Whale	9
Chimpanzee	1
Turtle	15
Tuna	21
Rabbit	4

   Part A

   (i) **Explain** how amino acid differences in a conserved protein can help scientists determine degrees of relationships between organisms.

   Part B

   (i) Use the data to **construct** a cladogram of these organisms on the template provided.

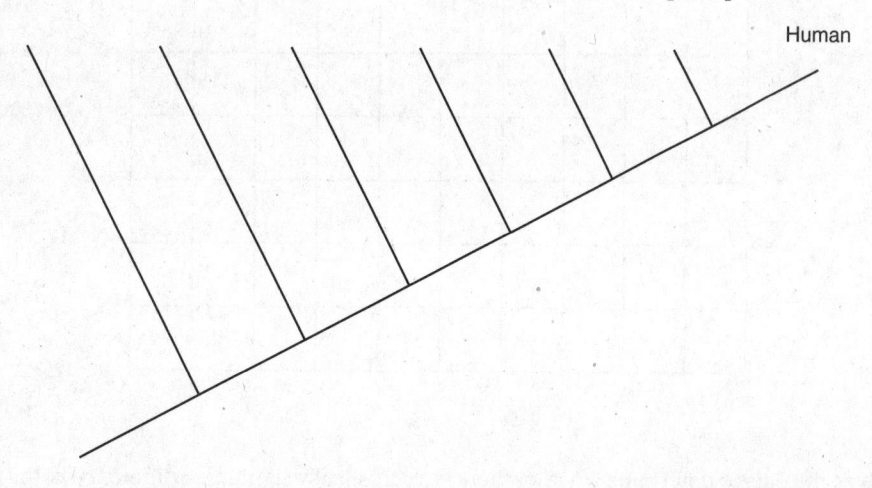

   Human

   Part C

   (i) **Justify** your placement of the organisms on the cladogram in your response to Part B.

Part D

A claim is made that chickens share a more recent common ancestor with humans than turtles do.

(i) **Identify** at least one piece of experimental evidence that might be used to support this claim.

2. A botanist measures the stomatal density on both the upper surface (facing the Sun) and lower surface (shaded from the Sun) of the leaves from two plant species. Purple passionflower, *Passiflora incarnata,* is native to Florida and Texas and is found in habitats with a humid climate. White oak, *Quercus alba*, is native to the central and northern United States and prefers a drier climate with seasons. The mean stomatal density and the standard errors of the mean are shown in the following table.

Plant Species	Mean Stomatal Density of Upper Surface (Number of Stomata per mm^2) $\pm$ SEM*	Mean Stomatal Density of Lower Surface (Number of Stomata per mm^2) $\pm$ SEM
Passiflora incarnata	90 $\pm$ 10	345 $\pm$ 10
Quercus alba	25 $\pm$ 3	375 $\pm$ 12

*Standard Error of the Mean

Part A

(i) **Describe** the role of stomata in water regulation and photosynthesis in plants.

Part B

(i) **Construct** a graph that shows the mean stomatal density for both the upper and lower leaf surfaces of both plants. Be sure to include 95% confidence intervals.

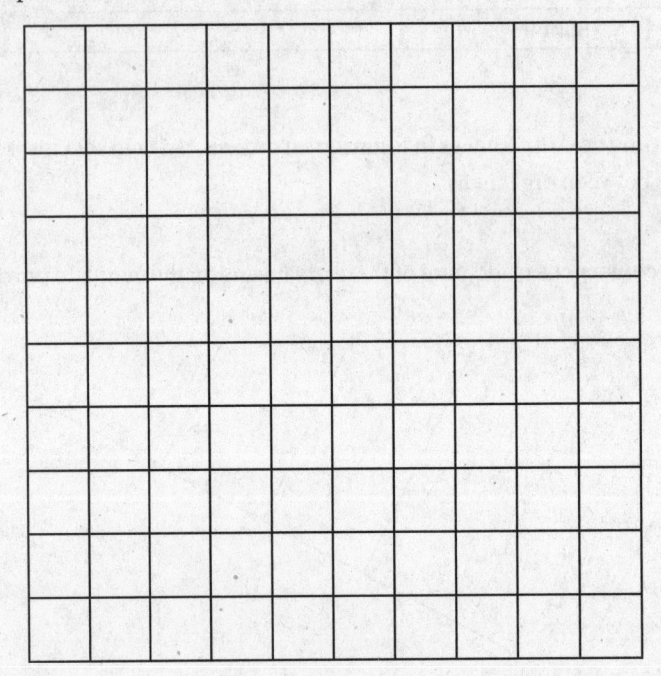

Part C

(i) **Analyze** the data to determine whether there is a statistically significant difference in the mean stomatal densities between the two plants.

Part D

(i) **Make a claim** explaining why the mean stomatal density on the upper surface of the leaves of *Passiflora incarnata* is different from the mean stomatal density on the upper surface of the leaves of *Quercus alba*.

(ii) **Justify** your claim with your knowledge of the role stomata play in water regulation and photosynthesis in plants.

3. A student conducts an investigation to measure the amount of gas produced by the aquatic plant *Elodea*. Equal amounts of *Elodea* are placed in inverted test tubes in four different water baths (at different temperatures) under lights (of the same intensity), as shown in the figure.

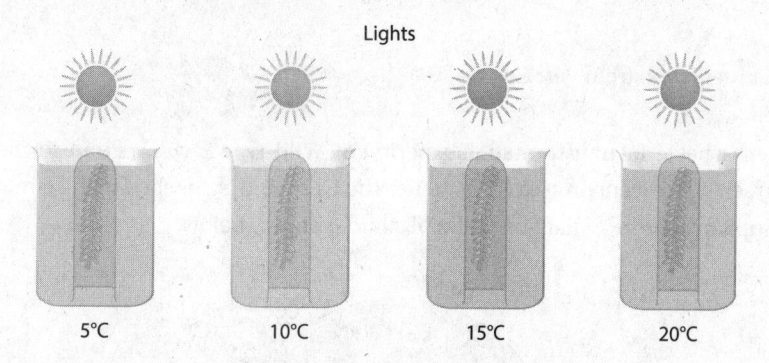

The size of the gas bubble at the top of the inverted test tube is measured after 30 minutes, and that data are displayed in the following table.

Temperature	Size of Gas Bubble After 30 Minutes (mm)
5°C	3.5
10°C	5.2
15°C	7.1
20°C	10.4

Part A

(i) **Describe** why a gas bubble would form in each tube.

Part B

(i) **Identify** the independent variable in this experiment.

(ii) **Identify** the dependent variable in this experiment.

(iii) **Identify** one experimental control in this experiment.

Part C

(i) **Predict** the size of the gas bubble if the experiment were repeated with a water bath at a temperature of 25°C.

Part D

(i) **Justify** your prediction from Part C.

4. Acetylcholine (ACh) is a neurotransmitter. The binding of acetylcholine to membrane receptors starts a cell signaling cascade in neurons.

Part A

(i) **Describe** the three steps in cell signaling.

Part B

(i) **Explain** how cell signals are amplified in the cell.

Part C

An autoimmune disorder produces antibodies that bind to the acetylcholine receptors, blocking them.

(i) **Predict** the effect this would have on the cell signaling pathway that involves acetylcholine.

Part D

(i) **Justify** your prediction from Part C.

5. A student performs a bacterial transformation experiment. Wild-type *E. coli* are transformed with a plasmid containing the green fluorescent protein (*GFP*) gene. The *GFP* gene is attached to the control elements of the repressible tryptophan operon. A diagram of the plasmid is shown below.

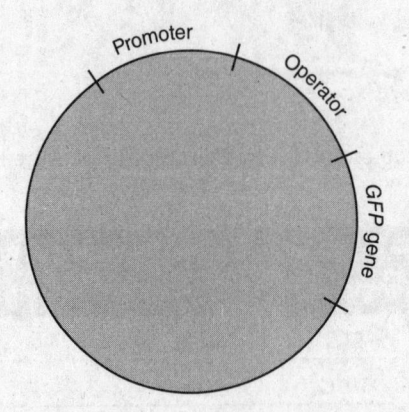

Part A

(i) **Describe** the roles of the promoter and operator in operons.

Part B

(i) **Explain** the effect, if any, of adding tryptophan to the medium in which the transformed bacteria are growing.

Part C

Tryptophan is added to the medium in which the bacteria are growing.

(i) Using an "X" for tryptophan and an "O" for the repressor protein, **represent** the likely locations of tryptophan and the repressor protein on or around the operon in the figure above.

Part D

(i) **Explain** how the regulation of the expression of the tryptophan operon relates to negative feedback.

6. In pea plants, round peas are dominant over wrinkled peas, and a purple flower color is dominant over a white flower color. Two pea plants (both of which are heterozygous for round peas and heterozygous for purple flowers) were crossed and produced 1,600 offspring. The number of offspring with each phenotype is shown in the following table.

Phenotype	Number of Offspring
Round peas, purple flowers	860
Round peas, white flowers	320
Wrinkled peas, purple flowers	310
Wrinkled peas, white flowers	110

Part A

(i) **Predict** the expected numbers of each phenotype. (Assume the genes for pea shape and flower color are on different chromosomes.)

Part B

(i) Using your predicted numbers from Part A, **calculate** the chi-square value for this experiment.

Part C

(i) **State** the null hypothesis for this experiment.

(ii) **Evaluate** the null hypothesis using a p-value of 0.05 and the chi-square value you calculated in Part B.

Part D

The genes for pea shape and flower color are located on different chromosomes.

(i) **Explain** how the data might have differed if the two genes had been located close together on the same chromosome.

Answer Explanations

Section I

1. **(B)** Ionic compounds have both positive and negative ions. Water is polar, and it has a partially negative end and a partially positive end. When ionic compounds dissolve in water, the partially negative ends of the water molecules are attracted to the positive ions of the ionic compound, and the partially positive ends of the water molecules are attracted to the negative ions of the ionic compound. Choices (A) and (C) are both incorrect because water is polar, not nonpolar. While hydrogen bonds will form between water molecules, hydrogen bonds do not form between water molecules and ions, so choice (D) is incorrect.

2. **(A)** The hydrophilic phosphate heads will be attracted to the surrounding water molecules. The hydrophobic lipid tails will be repelled by the water and orient themselves in the center of the micelle. Thus, part A represents the hydrophilic portions of the micelle, and part B represents the hydrophobic portions of the micelle. Choice (B) is incorrect because the phosphate heads are hydrophilic, not hydrophobic, and the tails are hydrophobic, not hydrophilic. Although choices (C) and (D) do refer to fats and associated fatty molecules that may be found in a plasma membrane of a cell, that is not what is represented in this image. Thus, both (C) and (D) are incorrect.

3. **(B)** This graph shows the means correctly plotted with 95% confidence intervals. The upper limit of the 95% confidence interval is the mean plus 2 $SE_{\bar{x}}$, and the lower limit is the mean minus 2 $SE_{\bar{x}}$. Choice (A) is incorrect because the upper and lower limits were calculated with only 1 $SE_{\bar{x}}$, not 2 $SE_{\bar{x}}$. Choice (C) is incorrect because even though it correctly plots the means, it does not correctly plot the 95% confidence intervals. Also, it is a line graph, which is inappropriate for categorical data, such as the data in this question. Choice (D) plots the means and confidence intervals correctly, but it is a line graph, which is inappropriate for the categorical data given in the question, so it is incorrect.

4. **(D)** The 95% confidence intervals of rubbing alcohol and olive oil do overlap, so these two liquids are the least likely to have a statistically significant difference. Choice (A) is incorrect because the confidence intervals for water and rubbing alcohol do *not* overlap. Since the confidence intervals for water and olive oil do not overlap, they are *more* likely to have a statistically significant difference, so choice (B) is incorrect. Similarly, the confidence intervals for water and rubbing alcohol do not overlap, so they also are likely to have a statistically significant difference, making choice (C) incorrect as well.

5. **(A)** Surfactants, like soap for example, are amphipathic molecules, which means they have both polar and nonpolar components. So the surfactant will disrupt the formation of hydrogen bonds between the water molecules, reducing the surface tension of the water. So fewer drops of water will be able to stay on the surface of the penny. Choices (B) and (C) are incorrect because the surfactant will neither disrupt nor strengthen the formation of polar ends, or covalent bonds, within the water molecules. The surfactant will *decrease* the surface tension of water, so choice (D) is incorrect.

6. **(C)** The nucleic acids, DNA and RNA, are the carriers of genetic information. Choices (A), (B), and (D) are incorrect because carbohydrates, lipids, and proteins do not carry genetic information.

7. **(D)** A polypeptide chain with 15 amino acids would have 14 peptide bonds. Breaking each peptide bond would require a hydrolysis reaction that would consume 1 molecule of water per bond, so 14 molecules of water would be needed. Choices (A) and (B) are incorrect because dehydration reactions form bonds, not break them. Choice (C) is incorrect because only 14 water molecules would be needed, not 15, and those water molecules would be consumed, not produced.

8. **(B)** Secondary structure in proteins is formed by hydrogen bonds between the amino acids in the polypeptide chain. Choice (A) is incorrect because primary structure is held together by

peptide bonds. Tertiary structure is driven primarily by hydrophobic and hydrophilic interactions and is not formed by peptide bonds, so choice (C) is incorrect. Choice (D) is incorrect because quaternary structure is only found in proteins that are composed of more than one subunit.

9. **(A)** Solutions with lower solute concentrations have higher water potentials; side A initially has a lower solute concentration, so it has a higher water potential. Choice (B) is incorrect because solutions with higher solute concentrations would have lower, not higher, water potentials. Aqueous solutions with different solute concentrations would have different water potentials, so choice (C) is incorrect. While sides A and B do contain the same solutes, they have different concentrations of each solute, so their water potentials are different, making choice (D) incorrect.

10. **(C)** As stated in the question, glucose can cross the membrane. Since side B has an initially higher concentration of glucose than side A, glucose will move down its concentration gradient from side B to side A, and the concentration of glucose on side A will increase. Choice (A) is incorrect because the concentration of glucose on side B will decrease, not increase, as glucose moves down its concentration gradient. Choices (B) and (D) are incorrect because albumin cannot cross the membrane according to the question.

11. **(A)** Side A has a lower solute concentration and a higher water potential than side B, so water will move from the higher water potential on side A to the lower water potential on side B. Since albumin cannot cross the barrier in order to satisfy equilibrium, water will flow toward side B in order to dilute the solution. Choice (B) correctly states that water will move from side A to side B, but it incorrectly explains the reason for the movement of water in this direction. Choices (C) and (D) are incorrect because water will move from side A to side B, not from side B to side A.

12. **(A)** Pepsin is found in the stomach, which has a pH of 2.5, so it is most likely to have its optimum activity at a pH of 2.5. Choice (B) shows the optimum activity at a pH of 7.0, so it is incorrect based on the information in the table. The graph in choice (C) shows the optimum activity at a basic pH of 10.0, so it is incorrect. Choice (D) shows a constant rate of increasing enzyme activity as the pH increases, so it is also incorrect.

13. **(A)** According to the table, amylase is found in the mouth and pepsin is found in the stomach. So changing the pH of those locations would most likely affect the activity of those enzymes. Lipase and trypsin are found in the small intestine, so changes in the pH of the mouth and the stomach acid would not likely affect their activity. Thus, choices (B), (C), and (D) are incorrect.

14. **(C)** Lipase and trypsin are found in the relatively neutral pH of the small intestine. So if the acidic contents of the stomach were added to the small intestine in an uncontrolled manner, the pH of the small intestine would most likely change. This would affect the activity of lipase and trypsin. Amylase is found in the mouth, so a change in the pH of the small intestine would not affect its activity. This rules out choices (A) and (D). Choice (B) is incorrect because pepsin is most active in the stomach and a change in the pH of the small intestine would not affect its activity.

15. **(A)** Anaerobic prokaryotes do not have mitochondria nor do they have access to oxygen, so they are limited to performing glycolysis. Choices (B), (C), and (D) would require the presence of oxygen and mitochondria, so those choices are all incorrect.

16. **(D)** Oxygen is a product of photosynthesis, so the areas where the bacteria congregated would correspond with the areas that carried out the most photosynthesis. According to the table, wavelengths of 420–470 nm and 660–700 nm correspond to blue-violet and red light, respectively, so those colors of light resulted in the greatest amount of photosynthesis. Photosynthetic pigments absorb light of specific wavelengths; not every wavelength of light can drive the process of photosynthesis, making choice (A) incorrect. Choice (B) is incorrect because aerobic bacteria are not required for the algae to produce oxygen; the concentration of aerobic bacteria are used as an indirect measure of the amount of oxygen produced. According to the table, green light has

a wavelength in the range of 495–570 nm, which does not correspond to the areas where the greatest numbers of bacteria congregated. Therefore, less oxygen was present in that area and less photosynthesis occurred under green light, so choice (C) is incorrect.

17. **(B)** Green light has wavelengths between 495–570 nm based on the table. Choices (A), (C), and (D) are incorrect because violet, orange, and red light have wavelengths in the ranges of 380–450 nm, 590–620 nm, and 620–750 nm, respectively.

18. **(C)** The role of photons of light energy in photosynthesis is to excite the electrons in the photosystems. Light does not activate the genes used in photosynthesis, so choice (A) is incorrect. Choice (B) is incorrect because carbon dioxide is consumed, not produced, during photosynthesis. Carbon dioxide is fixed in the light-independent reactions, not during the Krebs (citric acid) cycle, so choice (D) is incorrect.

19. **(B)** The temperature of the chambers is the independent variable. The number of germinating mung beans was kept constant, so it is not an independent variable. Thus, choice (A) is incorrect. The amounts of carbon dioxide produced and oxygen consumed are the dependent variables, so choices (C) and (D) are incorrect.

20. **(C)** The mung beans at 25° Celsius increased the level of carbon dioxide from 400 to 600 ppm over the 30-minute period, so the rate is $\frac{600 - 400 \text{ ppm}}{30 \text{ minutes}} = \frac{200 \text{ ppm}}{30 \text{ minutes}} = 6.67$ ppm CO_2 per minute. The mung beans at 4° Celsius increased the level of carbon dioxide from 400 to 450 ppm over the 30-minute period, so that rate is $\frac{450 - 400 \text{ ppm}}{30 \text{ minutes}} = \frac{50 \text{ ppm}}{30 \text{ minutes}} = 1.67$ ppm CO_2 per minute. Based on this data, the mung beans at 25° Celsius have a higher respiration rate than the mung beans at 4° Celsius. Thus, choice (C) is the only correct statement, and choices (A), (B), and (D) are incorrect.

21. **(C)** Paracrine signaling stimulates nearby cells. Autocrine signaling stimulates the same cell that produces the signal, so choice (A) is incorrect. Choice (B) is incorrect because juxtacrine signaling occurs between cells that are in direct contact (touching). Endocrine signals travel long distances, either through circulatory systems or through the air, so choice (D) is incorrect.

22. **(C)** Quorum sensing in bacteria affects nearby bacteria cells and is an example of paracrine signaling. Choice (A) is incorrect because synaptic signaling is specific to neurons that secrete neurochemicals across a synaptic gap. Cytokine signaling occurs between cells that are in direct contact, so choice (B) is incorrect. Endocrine signaling occurs over long distances, whereas quorum sensing occurs over short distances, so choice (D) is incorrect.

23. **(C)** Steroid hormones can cross the cell membrane unassisted and bind to receptors in the cell's cytoplasm. G-protein-linked receptors, tyrosine kinase receptors, and membrane ion channel receptors are all types of membrane-bound receptors, which bind to hydrophilic ligands, so choices (A), (B), and (D) are incorrect.

24. **(D)** Cyclic AMP (cAMP) and the calcium ion are both secondary messengers. Choices (A) and (C) are incorrect because cAMP and the calcium ion do not function as receptors. cAMP and the calcium ion are not ligands, so choice (B) is also incorrect.

25. **(B)** Gap junctions and plasmodesmata are channels that allow direct contact between cells and an exchange of materials, so they are examples of juxtacrine signaling. Choices (A), (C), and (D) do not describe the correct type of signaling.

26. **(B)** DNA is replicated during the S stage of the cell cycle. At the start of the S stage, each chromosome has one chromatid; at the end of the S stage, each chromosome has two identical chromatids, with twice the amount of DNA as they had at the beginning of the S stage. Choices (A), (C), and (D) are all incorrect because they list incorrect amounts of DNA and/or incorrect numbers of chromosomes.

27. **(A)** Cells that are not actively dividing and are fully differentiated will leave the cell cycle and enter the G0 stage. Choice (B) is incorrect because during G1, the cell is growing and getting ready

for cell division. During the S stage, the DNA is replicated in preparation for cell division, so choice (C) is also incorrect. Choice (D) is incorrect because during G2, the cell is making final preparations for mitosis.

28. **(C)** Cyclin-dependent kinases (CDKs) are present at near-constant levels throughout the cell cycle but do not become active until they are bound to cyclins. The levels of cyclins rise and fall during the cell cycle. Choices (A) and (B) are incorrect because cyclins are not present at near-constant levels during the cell cycle. Phosphatases remove phosphate groups from compounds, which usually inactivates them, so choice (D) is incorrect.

29. **(B)** Replication of the genetic material occurs during the S stage. Chromosomes align along the center of the cell during metaphase. The chromatids of chromosomes separate during anaphase. Choice (A) is incorrect because G1 and G2 are not the stages during which chromosomes align or separate. G2 is not when replication of the genetic material occurs, and telophase involves the formation of two new nuclei. So choice (C) is incorrect. Choice (D) is incorrect because prophase involves the condensation of the genetic material into visible chromosomes and the dissolution of the nuclear membrane.

30. **(A)** The first step in cell signaling is the binding of the ligand to the receptor. So if an antibody irreversibly bound to a receptor, the cell signaling process could not start. Choice (B) would affect the secondary messenger cyclic AMP (cAMP) because adenylyl cyclase catalyzes the formation of cAMP from ATP. Choices (C) and (D) are incorrect because protein kinases and protein phosphatases are involved in the transduction of the signal, which occurs after the binding of the ligand to the receptor.

31. **(A)** In this problem, it is stated that the population is NOT in Hardy-Weinberg equilibrium, nor is this population under the conditions necessary for Hardy-Weinberg equilibrium. So the Hardy-Weinberg equations cannot be used. To find the frequency of the nontaster allele, total up how many recessive alleles are in the population and divide by the total number of alleles in the population. There are a total of 1,000 individuals (520 + 400 + 80 = 1,000), so there are 2,000 alleles in total. Each of the 400 heterozygous individuals contributes one nontaster allele (400 nontaster alleles), and each of the 80 nontaster individuals contributes two nontaster alleles (160 nontaster alleles) for a total of 560 nontaster alleles: $\frac{560}{2,000} = 0.28$. Choices (B), (C), and (D) do not list the correct frequency.

32. **(B)** Carriers of the nontasting allele are heterozygous. There are 400 heterozygous individuals in a total population of 1,000 individuals, so the frequency of carriers is $\frac{400}{1,000} = 0.40$. Choices (A), (C), and (D) do not list the correct frequency.

33. **(B)** Prokaryotic cells do not contain mitochondria or chloroplasts, and eukaryotic cells have linear chromosomes, so cell B has characteristics of both prokaryotic and eukaryotic cells. Choice (A) is incorrect because circular chromosomes and the lack of mitochondria and chloroplasts are all characteristics of prokaryotic cells. Choices (C) and (D) are eukaryotic cells because they both have linear chromosomes and membrane-bound organelles.

34. **(D)** Cell D has both linear chromosomes and chloroplasts. Thus, of the four cells, cell D is most closely related to modern plants. Choice (A) is incorrect because it has circular chromosomes, which are a characteristic of prokaryotic cells. Choices (B) and (C) both do not contain chloroplasts, so neither is closely related to modern plants.

35. **(A)** Two species that occupy different habitats cannot interbreed and will remain separate species. Choice (B) is incorrect because temporal isolation occurs when organisms are active at different times of the day or have breeding seasons at different times of the year. Behavioral isolation involves different mating behaviors, for example mating dances or bird calls, so choice (C) is incorrect. Mechanical isolation occurs when morphological differences prevent the meeting of gametes, so choice (D) is incorrect.

36. **(B)** Gametic isolation occurs when the gametes of different species are incompatible, leading to reproductive isolation. Choice (A) is incorrect because mechanical isolation occurs when morphological differences prevent the meeting of gametes. Since both species of sea urchins can release their gametes into the water and their gametes can come into contact, there is no mechanical isolation in this instance. Reduced hybrid fertility occurs when a hybrid of the two organisms is produced but that hybrid is sterile, so choice (C) is incorrect. Choice (D) is incorrect because hybrid breakdown occurs when a fertile hybrid is produced but with each successive generation, the offspring of the hybrid are weaker or less fertile.

37. **(B)** The two strands of DNA are antiparallel, so the new strand would start with a 5′ end. Following the pairing rules for DNA (A-T and C-G), this leaves choice (B) as the correct answer. Choice (A) is incorrect because the new DNA strand is not antiparallel to the original strand. Choices (C) and (D) are incorrect because DNA replication does not incorporate U's into the new strand; the base pair uracil is only found in RNA.

38. **(B)** DNA replication is semiconservative, because each strand of DNA produced during DNA replication contains one original strand from the template DNA and one newly synthesized DNA strand. Choice (A) is incorrect because the entire DNA is copied during DNA replication. While newly synthesized DNA does contain parent and daughter DNA, it is not a random mixture of the two (one strand is from the parent molecule and one strand is newly synthesized). Thus, choice (C) is incorrect. Choice (D) is incorrect because the original DNA molecule is separated into two strands during DNA replication and each of those strands serves as a template for a new molecule of DNA.

39. **(A)** On the lagging strand, short fragments of new DNA are synthesized discontinuously during DNA replication. Choice (B) is incorrect because the leading strand of DNA is synthesized continuously, not discontinuously. DNA polymerase is used on both strands to synthesize DNA, so choice (C) is incorrect. Lagging strand and leading strand synthesis occur simultaneously, so choice (D) is incorrect.

40. **(D)** RNA polymerase is the enzyme that transcribes RNA from a DNA template. Choice (A) is incorrect because helicase is the enzyme that unwinds DNA at the start of DNA replication. Ligase is used to join DNA fragments (for example, the Okazaki fragments created during the discontinuous replication of the lagging strand of DNA), so choice (B) is incorrect. DNA polymerase synthesizes DNA during DNA replication, so choice (C) is incorrect.

41. **(C)** Introns are found in eukaryotes (and in archaea). Introns are not translated, so choice (A) is incorrect. Although introns are not directly converted into proteins, some introns may have a function in gene regulation, so choice (B) is incorrect. Eukaryotes do contain introns, so choice (D) is incorrect.

42. **(D)** Ribosomes are involved in translation, so an antibiotic that interferes with ribosomes would negatively impact the process of translation. Choices (A), (B), and (C) are incorrect because none of these processes (DNA replication, posttranscriptional modification of mRNA, or transcription) involve ribosomes.

43. **(B)** Eukaryotic mRNA contains introns, which are removed before the mRNA exits the nucleus and enters the cytoplasm, so eukaryotic mRNA is shorter than the DNA that codes for it. Choice (A) is incorrect because there is no evidence in the question that the scientist made any errors. While RNA is a less stable molecule than DNA due to the extra oxygen atom in RNA's five-carbon sugar, that is not the reason why eukaryotic mRNAs are shorter than the genes that code for them. Thus, choice (C) is incorrect. The poly-A tail added to eukaryotic mRNAs during posttranscriptional processing would make the mRNA longer, not shorter, so choice (D) is incorrect.

44. **(B)** The best way to solve this type of problem is to treat each gene separately and then combine the probabilities. The probability that the cross $AA \times Aa$ would produce Aa is $\frac{1}{2}$. Similarly, the

probability that the cross $Bb \times Bb$ would produce Bb is also $\frac{1}{2}$. The probability that the cross $Cc \times CC$ would produce Cc is also $\frac{1}{2}$. Therefore, the probability of all three events occurring is the product of their individual probabilities: $\frac{1}{2} \times \frac{1}{2} \times \frac{1}{2} = \frac{1}{8}$. Choices (A), (C), and (D) do not list the correct probability.

45. **(A)** The trait occurs in every generation and affects both males and females, so it is most likely an autosomal dominant trait. The offspring of a female with a sex-linked dominant trait would have a 50% chance of inheriting the trait, and males with the trait would pass it on to *all* of their female offspring. That's not what's shown here, so choice (B) is incorrect. Choice (C) is incorrect because sex-linked recessive traits are more often expressed in males than females because males have only one X chromosome. Mitochondrial inheritance occurs when a trait is passed from mother to offspring; males do not pass on mitochondrial genes to their offspring, so choice (D) is incorrect.

46. **(C)** Females have two X chromosomes, and the genes for color blindness are located on the X chromosome. Females inherit one X chromosome from each parent. If the female in the question stem does not express color blindness, one of her X chromosomes does not contain the color blindness allele. Her father was color blind, so the X chromosome she inherited from him must have the color blindness allele, and therefore she must be heterozygous (X^BX^b). If she has a child with a man who is not color blind (X^BY) and that child is a son, the likelihood the son would be color blind would be 50%, as shown in the following Punnett square.

	X^B	X^b
X^B	X^BX^B	X^BX^b
Y	X^BY	X^bY

Choices (A), (B), and (D) do not list the correct percentage.

47. **(A)** The color blindness trait is recessive. In order for the daughter to be color blind, she would have to inherit the allele from each of her parents. Since her father is not color blind, she cannot inherit the allele from him, so there is no probability for her to be color blind. Choices (B), (C), and (D) do not list the correct percentage.

48. **(D)** The woman in generation II is not color blind but her father in generation I was color blind, so the pedigree shown in choice (D) matches the text description. Choice (A) is incorrect because the male in generation II is not color blind (according to the text), so his square should not be shaded. The woman in generation II is not color blind (according to the text), so her circle should not be shaded. Therefore, choice (B) is incorrect. Choice (C) is incorrect for two reasons. First, the male in generation I was color blind, so his square should be shaded. Second, the male in generation II is not color blind, so his square should not be shaded.

49. **(D)** Weather is a density-independent factor that is not affected by population density. Choices (A), (B), and (C) all do depend on population density and are therefore incorrect.

50. **(D)** The number of individuals is reaching its maximum at point D, which represents the carrying capacity of the environment. Choices (A) and (B) represent the lag phase of growth and are therefore incorrect. Choice (C) is when the rate of population growth is the greatest, which is exponential growth. Thus, choice (C) does not represent the carrying capacity and is therefore incorrect.

51. **(C)** Exponential growth is when the rate of population growth is the greatest. On this graph, exponential growth is shown at point C. Choices (A) and (B) represent points during the lag phase of growth, so those choices are incorrect. The carrying capacity of the environment is represented by point D, so choice (D) is incorrect.

52. **(C)** Mutualism is a type of symbiotic relationship that results when both organisms benefit: the clown fish is protected from predators, and the sea anemone gains access to prey that is attracted to the clown fish. Commensalism is when one member benefits and the other member is neither helped nor harmed; that is not the case in this scenario, so choice (A) is incorrect. Competition is when two organisms are competing for the same limited resource; that is not what's happening in this situation, so choice (B) is incorrect. Parasitism is when one member benefits and the other member is harmed; no member is harmed in this relationship, so choice (D) is also incorrect.

53. **(C)** Keystone species have a disproportionately large effect on their environment relative to their abundance. When the sea otter populations declined in numbers, the biodiversity of the entire ecosystem was affected, so sea otters are a keystone species in this ecosystem. Autotrophs produce their own nutrition, usually through photosynthesis. Sea otters are not autotrophs, so choice (A) is incorrect. Detritivores eat dead organic material, allowing nutrients to be recycled through the ecosystem. This description does not match the role of sea otters in the question, so choice (B) is incorrect. Sea otters are a predator species, not a prey species, in this ecosystem, so choice (D) is also incorrect.

54. **(C)** Ecosystems can be disrupted and thrown out of balance if energy availability changes. A net gain in energy can result in the *gain* of mass by an organism; it would not result in the death of the organism, so choice (A) is incorrect. Energy is not recycled through ecosystems; each change in trophic level results in a loss of energy (from the consumer to the environment) as heat. So choice (B) is also incorrect. Choice (D) is incorrect because autotrophs, not heterotrophs, capture the energy in sunlight.

55. **(D)** *K*-selected populations do *not* far exceed their environment's carrying capacity; once a *K*-selected population reaches the carrying capacity, its numbers fluctuate moderately around the environment's carrying capacity. *r*-selected populations will fluctuate greatly around their environment's carrying capacity and far exceed it on occasion. Choices (A), (B), and (C) are all true statements about population growth. Without limiting factors, organisms will exhibit exponential growth. Density-dependent limits will cause populations to grow in a logistic manner. The resources available determine an environment's carrying capacity.

56. **(C)** In repressible operons, the repressor protein needs to bind to a corepressor to be capable of binding to the operator. The corepressor is often the product produced by the operon. In the *trp* operon, the amino acid tryptophan serves as the corepressor. When excess amounts of tryptophan are available, tryptophan binds to the repressor, which then binds to the operator. This prevents transcription of the operon by RNA polymerase and prevents the production of more tryptophan. Choice (A) is incorrect because while it does show tryptophan bound to the repressor, the repressor is not bound to the operator. In choice (B), the repressor is bound to the operator but without its corepressor, so choice (B) is incorrect. Choice (D) is incorrect because it shows tryptophan bound to the operator, but tryptophan cannot bind to the operator on its own.

57. **(A)** In inducible operons, the repressor protein can bind to the operator on its own, shutting down the operon. Inducible operons usually produce enzymes that are involved in digesting a compound. The inducer molecule is often the compound that is digested by the products of the inducible operon. When the inducer molecule is present, the inducer binds to the repressor and the repressor undergoes a shape change, which causes it to be released from the operator. RNA polymerase can then begin transcription of the operon. Choice (B) is incorrect because lactose is present but is not bound to the repressor protein (and the repressor is bound to the operator). In choice (C), lactose is bound to the repressor but the repressor is still bound to the operator, so this choice is incorrect. Choice (D) is incorrect because it shows lactose bound to the operator, but only repressors can bind directly to the operator.

58. **(C)** The cell membrane is composed of a phospholipid bilayer in which the polar phosphate heads are on both the inner and outer surfaces of the membrane and the hydrophobic lipid tails are in the interior of the membrane. The hydrophobic portions of transmembrane proteins would most likely be found in contact with the hydrophobic (nonpolar) lipids in the cell membrane. The hydrophobic portions of transmembrane proteins would be repelled by the polar phosphates of the phospholipids, so choice (A) is incorrect. Both the aqueous surroundings of the cell and the cell's cytosol are largely polar, so the hydrophobic portions of transmembrane proteins would be repelled by those environments, making choices (B) and (D) incorrect.

59. **(D)** Ribosomes produce proteins, so a cell that produced large amounts of a peptide (protein) hormone would likely have a high concentration of ribosomes. Lysosomes are involved in digesting compounds; they are not involved in protein production, so choice (A) is incorrect. Choice (B) is incorrect because centrioles are involved in organizing microtubules during cell division. Smooth endoplasmic reticulum detoxifies poisons and is involved in lipid production, not the production of peptides, so choice (C) is also incorrect.

60. **(B)** During fertilization, the vast majority of mitochondria are passed on to offspring from the mother because the eggs are thousands of times larger than the sperm. Thus, mitochondrial DNA is inherited from the mother. Males do not pass on their mitochondrial DNA to their offspring. Choice (A) is incorrect because sex-linked recessive genes can be passed on to offspring by either the mother or the father. Autosomal genes can be passed on to offspring by either parent, so choices (C) and (D) are incorrect.

Self-Analysis Chart for Section I

Use this chart to help identify areas where you need to focus your review. After completing the multiple-choice section and reviewing the answer explanations, circle the questions in this chart that you answered incorrectly. Note the units that you scored well in as well as the units that might require further study. Be sure to also review the answer explanations for Section II as all of those questions cover a wide variety of units and topics.

Unit	Questions
Unit 1: Chemistry of Life	1, 2, 3, 4, 5, 6, 7, 8
Unit 2: Cell Structure and Function	9, 10, 11, 33, 34, 58, 59
Unit 3: Cellular Energetics	12, 13, 14, 15, 16, 17, 18, 19, 20
Unit 4: Cell Communication and Cell Cycle	21, 22, 23, 24, 25, 26, 27, 28, 29, 30
Unit 5: Heredity	31, 32, 44, 45, 46, 47, 48, 60
Unit 6: Gene Expression and Regulation	37, 38, 39, 40, 41, 42, 43, 56, 57
Unit 7: Natural Selection	35, 36
Unit 8: Ecology	49, 50, 51, 52, 53, 54, 55

Section II

1. (A-i) Organisms that share a more recent common ancestor have had less time during which mutations can accumulate, so their proteins will have fewer amino acid differences than organisms that share a more distant common ancestor.

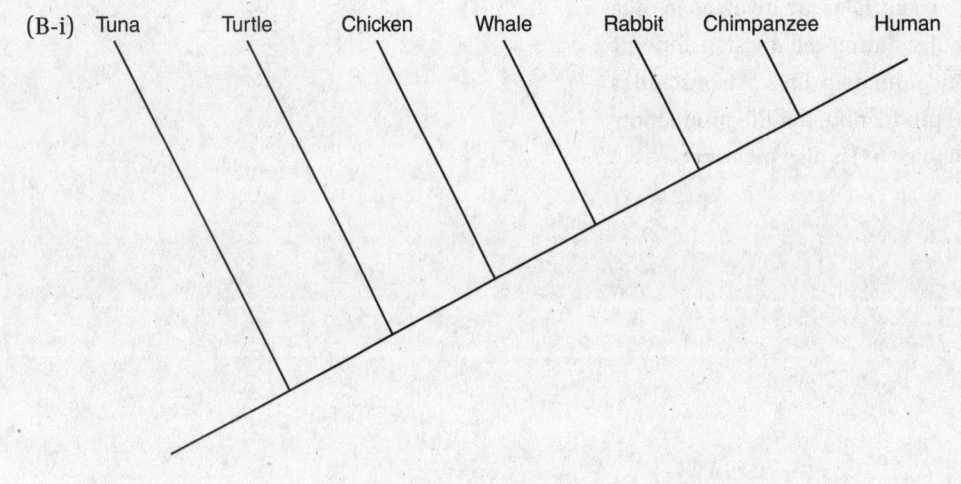

    (B-i) Tuna   Turtle   Chicken   Whale   Rabbit   Chimpanzee   Human

    (C-i) Chimpanzees have the fewest number of amino acid differences from humans in their cytochrome c protein, so they are placed closest to humans on the cladogram as the most recent common ancestor. Tuna have the greatest number of amino acid differences from humans in their cytochrome c protein, so they are placed farthest from humans on the cladogram as the most distant common ancestor. The other organisms are placed in order of the number of amino acid differences from humans. Those with fewer differences are closer to humans, and those with more differences are closer to tuna.

    (D-i) One could look at the number of amino acid differences in a different protein to see if chickens still have fewer amino acid differences from humans than turtles do. One could also look at DNA sequences of the three species or look for homologous structures.

2. (A-i) When stomata open, the plant can take in the carbon dioxide it needs for photosynthesis. However, when stomata open, the plant can also lose water through the process of transpiration. The lower the water potential in the plant's surroundings, the more likely the plant is to lose water when its stomata are open.

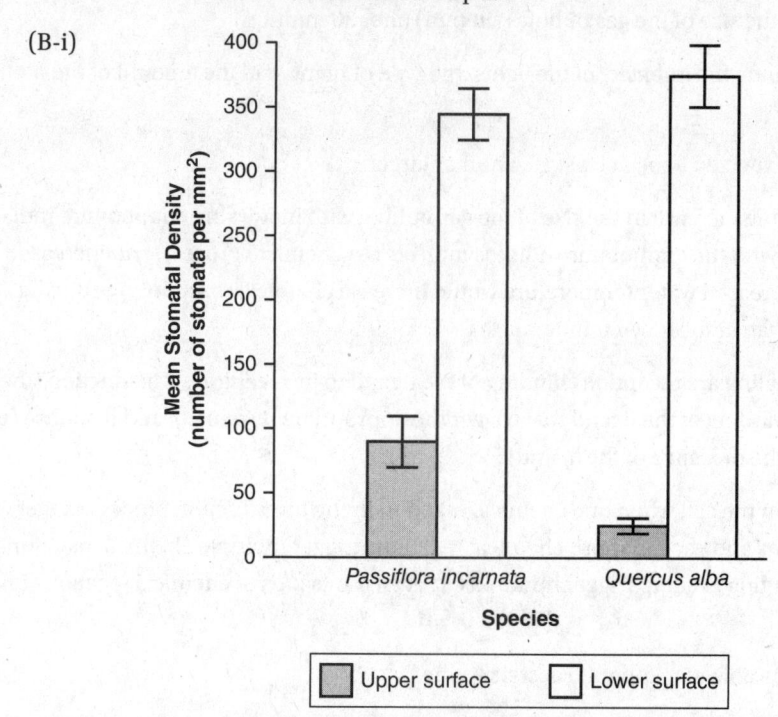

(B-i)

[Bar graph showing Mean Stomatal Density (number of stomata per mm^2) on the y-axis (0 to 400) for two species on the x-axis: Passiflora incarnata and Quercus alba. For Passiflora incarnata: upper surface ~90, lower surface ~345. For Quercus alba: upper surface ~25, lower surface ~375. Error bars shown. Legend: Upper surface, Lower surface.]

(C-i) In looking at the data for the lower surfaces of the leaves of the two species, the 95% confidence intervals overlap. So one cannot be sure if there is a statistically significant difference between the mean stomatal densities on the lower surfaces of the leaves of the *Passiflora incarnata* and *Quercus alba*. However, the 95% confidence intervals for the upper surfaces of the two species do not overlap. (The lower limit of the 95% confidence interval of *Passiflora incarnata* is 70 stomata per mm^2, and the upper limit of the 95% confidence interval of *Quercus alba* is 31 stomata per mm^2.) Therefore, there is likely a statistically significant difference in the mean stomatal densities found on the upper surfaces of the leaves of the two species.

(D-i and ii) The mean stomatal density on the upper surface of the leaves of *Passiflora incarnata* is greater than the mean stomatal density on the upper surface of the leaves of *Quercus alba* because *Passiflora incarnata* typically grows in more humid environments than does *Quercus alba*. If the environment surrounding a plant is more humid, the difference in water potential between the plant and its surroundings will be less, and the plant will likely lose less water to its environment when it opens its stomata to take in the carbon dioxide needed for photosynthesis. In drier environments, like the ones in which *Quercus alba* is typically found, the difference in water potential between the plant and its environment is greater. When the plant opens its stomata to take in the carbon dioxide needed for photosynthesis, it is more likely to lose water to its environment.

3. (A-i) Plants produce oxygen during photosynthesis, and the oxygen produced by the *Elodea* during photosynthesis would accumulate in the inverted tubes.

   (B-i) The independent variable is the temperature of the water baths.

   (B-ii) The dependent variable is the size of the gas bubbles (in mm) after 30 minutes.

   (B-iii) Experimental controls include the intensity of the lights, the type of plant, and the amount of the plant in the tubes.

   (C-i) The gas bubble would be expected to be at least 10.4 mm or larger.

   (D-i) Based on the data shown, the increase in the size of the gas bubble with increasing temperature indicates a direct relationship with the temperature. Photosynthesis is a chemical process, and increased kinetic energy from the increased water temperature would increase chemical reactions, producing more oxygen gas at 25°C than at the lower temperatures.

4. (A-i) The three steps in cell signaling are reception (binding of the ligand to its receptor), transduction (the steps that occur after the binding of the ligand that trigger the appropriate response), and response (the cell's desired response to the presence of the ligand).

   (B-i) Cell signals are amplified in the cell when one chemical reaction in the transduction process triggers multiple chemical reactions and each of those chemical reactions trigger multiple chemical reactions. This allows one ligand-binding event to trigger hundreds or even thousands of chemical reactions in the cell.

   (C-i) The cell would no longer be able to respond to acetylcholine.

   (D-i) If an antibody irreversibly binds to the acetylcholine receptors, acetylcholine cannot bind to the receptors and can no longer trigger the appropriate response in the cell.

5. (A-i) The promoter serves as a binding site for RNA polymerase. The operator serves as a binding site for the repressor protein.

   (B-i) Since the tryptophan operon is a repressible operon, the presence of tryptophan would turn the operon off. Tryptophan serves as a corepressor. It would bind to the repressor protein, and then the repressor protein would bind to the operator, blocking RNA polymerase's ability to transcribe the genes on the plasmid. Green fluorescent protein would not be expressed.

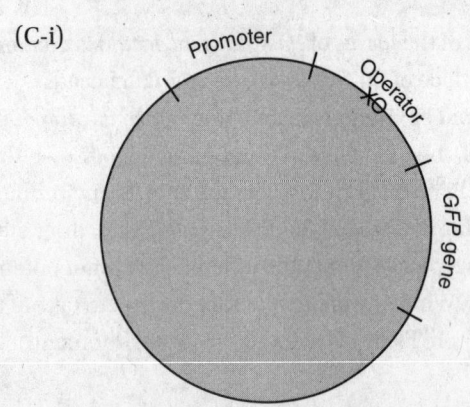

   (C-i)

   (D-i) Negative feedback helps maintain homeostasis and near-constant levels of different molecules in a cell. When there is excess tryptophan in a cell, the tryptophan binds to the repressor protein, which then binds to the operator of the tryptophan operon, preventing the production of more tryptophan. This prevents the cell from wasting energy and resources on producing more tryptophan than it needs.

6. (A-i) If the pea plants are heterozygous for both traits, the cross would be $RrPp \times RrPp$. So $\frac{3}{4}$ of the offspring would be expected to have round peas, and $\frac{1}{4}$ of the offspring would be expected to have wrinkled peas. Additionally, $\frac{3}{4}$ of the offspring would be expected to have purple flowers, and $\frac{1}{4}$ of the offspring would be expected to have white flowers. So the expected numbers of offspring would be:

Round peas and purple flowers $= \frac{3}{4} \times \frac{3}{4} \times 1{,}600 = 900$

Round peas and white flowers $= \frac{3}{4} \times \frac{1}{4} \times 1{,}600 = 300$

Wrinkled peas and purple flowers $= \frac{1}{4} \times \frac{3}{4} \times 1{,}600 = 300$

Wrinkled peas and white flowers $= \frac{1}{4} \times \frac{1}{4} \times 1{,}600 = 100$

(B-i) Using the chi-square formula:

$$\chi^2 = \sum \frac{(\text{observed} - \text{expected})^2}{\text{expected}}$$

$$\frac{(860-900)^2}{900} + \frac{(320-300)^2}{300} + \frac{(310-300)^2}{300} + \frac{(110-100)^2}{100} = 4.44$$

(C-i and ii) The null hypothesis is that there is no statistically significant difference between the observed and expected data. Since there are four possible phenotypic outcomes in the experiment, there are 3 degrees of freedom. Using a p-value of 0.05 and 3 degrees of freedom, the critical value from the chi-square table is 7.81. Since the calculated chi-square value is less than the critical value, fail to reject the null hypothesis.

(D-i) If the genes had been located on the same chromosome, they would be linked genes and would not assort independently. Genes that are linked would not follow the Mendelian ratios predicted by Punnett squares.

ANSWER SHEET
Practice Test 2

1. Ⓐ Ⓑ Ⓒ Ⓓ
2. Ⓐ Ⓑ Ⓒ Ⓓ
3. Ⓐ Ⓑ Ⓒ Ⓓ
4. Ⓐ Ⓑ Ⓒ Ⓓ
5. Ⓐ Ⓑ Ⓒ Ⓓ
6. Ⓐ Ⓑ Ⓒ Ⓓ
7. Ⓐ Ⓑ Ⓒ Ⓓ
8. Ⓐ Ⓑ Ⓒ Ⓓ
9. Ⓐ Ⓑ Ⓒ Ⓓ
10. Ⓐ Ⓑ Ⓒ Ⓓ
11. Ⓐ Ⓑ Ⓒ Ⓓ
12. Ⓐ Ⓑ Ⓒ Ⓓ
13. Ⓐ Ⓑ Ⓒ Ⓓ
14. Ⓐ Ⓑ Ⓒ Ⓓ
15. Ⓐ Ⓑ Ⓒ Ⓓ
16. Ⓐ Ⓑ Ⓒ Ⓓ
17. Ⓐ Ⓑ Ⓒ Ⓓ
18. Ⓐ Ⓑ Ⓒ Ⓓ
19. Ⓐ Ⓑ Ⓒ Ⓓ
20. Ⓐ Ⓑ Ⓒ Ⓓ
21. Ⓐ Ⓑ Ⓒ Ⓓ
22. Ⓐ Ⓑ Ⓒ Ⓓ
23. Ⓐ Ⓑ Ⓒ Ⓓ
24. Ⓐ Ⓑ Ⓒ Ⓓ
25. Ⓐ Ⓑ Ⓒ Ⓓ
26. Ⓐ Ⓑ Ⓒ Ⓓ
27. Ⓐ Ⓑ Ⓒ Ⓓ
28. Ⓐ Ⓑ Ⓒ Ⓓ
29. Ⓐ Ⓑ Ⓒ Ⓓ
30. Ⓐ Ⓑ Ⓒ Ⓓ
31. Ⓐ Ⓑ Ⓒ Ⓓ
32. Ⓐ Ⓑ Ⓒ Ⓓ
33. Ⓐ Ⓑ Ⓒ Ⓓ
34. Ⓐ Ⓑ Ⓒ Ⓓ
35. Ⓐ Ⓑ Ⓒ Ⓓ
36. Ⓐ Ⓑ Ⓒ Ⓓ
37. Ⓐ Ⓑ Ⓒ Ⓓ
38. Ⓐ Ⓑ Ⓒ Ⓓ
39. Ⓐ Ⓑ Ⓒ Ⓓ
40. Ⓐ Ⓑ Ⓒ Ⓓ
41. Ⓐ Ⓑ Ⓒ Ⓓ
42. Ⓐ Ⓑ Ⓒ Ⓓ
43. Ⓐ Ⓑ Ⓒ Ⓓ
44. Ⓐ Ⓑ Ⓒ Ⓓ
45. Ⓐ Ⓑ Ⓒ Ⓓ
46. Ⓐ Ⓑ Ⓒ Ⓓ
47. Ⓐ Ⓑ Ⓒ Ⓓ
48. Ⓐ Ⓑ Ⓒ Ⓓ
49. Ⓐ Ⓑ Ⓒ Ⓓ
50. Ⓐ Ⓑ Ⓒ Ⓓ
51. Ⓐ Ⓑ Ⓒ Ⓓ
52. Ⓐ Ⓑ Ⓒ Ⓓ
53. Ⓐ Ⓑ Ⓒ Ⓓ
54. Ⓐ Ⓑ Ⓒ Ⓓ
55. Ⓐ Ⓑ Ⓒ Ⓓ
56. Ⓐ Ⓑ Ⓒ Ⓓ
57. Ⓐ Ⓑ Ⓒ Ⓓ
58. Ⓐ Ⓑ Ⓒ Ⓓ
59. Ⓐ Ⓑ Ⓒ Ⓓ
60. Ⓐ Ⓑ Ⓒ Ⓓ

PRACTICE TEST 2

Practice Test 2

Section I: Multiple-Choice

TIME: 90 MINUTES

DIRECTIONS: For each question or incomplete statement, select the choice that best answers the question or completes the statement.

Questions 1 and 2

Refer to the following table, which describes some of the birds known as "Darwin's finches," which live on the Galápagos Islands.

Finch Species	Type of Beak	Food Source	Tree or Ground Finch
Cactus Finch	Probing	Cactus	Ground
Medium Ground Finch	Crushing	Seeds	Ground
Vegetarian Tree Finch	Parrot	Fruit	Tree
Warbler Finch	Probing	Insects	Ground
Woodpecker Finch	Probing	Insects	Tree

1. A drought eliminates most insects and fleshy fruits from the islands, and the predominant food source is now hard-shelled seeds. Birds with which type of beak would be expected to increase in numbers?

    (A) parrot beak
    (B) grasping beak
    (C) probing beak
    (D) crushing beak

2. A hurricane eliminates most trees from the islands. Which type of birds would be expected to decrease in numbers?

    (A) cactus finch
    (B) warbler finch
    (C) woodpecker finch
    (D) medium ground finch

3. The lyrebird (*Menura novaehollandiae*) is an Australian songbird and is known for its ability to mimic other sounds and produce bird songs with great complexity. The more complex the bird song, the more female lyrebirds are attracted. This is an example of which type of selection?

    (A) artificial
    (B) directional
    (C) disruptive
    (D) sexual

Questions 4–6

Drosophila body color can be brown or black, with the *B* (brown) allele dominant to the *b* (black) allele. A large population of *Drosophila* in Hawaii was studied over a period of 10 years, and the frequency of *Drosophila* with black bodies is shown over time in the figure.

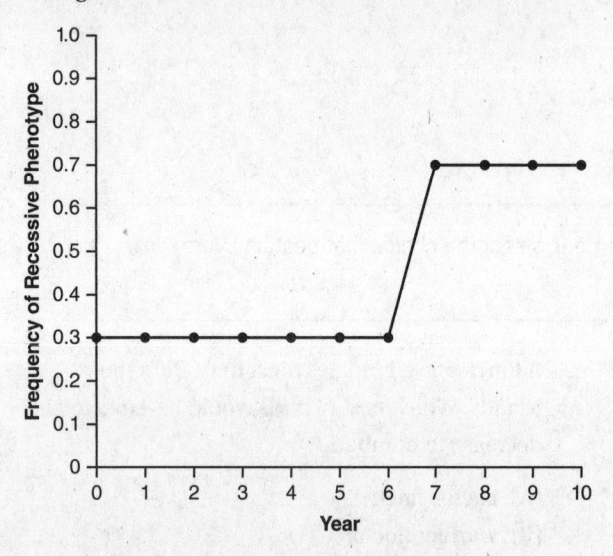

4. Assuming the population was in Hardy-Weinberg equilibrium during year 4, what percentage of the population was heterozygous during year 4 of the study?

   (A) 30.0%
   (B) 49.5%
   (C) 54.8%
   (D) 70.0%

5. Assuming the population was in Hardy-Weinberg equilibrium during year 10, what would be a characteristic of the population during year 10 of the study?

   (A) small population size
   (B) random mating
   (C) extensive migration
   (D) heterozygote advantage

6. Which of the following statements best describes the change in the population between year 6 and year 7 and a possible reason for the change?

   (A) The frequency of the recessive phenotype decreased because the black body phenotype was 100% lethal.
   (B) The frequency of the recessive phenotype increased because a change in the environment provided a selective advantage to flies with the black body phenotype.
   (C) The frequency of the recessive phenotype decreased because the dominant phenotype is always the most common phenotype.
   (D) The frequency of the recessive phenotype increased because individual flies adapted to a new, darker environment.

7. In a rural area of the country, there is a high incidence of a rare condition called methemoglobinemia, in which a mutation causes the skin of affected individuals to take on a bluish tint when under low oxygen conditions. All of the individuals with methemoglobinemia are descendants of one individual who immigrated to the area from another country. This is an example of

   (A) natural selection.
   (B) heterozygote advantage.
   (C) bottleneck effect.
   (D) founder effect.

Questions 8 and 9

The table shows the presence (+) or absence (−) of five traits in species L, M, N, O, and P.

Species	Trait 1	2	3	4	5
L	+	+	+	+	−
M	+	+	+	+	+
N	−	+	−	−	−
O	−	+	+	−	−
P	−	+	+	+	−

8. Which of the following cladograms best represents the data in the table?

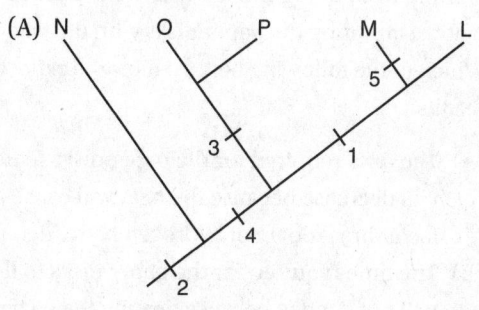

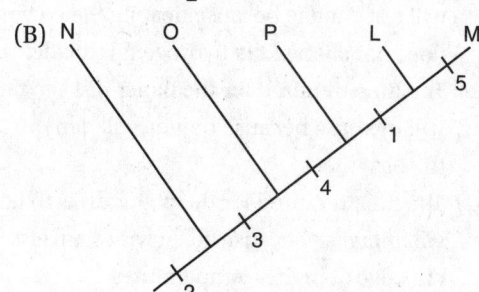

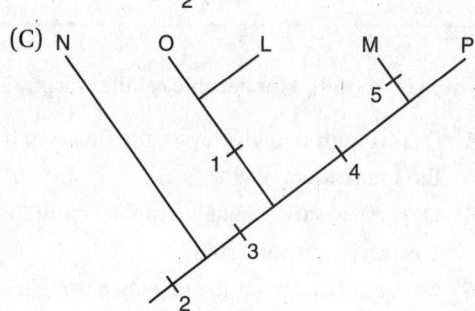

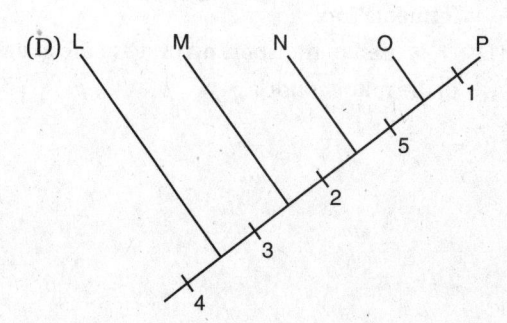

   (A) [cladogram with N, O, P, M, L; nodes 2, 4, 3, 1, 5]
   (B) [cladogram with N, O, P, L, M; nodes 2, 3, 4, 1, 5]
   (C) [cladogram with N, O, L, M, P; nodes 2, 3, 1, 4, 5]
   (D) [cladogram with L, M, N, O, P; nodes 4, 3, 2, 5, 1]

9. Which species represents the outgroup in the cladogram?

   (A) M
   (B) N
   (C) O
   (D) P

10. Two species of blind mole rats, *Spalax ehrenbergi* and *Spalax galili*, live in the same habitat and breed during the same season. However, *Spalax ehrenbergi* and *Spalax galili* do not interbreed. Studies have revealed that while they share a common ancestor, the DNA of the two species is significantly different. This is an example of

    (A) allopatric speciation.
    (B) sympatric speciation.
    (C) polygenic inheritance.
    (D) temporal isolation.

11. Fossil evidence shows that throughout Earth's history, there were long periods of time in which the number and types of species found on Earth were relatively constant, followed by short periods of rapid change in the number and types of species found on Earth. Which of the following best describes this observation?

    (A) gradualism
    (B) punctuated equilibrium
    (C) convergent evolution
    (D) coevolution

12. Which of the following is an example of natural selection?

    (A) A farmer breeds her two biggest cattle in an effort to obtain the largest offspring.
    (B) A person spends time in the Sun and his skin tans and becomes darker.
    (C) A horse breeder breeds two winners of the Kentucky Derby in order to produce a faster horse.
    (D) A mouse with lighter-colored fur is more likely to survive and reproduce in a sandy environment than a mouse with darker-colored fur.

Questions 13–16

Catalase is an enzyme that breaks down hydrogen peroxide (H_2O_2) as shown in the following reaction:

$$2H_2O_{2(l)} \xrightarrow{\text{Catalase}} 2H_2O_{(l)} + O_{2(g)}$$

A student isolates catalase from potato tissue and places it on ice. The student uses a hole punch to make equally sized paper disks, saturates each disk in the catalase solution, and then places them in beakers with different concentrations of H_2O_2. As the enzyme-catalyzed reaction proceeds, oxygen bubbles are produced and they cling to the paper disks, which then float to the surface. The time required for the catalase-saturated filter paper to float is recorded. The experiment is repeated five times for each concentration of hydrogen peroxide. The data are shown in the table.

Concentration of $H_2O_{2(l)}$	Mean Time (sec) $\pm$ 1 SEM*
1%	100 ± 7.0
3%	75 ± 5.2
6%	45 ± 4.3
9%	30 ± 3.9

*Standard Error of the Mean

13. Which of the following would be a valid control for the experiment?

    (A) changing the size of the paper disk
    (B) saturating the paper disk in distilled water
    (C) reducing the temperature of the hydrogen peroxide
    (D) pureeing the potatoes with a pH 3 buffer instead of water

14. Based on the data obtained, there is least likely to be a statistically significant difference between which two concentrations of hydrogen peroxide?

    (A) 1% and 3%
    (B) 3% and 6%
    (C) 6% and 9%
    (D) 1% and 6%

15. What is the most likely explanation for the relationship between time for the disk to float and concentration of hydrogen peroxide?

    (A) Higher hydrogen peroxide concentrations generate more heat, causing disks to rise faster.
    (B) Increasing the concentration of hydrogen peroxide changes the active site on the catalase, increasing its reaction rate.
    (C) The increasing levels of oxygen produced in the reaction functioned as an allosteric inhibitor of the enzyme.
    (D) More substrate was available to the enzyme.

16. In a follow-up experiment, the student placed the enzyme in a boiling water bath for one minute before saturating the paper disks with the enzyme. Which of the following best predicts the expected results?

    (A) The time required for the paper disks to float will decrease because the heat will cause a faster breakdown of hydrogen peroxide.
    (B) The time required for the paper disks to float will not change because heating the enzyme does not change the hydrogen peroxide.
    (C) The time required for the paper disks to float will increase because boiling will denature the enzyme.
    (D) The time required for the paper disks to float will increase because all enzymes are less effective at higher temperatures.

17. Why is O_2 required for aerobic cellular respiration?

    (A) O_2 is required in glycolysis, the first step in the breakdown of glucose.
    (B) O_2 serves as the final electron acceptor in the electron transport chain.
    (C) O_2 is a substrate for the enzymes used in fermentation.
    (D) O_2 is used to transport pyruvate to the matrix of the mitochondria.

18. Which of the following questions is most relevant in understanding the Krebs cycle?

    (A) How is glucose broken down into pyruvate?
    (B) How is the proton gradient in the mitochondria formed?
    (C) How is NAD⁺ reduced to NADH?
    (D) How does ATP synthase produce ATP?

19. Which of the following is a correct statement about fermentation and the electron transport chain?

    (A) Both processes require O_2.
    (B) Both processes oxidize NADH to NAD⁺.
    (C) Both processes occur in the mitochondria.
    (D) Both processes result in the formation of metabolic water.

20. Telomeres are repetitive sequences of DNA found on the ends of linear chromosomes. With each cycle of DNA replication, a few bases are lost from the telomeres and the length of telomeres decrease. What is the function of telomeres?

    (A) to assist in the synthesis of proteins
    (B) to protect the genetic information in chromosomes during cell division
    (C) to correct mutations in DNA
    (D) to inhibit cell division

21. A celery stalk placed in distilled water for a few hours becomes stiff while a celery stalk placed in a 1.0 molar sodium chloride solution wilts. Which of the following best explains these observations?

    (A) Celery cells have a higher water potential than both the distilled water and the 1.0 molar sodium chloride solution.
    (B) Celery cells have a lower water potential than both the distilled water and the 1.0 molar sodium chloride solution.
    (C) Celery cells have a higher water potential than the distilled water and a lower water potential than the 1.0 molar sodium chloride solution.
    (D) Celery cells have a lower water potential than the distilled water and a higher water potential than the 1.0 molar sodium chloride solution.

22. When carbon dioxide is added to water, carbonic acid forms and the pH of the solution decreases. Ten alginate beads of the photosynthetic algae *Chlorella vulgaris* are placed in clear glass vials, along with a pH indicator dye. One vial is wrapped in foil to exclude all light. Both vials are placed under a light source for 40 minutes. The pH is recorded every 10 minutes. The data are shown in the table.

Time (minutes)	pH of Vial with Foil	pH of Vial Without Foil
0	7	7
10	6.5	7.2
20	6.0	7.7
30	5.5	8.2
40	5.0	8.5

Which of the following best explains the changes in pH seen in the vials?

(A) The vial covered with foil only performs cellular respiration, and the vial not covered with foil only performs photosynthesis.
(B) The vial covered with foil only performs cellular respiration, and the vial not covered with foil performs both photosynthesis and cellular respiration.
(C) The vial covered with foil only performs photosynthesis, and the vial not covered with foil performs both photosynthesis and cellular respiration.
(D) The vial covered with foil only performs photosynthesis, and the vial not covered with foil only performs cellular respiration.

Questions 23 and 24

A study compared cancer rates in five species with their different body masses. The results are shown in the table.

Species	Cancer Rate (%)	Body Mass (kg)
Elephant	3	5,000
Marmoset	16	0.4
Prairie Dog	18	2.0
Tasmanian Devil	50	10
Tiger	12	150

23. A scientist proposes that since larger animals experience more cell divisions over their lifetimes, they should have a greater cancer rate. Do the data shown support this theory?

    (A) yes, because the largest animals have the greatest cancer rates
    (B) yes, because the smallest animals have the lowest cancer rates
    (C) no, because the largest animals have the greatest cancer rates
    (D) no, because the largest animals have the lowest cancer rates

24. The gene *TP53* is a tumor suppressor gene, which produces a gene product that lowers cancer rates. Which animal is most likely to have the highest number of copies of the cancer-reducing *TP53* gene?

    (A) elephant
    (B) marmoset
    (C) Tasmanian devil
    (D) tiger

25. In *Drosophila melanogaster*, the gene for eye color is on the X chromosome. Female *Drosophila* have two X chromosomes, and male *Drosophila* have one X chromosome. The allele for red eye color is dominant, and the allele for white eye color is recessive. A female with white eyes is mated with a male with red eyes. Which of the following best predicts their probable offspring?

    (A) Half of the male offspring and half of the female offspring will have white eyes, and half of the male offspring and half of the female offspring will have red eyes.
    (B) Half of the male offspring will have white eyes, half of the male offspring will have red eyes, and all of the female offspring will have white eyes.
    (C) All of the male offspring will have red eyes, and all of the female offspring will have white eyes.
    (D) All of the male offspring will have white eyes, and all of the female offspring will have red eyes.

26. The bacteria *Streptococcus pneumonia* has a rough strain (R) and a smooth strain (S). The S strain is covered in a capsule, which makes the virulent (disease-causing) bacteria more difficult to detect by the immune system. In a classic experiment, Frederick Griffith injected different combinations of live and heat-killed R strain and S strain *Streptococcus pneumonia* bacteria into mice and obtained the following results.

Trial	Contents of Injection	Nonvirulent	Virulent
1	Live R bacteria	X	
2	Live S bacteria		X
3	Heat-killed R bacteria	X	
4	Heat-killed S bacteria	X	
5	Live R bacteria plus heat-killed S bacteria		X

Which process most likely occurred in trial 5?

(A) transcription
(B) transduction
(C) transformation
(D) translocation

Questions 27–29

Sickle cell disease (SCD) is an autosomal recessive disorder. If an individual has just one copy of the sickle cell allele, the person has sickle cell trait (SCT) and is more resistant to malaria than individuals with no copies of the sickle cell allele.

27. If a person with sickle cell disease has a child with a person who has the sickle cell trait, what is the likelihood the child will have sickle cell disease?

(A) 0%
(B) 25%
(C) 50%
(D) 75%

28. The fact that individuals with sickle cell trait (SCT) are more resistant to malaria than the general population is an example of which phenomenon?

(A) polygenic inheritance
(B) sexual selection
(C) sex-linked inheritance
(D) heterozygote advantage

29. If an individual who has the sickle cell trait (SCT) has a child with a person who does not have sickle cell disease (SCD) but whose father did have SCD, what is the likelihood the child will have the SCT?

(A) 0%
(B) 25%
(C) 50%
(D) 75%

30. Research by climate scientists indicates that rising ocean temperatures may be causing El Niño storms to increase in frequency. These El Niño storms bring more rainfall to the Galápagos Islands. Increased water availability allows an invasive species of the fly *Philornis downsi* to flourish on the islands. This fly feeds on the blood of finch nestlings. What is the ecological relationship between the fly *Philornis downsi* and these finches?

(A) commensalism
(B) competition
(C) mutualism
(D) parasitism

31. Since the early 20th century, over 90% of the seagrass plants off of the coast of the United Kingdom have disappeared. Over one million seagrass seeds are being planted in the area to offset this loss. Which of the following would be the most likely benefit of this effort?

(A) reduction of habitats for some invasive species
(B) removal of oxygen from the atmosphere as seagrass performs photosynthesis
(C) absorption of carbon dioxide in the atmosphere as seagrass performs photosynthesis
(D) increased soil erosion in the area where seagrass is planted

32. A student performed an experiment to investigate how many drops of different liquids could be placed on a penny before the liquids would overflow off of the surface of the penny. The table shows the data obtained.

Liquid	Mean Number of Drops ($\pm$2 SEM*)
Ethanol	22 ± 1.26
Vegetable Oil	6 ± 1.16
Distilled Water	28 ± 1.26
0.5 Molar NaCl	16 ± 1.08

*Standard Error of the Mean

Which of the following best explains the experimental results?

(A) Vegetable oil forms more hydrogen bonds than the other liquids tested.
(B) Distilled water forms more hydrogen bonds than the other liquids tested.
(C) The ions in NaCl are repelled by the metal in the penny.
(D) Ethanol is a larger molecule than water, so ethanol forms more hydrogen bonds than water.

33. A U-tube apparatus is separated into two chambers by a semipermeable membrane, as shown in the figure. Water and iodine molecules can pass through the membrane, but starch molecules cannot pass through the membrane. Starch molecules are stained a dark blue or black color when in direct contact with iodine.

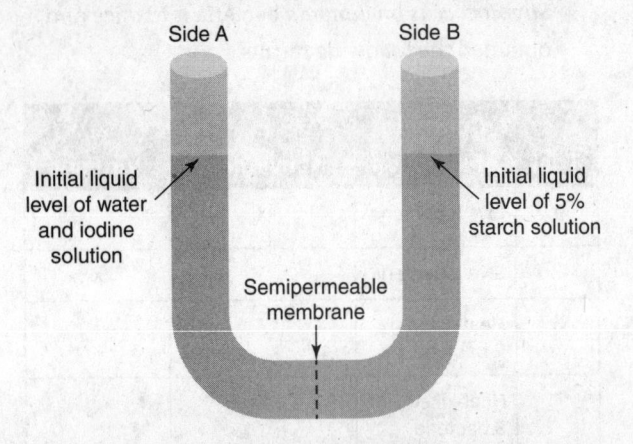

One hundred milliliters of an aqueous iodine solution are placed in the chamber on the left side of the U-tube apparatus (side A). One hundred milliliters of a 5% starch solution are placed in the chamber on the right side of the U-tube apparatus (side B). Which of the following is the most likely result after allowing the liquids to equilibrate for six hours?

(A) The water level on side A will rise, and side A will turn dark blue when the iodine and starch interact.
(B) The water level on side A will rise, and side B will turn dark blue when the iodine and starch interact.
(C) The water level on side B will rise, and side A will turn dark blue when the iodine and starch interact.
(D) The water level on side B will rise, and side B will turn dark blue when the iodine and starch interact.

34. A translocation between human chromosomes 9 and 22 results in the creation of a fusion gene called *BCR-ABL*. The *BCR-ABL* fusion can lead to an abnormally high cell division rate and can result in chronic myelogenous leukemia (CML). Which of the following would most likely slow the rate of cell division in cells that contain the BCR-ABL kinase?

    (A) addition of a cofactor
    (B) addition of a coenzyme
    (C) addition of a competitive inhibitor
    (D) addition of a transcription factor

35. Red blood cells are placed in distilled water. Which of the following would be the most likely result?

    (A) The red blood cells will shrivel as water moves from the area of higher water potential inside the red blood cells into the area of lower water potential in the beaker.
    (B) The red blood cells will swell and lyse as water moves from the area of higher water potential inside the red blood cells into the area of lower water potential in the beaker.
    (C) The red blood cells will shrivel as water moves from the area of higher water potential in the beaker to the area of lower water potential inside the red blood cells.
    (D) The red blood cells will swell and lyse as water moves from the area of higher water potential in the beaker to the area of lower water potential inside the red blood cells.

36. Rinderpest is a viral disease that affects cattle and wildebeests on the plains of the Serengeti in Africa. In the 1950s, a vaccine against rinderpest became available, and the number of wildebeests on the plains of the Serengeti increased. As the number of wildebeests increased, park rangers found that more grass was eaten by the wildebeests, and the number of wildfires in the area decreased. The decrease in frequency of wildfires led to increased tree density in the area. This increased tree density led to greater biodiversity of species in the area. Based on this information, what role in this ecosystem do the wildebeests play?

    (A) apex predator
    (B) keystone species
    (C) producer
    (D) secondary consumer

Questions 37–39

The chart indicates the presence (+) or absence (−) of components in different cells.

Cellular Component	Cell I	Cell II	Cell III	Cell IV
Cell Membrane	+	+	+	+
Nucleus	+	+	−	+
Mitochondria	+	+	−	+
Chloroplasts	+	−	−	−
Cell Wall	+	−	+	+

37. Which cell is most likely a bacterium?

    (A) I
    (B) II
    (C) III
    (D) IV

38. Which cell comes from an organism that is more likely to be dependent on pollinators for reproduction?

    (A) I
    (B) II
    (C) III
    (D) IV

39. Which cell is most likely from human muscle tissue?

    (A) I
    (B) II
    (C) III
    (D) IV

40. Which of the following best summarizes the relationship between hydrolysis and dehydration reactions?

    (A) Hydrolysis builds monomers, and dehydration reactions break down monomers.
    (B) Hydrolysis breaks down polymers into monomers, and dehydration reactions link monomers to form polymers.
    (C) Hydrolysis reactions release water into the environment, and dehydration reactions consume water from the environment.
    (D) Hydrolysis reactions only occur in carbohydrates and proteins, and dehydration reactions only occur in lipids and nucleic acids.

41. Which of the following contributes to the high specific heat of water?

    (A) covalent bonds between oxygen atoms of two adjacent water molecules
    (B) hydrogen bonds between oxygen atoms of two adjacent water molecules
    (C) covalent bonds between a hydrogen atom on one water molecule and the oxygen atom on an adjacent water molecule
    (D) hydrogen bonds between a hydrogen atom on one water molecule and the oxygen atom on an adjacent water molecule

42. Alpha helixes and beta sheets are examples of a protein's _____ structure and are formed by _____ bonds.

    (A) primary; peptide
    (B) primary; hydrogen
    (C) secondary; peptide
    (D) secondary; hydrogen

43. A molecule is analyzed and found to contain carbon, hydrogen, oxygen, nitrogen, and phosphorus. This molecule is most likely to be a _____.

    (A) carbohydrate
    (B) lipid
    (C) nucleic acid
    (D) protein

44. Which part of an amino acid is unique and would most likely determine how the amino acid would interact with the aqueous environment of a cell?

    (A) the carboxyl group attached to the central carbon
    (B) the amino group attached to the central carbon
    (C) the R-group attached to the central carbon
    (D) the hydrogen atom attached to the central carbon

45. Which of the following is an example of signal amplification?

    (A) a steroid hormone binding to a gene and activating its transcription
    (B) a kinase adding a phosphate group to an enzyme and activating the enzyme
    (C) opening a membrane ion channel, allowing hundreds of ions to enter a cell
    (D) a single ligand binding to a single receptor, triggering activation of dozens of enzymes

46. During which phase of the cell cycle does DNA replication occur?

    (A) G1
    (B) G2
    (C) S
    (D) M

47. Which of the following is an example of a positive feedback mechanism?

   (A) Adenylyl cyclase catalyzes the formation of cyclic AMP from ATP.
   (B) Ripening fruits release ethylene, and ethylene stimulates fruit ripening.
   (C) When a cell has an adequate supply of tryptophan, the operon containing the genes that code for the production of tryptophan is shut down.
   (D) When blood glucose levels rise, the pancreas releases insulin, which lowers blood glucose levels.

Questions 48–51

Cystic fibrosis is an autosomal recessive disorder caused by mutations in the cystic fibrosis transmembrane conductance regulator (*CFTR*) gene. The *CFTR* gene codes for a protein that functions as a gated membrane channel that helps control the flow of chloride ions from inside the cytosol of the cell to the extracellular fluid outside the cell. When the CFTR protein is not functioning properly, not enough chloride ions exit the cell and the mucus in the extracellular fluid becomes dehydrated, thick, and viscous.

The fully functional form of the CFTR protein is translated from 27 exons in the *CFTR* gene. Not everyone with cystic fibrosis has the same mutation in the *CFTR* gene; currently five classes of mutations that can cause cystic fibrosis have been identified.

48. In one type of cystic fibrosis, the sections of the protein coded by exon 10 and exon 17 are missing from CFTR protein. Which type of mutation is most likely the cause of this?

   (A) nonsense mutation
   (B) translocation between two nonhomologous chromosomes
   (C) mutation in a splice site
   (D) point mutation, which changes a codon for a hydrophilic amino acid to a codon for a hydrophobic amino acid

49. The most common mutation, F508del, deletes a single amino acid, resulting in slight misfolding of the protein. The cell detects and destroys these atypical proteins. A combination of two specific medications can help reduce some of the symptoms in patients with this mutation. What is the most likely mechanism of action of the two medications?

   (A) activation of *CFTR* gene transcription in patients with the F508del mutation
   (B) repression of the number of ribosomes in patients with the F508del mutation
   (C) correction of a premature stop codon in the middle of the F508del mutation
   (D) help for the CFTR protein to fold into a more correct shape

50. Another type of mutation causes reduced levels of protein production. Which of the following medications would be effective?

   (A) a drug that keeps the gate on the CFTR ion channel open for longer periods of time
   (B) a drug that suppresses expression of the *CFTR* gene
   (C) a drug that decreases the number of chloride ions exiting the cell
   (D) a drug that stimulates the Na^+/K^+ pump in the cell membrane

51. A person with cystic fibrosis has a child with a person who does not have the cystic fibrosis allele. What is the probability that the child will have cystic fibrosis?

   (A) 0%
   (B) 25%
   (C) 50%
   (D) 75%

52. The arginine (*arg*) operon is a repressible operon in bacteria. What is the most likely result when excess arginine is added to the environment of bacteria?

    (A) Arginine acts as an inducer, binding to the repressor protein and changing its shape so that transcription of the arginine operon starts.
    (B) Arginine binds to RNA polymerase, stimulating transcription of the arginine operon.
    (C) Arginine acts as a corepressor, helping the repressor bind to the operator and stopping transcription of the operon.
    (D) Arginine stimulates the enzyme adenylyl cyclase, causing the production of cyclic AMP.

53. Which of the following biotechnology techniques amplifies the number of copies of a specific DNA sequence?

    (A) gel electrophoresis
    (B) polymerase chain reaction
    (C) CRISPR-Cas9
    (D) DNA sequencing

54. Conjugation, transformation, and transduction are all examples of which of the following?

    (A) horizontal gene transfer
    (B) mutations caused by an atypical number of chromosomes
    (C) processes that occur only in eukaryotic cells
    (D) methods of reducing genetic diversity

55. Ecologists studying a population of 250 deer in the hills north of San Francisco, California, find that the birth rate for the population is 0.25 and the death rate is 0.15. Which of the following is the most likely prediction regarding the size of this population the following year?

    (A) The population size will decrease because the birth rate is less than 1.0.
    (B) The population size will decrease because the death rate is greater than 0.
    (C) The population size will increase because the deer population in the area has an unlimited food source.
    (D) The population size will increase because the birth rate is greater than the death rate.

56. Which of the following is a density-independent factor that can limit population growth?

    (A) competition for mates
    (B) competition for food
    (C) predation
    (D) a natural disaster

Questions 57 and 58

Fordinae geoica and *Fordinae formicaria* are two species of gall-forming aphids. By creating galls, nutrient-rich sap from the phloem of the host plant is suctioned into these voids, providing food for the aphids.

F. geoica were introduced onto a rose bush upon which *F. formicaria* had already inhabited and built galls. Three months later, the number of *F. formicaria* was reduced by 84%, while the number of *F. geoica* increased by over 300%. Analysis of the gall contents of the two species of aphids revealed that galls formed by *F. geoica* had a much higher sap content than galls formed by *F. formicaria*.

57. The ecological relationship between *F. geoica* and *F. formicaria* is best described by which of the following?

    (A) intraspecies competition
    (B) interspecies competition
    (C) mutualism
    (D) parasitism

58. The relationship between *Fordinae* and the rose bush is best described by which of the following?

    (A) intraspecies competition
    (B) interspecies competition
    (C) mutualism
    (D) parasitism

Questions 59 and 60

An invasive species of carp is introduced into a lake in 1986. The species of fish present in the lake and the number of each species before and after the introduction of the carp are shown in the table.

Species	Number Present in 1985	Number Present in 1991
Trout	1,375	934
Bass	1,410	733
Catfish	501	45
Carp	0	2,003
Steelheads	662	238
Pikes	52	47

59. Which species had the greatest percent population decline after the introduction of carp to the lake?

   (A) bass
   (B) catfish
   (C) steelheads
   (D) trout

60. If an ecological disturbance occurred in the lake, which lake community (preintroduction or postintroduction) could be hypothesized to recover from the ecological disturbance?

   (A) There is no difference because the total number of individuals in 1985 and 1991 was the same, meaning that the Simpson's Diversity Index was equal in both communities.
   (B) The population in 1985 would have a greater ability to recover from an ecological disturbance because its Simpson's Diversity Index was higher.
   (C) The population in 1991 would have a lower ability to recover from an ecological disturbance because its Simpson's Diversity Index was higher.
   (D) The population in 1991 would have a greater ability to recover from an ecological disturbance because the number of species present in 1991 was higher.

STOP If there is still time remaining, you may review your answers.

Section II: Free-Response

TIME: 90 MINUTES

> **DIRECTIONS:** Answer each of the following six free-response questions using complete sentences. Allow approximately 20–25 minutes each for the long free-response questions (Questions 1 and 2) and approximately 5–10 minutes each for the short free-response questions (Questions 3, 4, 5, and 6).

1. Six equally sized cubes of turnips are massed and placed in six different sucrose solutions at 21°C. The turnip cubes are massed again after 24 hours, and the percent change in mass is calculated. A graph of the data is constructed.

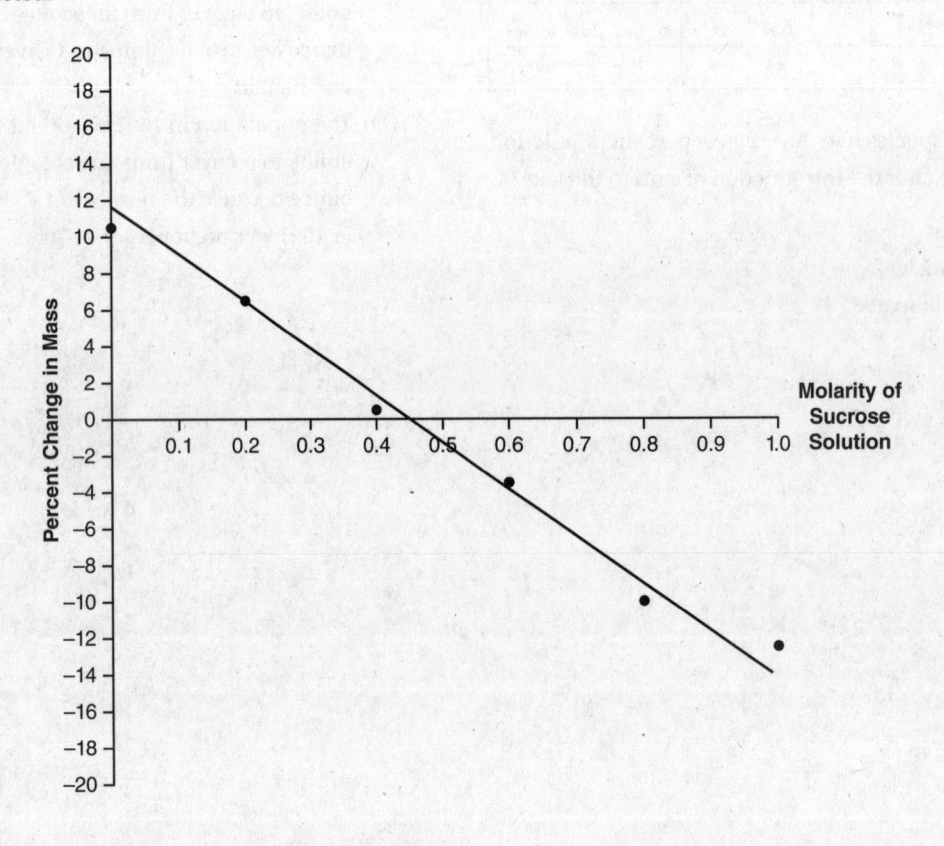

Part A

(i) For each of the cubes tested, **explain** the observed percent change in mass.

Part B

(i) **Predict** the molarity of a turnip cell.

(ii) **Justify** your prediction using evidence from the graph.

Part C

(i) **Calculate** the solute potential of a turnip cell.

Part D

A new turnip cube (of the same size and shape) is placed in a 0.5 M solution of sucrose.

(i) **Predict** the percent change in mass of the turnip cube after 24 hours.

(ii) **Justify** your prediction.

2. The transpiration rate was measured for three different plant species. Ten plants of each species were kept under the same experimental conditions for three days. Total water loss for each plant was measured during the three-day period using a potometer. At the end of the three-day period, the total surface area of the leaves of each plant was measured, and the water loss per square meter of leaf area per day was calculated. Data are shown in the table.

Plant Species	Mean Water Loss per Plant (mL/m^2/day) $\pm$ 2 SEM*
A	15.4 $\pm$ 2.1
B	21.6 $\pm$ 2.8
C	25.0 $\pm$ 3.1

*Standard Error of the Mean

Part A

(i) **Explain** how the relative amounts of transpiration and photosynthesis can affect water loss in a plant.

Part B

(i) On the axes provided, **construct** a graph of the data using 95% confidence intervals.

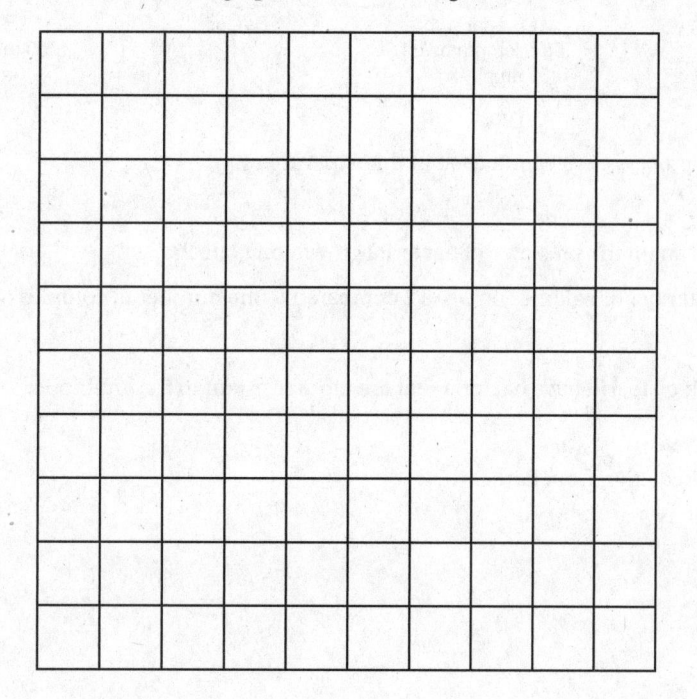

Part C

(i) **Analyze** the data to determine which, if any, plant species have a statistically significant difference between their amounts of water loss.

Part D

(i) **Predict** which plant species would be best suited for a dry climate.

(ii) **Justify** your prediction using the data from the experiment.

3. A sample of a culture of *E. coli* is plated onto two LB agar plates: plate 1 containing LB agar and no antibiotic (LB) and plate 2 containing LB agar with 50 micrograms/mL of the antibiotic ampicillin (LB + amp). Both plates are incubated at 37°C for 24 hours. Bacteria from plate 1 are plated onto two more plates: one with LB agar (plate 3) and the other containing LB agar with 50 micrograms/mL of the antibiotic ampicillin (plate 4). Bacteria from plate 2 are plated onto plate 5 (containing LB agar) and plate 6 (containing LB agar with 250 micrograms/mL of the antibiotic ampicillin). This second round of plates are all incubated at 37°C for 24 hours. A diagram of the experiment is shown below.

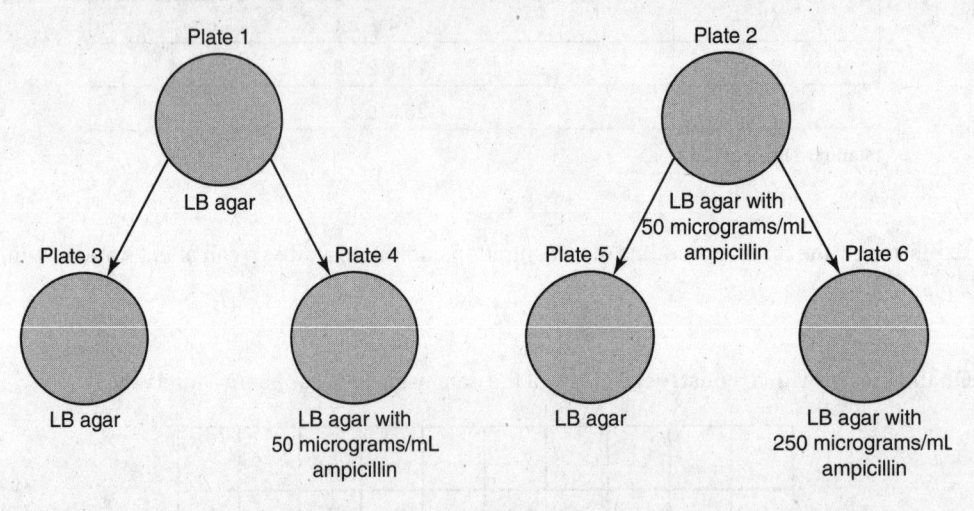

Part A

(i) **Describe** the process of selection used in this experiment.

Part B

(i) **Make a claim** about the presence of bacterial growth on plate 2.

(ii) **Predict** the number of colonies on plate 2 compared to the number of colonies on plate 1.

Part C

(i) **Predict** which plate will have bacteria that are more resistant to the antibiotic.

Part D

(i) **Justify** your prediction from Part C.

4. Below is a food web from an aquatic ecosystem.

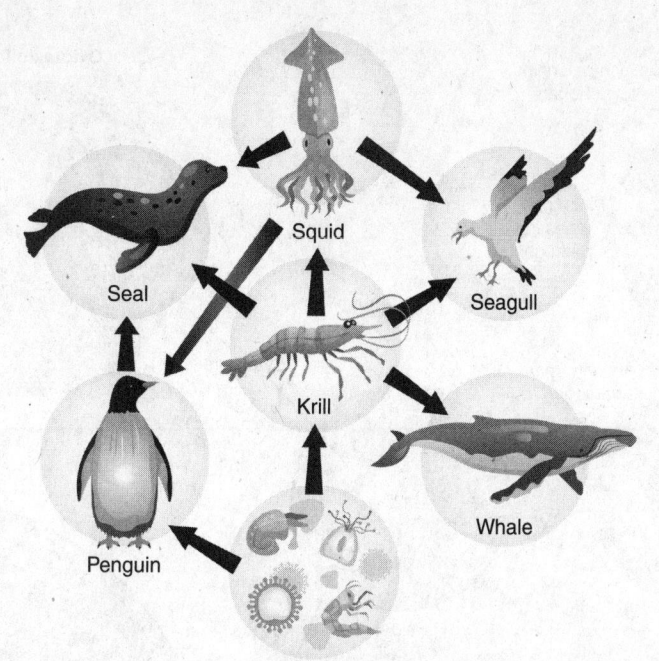

Part A

(i) **Describe** the roles of penguins and seagulls in this food web.

Part B

(i) **Identify** the primary consumer(s) in this food web.

(ii) **Explain** your reasoning.

Part C

Many species of phytoplankton are found in marine environments, but not all are digestible or desirable to eat. One species of phytoplankton, *Coccolithophore*, is not a food source.

(i) **Predict** the effect of a *Coccolithophore* bloom on this aquatic ecosystem.

Part D

(i) **Justify** your prediction from Part C.

5. A portion of the human immune system response to an invading pathogen is shown below.

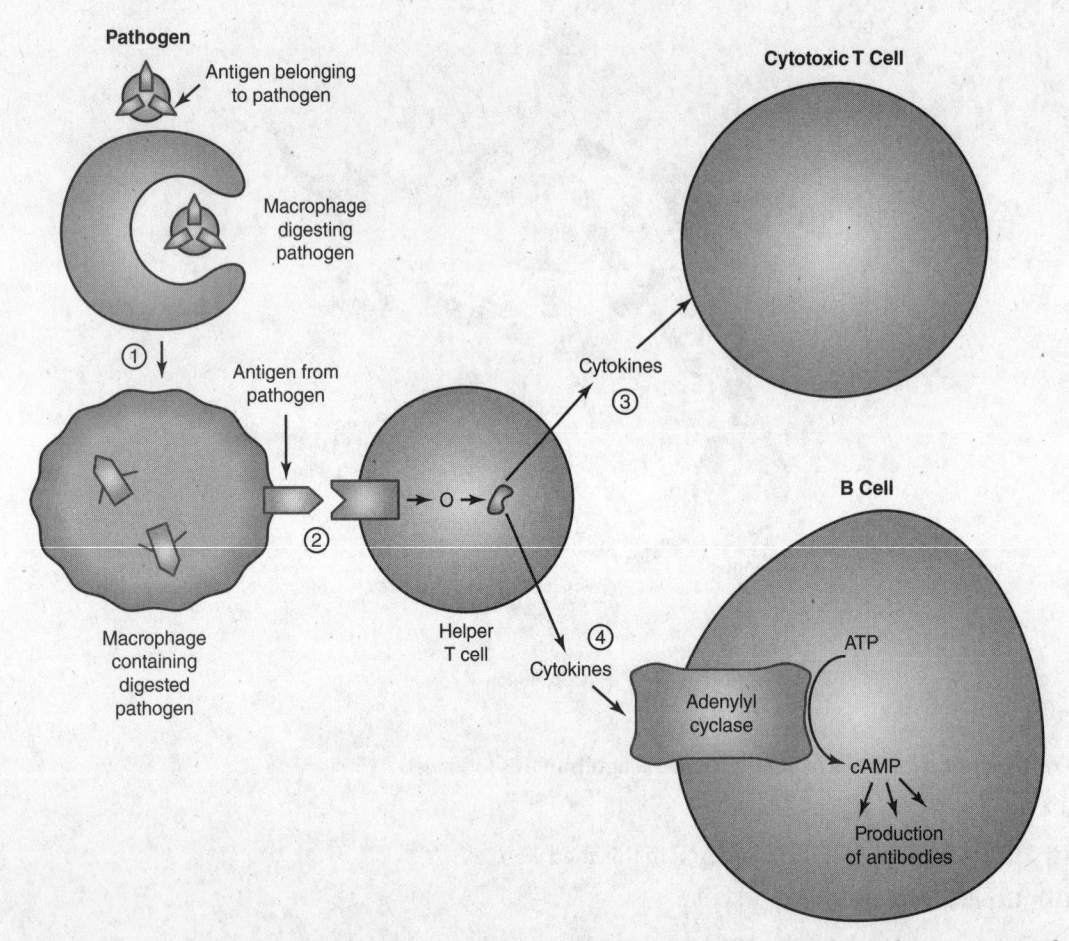

Macrophages are white blood cells that ingest invading pathogens. After digesting the pathogen, macrophages present the antigens from the pathogen to helper T cells in the immune system. These helper T cells then secrete chemical signals called cytokines. The cytokines can travel short distances to activate nearby cytotoxic T cells, which will kill infected cells. Some cytokines will activate the enzyme adenylyl cyclase in nearby B cells to produce cyclic AMP (cAMP). cAMP will then activate the production of antibodies in the B cells.

Part A

(i) **Describe** the characteristics of juxtacrine signaling.

(ii) **Draw** a square around the step (1, 2, 3, or 4) in the figure that illustrates an example of juxtacrine signaling.

Part B

(i) **Explain** the role cytokines play in the cell signaling process described above.

Part C

(i) **Draw** a circle around a molecule that functions as a secondary messenger in the pathway.

Part D

(i) **Explain** how secondary messengers can amplify a response in a cell.

6. A student investigated the effect of caffeine on the number of cells in mitosis and interphase in onion root tip cells. Onions were planted in sand in two groups. One group of onions received distilled water, and the second group received an aqueous solution of 0.002 M caffeine for two days. The tips of the roots were removed and stained, and the number of cells in mitosis and interphase was counted. The data appear in the table below.

Treatment	Phase of Cell	Observed	Expected
Distilled Water	Mitosis	20	22
	Interphase	94	92
Caffeine Solution	Mitosis	30	28
	Interphase	120	122

Part A

(i) **Calculate** the percentage of cells in mitosis and interphase observed in each group.

Part B

(i) **Identify** the control group in this experiment.

(ii) **Identify** the independent variable in this experiment.

(iii) **Identify** the dependent variable in this experiment.

Part C

The student states the following null hypothesis for this experiment: "There is no statistically significant difference between the number of cells in mitosis and in interphase between the onions grown in distilled water and the onions grown in 0.002 M caffeine."

(i) Using the data, the chi-square test, and a *p*-value of 0.05, **evaluate** the student's null hypothesis.

Part D

In cells, the start of mitosis is triggered by mitosis-promoting factor (MPF). MPF forms when cyclin protein binds to a cyclin-dependent kinase.

(i) **Explain** how a drug that irreversibly binds to the active site of a cyclin-dependent kinase would affect the cell cycle.

Answer Explanations

Section I

1. **(D)** Birds with crushing beaks would be better equipped to utilize hard-shelled seeds as a food source. The table indicates that birds with crushing beaks eat seeds, while birds with parrot beaks, grasping beaks, and probing beaks do not eat seeds, so choices (A), (B), and (C) are incorrect.

2. **(C)** The woodpecker finch is a tree finch, so if a hurricane eliminated most trees from the islands, it is expected that the number of woodpecker finches would decrease. The cactus finch, warbler finch, and medium ground finch are not tree finches and would be less likely to be affected by the elimination of trees from the islands, so choices (A), (B), and (D) are incorrect.

3. **(D)** Mate choice by females is an example of sexual selection. Choice (A) is incorrect because in artificial selection humans determine which individuals survive and reproduce at a greater rate. Directional selection occurs when one extreme phenotype is favored, which is not the case here, so choice (B) is incorrect. Disruptive selection occurs when both extremes of a phenotype are favored, so choice (C) is also incorrect.

4. **(B)** During year 4, 30% of the population had the recessive phenotype, so $q^2 = 0.30$ and $q = 0.548$. $p + q = 1$ so $p = 0.452$. The frequency of heterozygotes would be $2pq = 2 \times (0.452) \times (0.548) = 0.495$, or 49.5%.

5. **(B)** Random mating is a necessary condition for a population to be in Hardy-Weinberg equilibrium. Choice (A) is incorrect because large (not small) populations are required for Hardy-Weinberg equilibrium. There needs to be no migration into or out of a population for Hardy-Weinberg equilibrium, so choice (C) is incorrect. No individuals should have a selective advantage in a Hardy-Weinberg population, so choice (D) is incorrect.

6. **(B)** The frequency of the recessive phenotype (black body) increased, most likely due to a change in the environment that provided a selective advantage to flies with a black body. Choices (A) and (C) are incorrect because the frequency of the recessive phenotype increased between years 6 and 7, not decreased. Individuals do not evolve; populations evolve. So choice (D) is incorrect.

7. **(D)** Founder effect occurs when members of a population migrate to a new area and the frequency of one or more alleles in the migrating/founding population is higher than in the original population. Choice (A) is incorrect because there is no evidence that methemoglobinemia has a selective advantage, nor is there evidence that heterozygotes for the trait have a selective advantage, so choice (B) is also incorrect. The bottleneck effect occurs when a population undergoes a rapid and extreme decrease in population size, which is not the case in this example, so choice (C) is incorrect.

8. **(B)** Species M has all five traits, so it is on the far right of the cladogram. Species L has traits 1, 2, 3, and 4 but not 5, so trait 5 is placed after the branch point for L. Species P has traits 2, 3, and 4, so trait 1 is placed after the branch point for P but before the branch point for L. Species O has traits 2 and 3, so trait 4 is placed after the branch point for O but before the branch point for P. The only trait species N has is trait 2, so trait 3 is placed after the branch point for N but before the branch point for O. Trait 2 is placed before the branch point for N. Choices (A), (C), and (D) are incorrect in their placements of the traits and the species.

9. **(B)** Species N is the outgroup because it has the least in common with the other species, sharing only trait 2. Choices (A), (C), and (D) are incorrect because every other species shares two or more traits with the others in the cladogram.

10. **(B)** Sympatric speciation occurs when two species are separated by genetic differences. Choice (A) is incorrect because allopatric speciation requires geographic separation, which is not the case in this example. Polygenic inheritance is when more than one gene contributes to a trait, also not the case here, so choice (C) is incorrect. Choice (D) is incorrect because temporal isolation requires that the species breed at different times.

11. **(B)** Punctuated equilibrium describes long periods of stability during evolution interspersed with periods of rapid change. Choice (A) is incorrect because gradualism describes a slow, constant rate of evolutionary change. Convergent evolution occurs when species share similar traits because of similar environments, not common ancestry, so choice (C) is incorrect. Choice (D) is incorrect because coevolution occurs when two species influence each other's evolution.

12. **(D)** Natural selection occurs when the natural environment determines which individuals survive and reproduce at a greater rate. Choices (A) and (C) are incorrect because both are examples of artificial selection, when humans breed organisms for desired traits. The darkening of a person's skin when exposed to the Sun is an acquired trait, so choice (B) is also incorrect.

13. **(B)** Distilled water does not contain enzymes, so it would be an appropriate control. Choice (A) is incorrect because changing the size of the paper disk would add another variable. Reducing the temperature of the hydrogen peroxide would not be an appropriate control, so choice (C) is incorrect. Pureeing the potatoes with buffer instead of water would change the conditions of the experiment and would not be a control, so choice (D) is incorrect.

14. **(C)** The upper limit of the 95% confidence interval for 9% hydrogen peroxide $(30 + 2(3.9) = 37.8$ sec) overlaps with the lower limit of the 95% confidence interval for 6% hydrogen peroxide $(45 − 2(4.3) = 36.4$ sec). So it is not likely there is a statistically significant difference between those two concentrations of hydrogen peroxide. The 95% confidence intervals for all of the other pairs of hydrogen peroxide do not overlap and therefore are more likely to have a statistically significant difference between them.

15. **(D)** At higher concentrations of hydrogen peroxide, more substrate was available to the enzyme so more oxygen was produced and the paper disks floated faster. The amount of heat generated did not change with higher concentrations of hydrogen peroxide, so choice (A) is incorrect. There is no evidence that more concentrated hydrogen peroxide changed the shape of the active site of the enzyme, so choice (B) is incorrect. Oxygen is not an allosteric inhibitor of the enzyme, so choice (C) is also incorrect.

16. **(C)** Boiling denatures most enzymes, irreversibly changing their shape so they do not function. Choice (A) is incorrect because the enzyme is exposed to the boiling water bath, not the hydrogen peroxide. Boiling will change the time required for the paper disk to float, so choice (B) is incorrect. Not all enzymes are less effective at higher temperatures, so choice (D) is incorrect.

17. **(B)** Oxygen is the final electron acceptor in the electron transport chain in cellular respiration. Oxygen is not required in glycolysis, so choice (A) is incorrect. Oxygen is not required for fermentation nor is it a substrate for enzymes used in fermentation, so choice (C) is incorrect. Pyruvate does not enter the matrix of the mitochondria, so choice (D) is incorrect.

18. **(C)** During the Krebs cycle, NAD^+ is reduced to NADH, so this is a relevant question in understanding the Krebs cycle. The breakdown of glucose into pyruvate occurs during glycolysis, not the Krebs cycle, so choice (A) is incorrect. The proton gradient in the mitochondria is formed during oxidative phosphorylation, not the Krebs cycle, so choice (B) is incorrect. Choice (D) is incorrect because the formation of ATP by ATP synthase happens during chemiosmosis, not the Krebs cycle.

19. **(B)** Both fermentation and the electron transport chain oxidize NADH to NAD^+. Choice (A) is incorrect because fermentation does not require oxygen. Fermentation occurs in the cytoplasm, not the mitochondria, so choice (C) is incorrect. One of the end products of the electron transport chain is metabolic water, but fermentation does not form metabolic water. So choice (D) is incorrect.

20. **(B)** Since a few base pairs are lost each time a linear chromosome is replicated, the repetitive telomere sequences at the ends of linear chromosomes help protect the genetic information in chromosomes from degradation during cell division.

Choice (A) is incorrect because telomeres do not function in protein synthesis. Telomeres do not correct mutations in DNA, so choice (C) is incorrect. Choice (D) is incorrect because telomeres do not inhibit cell division.

21. **(D)** Water moves from areas of higher water potential to areas of lower water potential. Celery stalks become stiff after being placed in distilled water for a few hours because water will move from the area of higher water potential in the distilled water to the area of lower water potential in the celery stalks. Celery stalks placed in 1.0 molar sodium chloride will wilt because water will move from the area of higher water potential in the celery to the area of lower water potential in the 1.0 molar sodium chloride solution. Based on the fact that the celery cells became stiff in the distilled water, they must have a lower water potential than that of the distilled water, so choice (A) is incorrect. Since the celery cells wilted in the sodium chloride solution, the water potential in the celery cells must be higher than the water potential in the sodium chloride solution, so choice (B) is incorrect. Celery cells have solutes and therefore cannot have a higher water potential than that of distilled water, so choice (C) is incorrect.

22. **(B)** Photosynthesis requires light energy, so the vial covered in foil will only perform cellular respiration. Cellular respiration produces carbon dioxide, which forms an acid when dissolved in water and therefore lowers the pH. The vial without foil performs both cellular respiration and photosynthesis. Photosynthesis consumes carbon dioxide, so the vial without foil will not build up as much carbon dioxide and will have a higher pH than the vial covered in foil. Choice (A) is incorrect because the vial without foil will perform both cellular respiration and photosynthesis. The vial covered with foil will not perform photosynthesis, so choices (C) and (D) are incorrect.

23. **(D)** According to the table, the largest animals have lower cancer rates, so the data do not support the claim that larger animals have greater cancer rates. Choices (A) and (C) are incorrect because larger animals have lower cancer rates. Smaller animals have higher cancer rates than larger animals, so choice (B) is incorrect.

24. **(A)** Elephants have lower cancer rates than marmosets, Tasmanian devils, and tigers, so elephants are more likely to have more copies of the tumor-suppressing *TP53* gene. Choices (B), (C), and (D) are incorrect because marmosets, Tasmanian devils, and tigers all have higher cancer rates than elephants.

25. **(D)** Since males receive their sole X chromosome from their mother, all the males will have white eyes. All of the females will receive an X chromosome with the dominant red allele from their father, so all of the females will have red eyes. The following Punnett square shows the genotypes of the parent flies and the probable offspring from the cross.

	X^r	X^r
X^R	$X^R X^r$	$X^R X^r$
Y	$X^r Y$	$X^r Y$

26. **(C)** Transformation is the uptake of naked foreign DNA by a cell. Choice (A) is incorrect because transcription describes the process of creating a complementary RNA copy of a DNA sequence. Transduction is the introduction of foreign DNA into a cell by a virus, so choice (B) is incorrect. Choice (D) is incorrect because translocation describes the exchange of genetic material between nonhomologous chromosomes.

27. **(C)** A person with sickle cell disease will be homozygous recessive, and a person with the sickle cell trait will be heterozygous. Their offspring will have a 50% likelihood of having sickle cell disease, as shown in the figure.

	Sickle cell disease (ss)	
	s	s
Sickle cell trait Ss — S	Ss	Ss
s	ss	ss

28. **(D)** Heterozygote advantage occurs when the individuals with the heterozygous genotype have an advantage over homozygous dominant and

homozygous recessive genotypes. In this case, heterozygotes do not have sickle cell disease and are more resistant to malaria. Choice (A) is incorrect because polygenic inheritance describes when more than one gene contributes to a trait. Sexual selection occurs when mate choice determines reproductive success, so choice (B) is incorrect. Choice (C) is incorrect because sex-linked inheritance occurs when the gene for a trait is located on a sex chromosome.

29. **(C)** A person who does not have sickle cell disease but whose parent had sickle cell disease must be heterozygous. A person with the sickle cell trait is heterozygous, so the child of the two individuals would have a 50% likelihood of having the sickle cell trait, as shown in the figure.

	S	s
S	SS	Ss
s	Ss	ss

30. **(D)** Parasitism is a symbiotic relationship where one organism is harmed (the finch nestlings) and the other organism benefits (*Philornis downsi*). Choice (A) is incorrect because commensalism is a symbiotic relationship where one organism benefits and the other organism neither benefits nor is harmed. Competition is when two species are competing for the same resources, so choice (B) is incorrect. Choice (C) is incorrect because mutualism means both species in the symbiotic relationship benefit.

31. **(C)** Seagrass will absorb carbon dioxide as it performs photosynthesis. Choice (A) is incorrect because there is no evidence that planting seagrass would reduce habitats for invasive species. Photosynthesis adds oxygen to the atmosphere; it does not remove oxygen, so choice (B) is incorrect. There is no evidence that seagrass plantings would increase soil erosion, so choice (D) is incorrect.

32. **(B)** More drops of water can be placed on the penny than the other liquids, demonstrating that water has stronger cohesive properties than the other liquids, and therefore water forms more hydrogen bonds than the other liquids. Choice (A) is incorrect because vegetable oil is a nonpolar compound and does not form hydrogen bonds. Ions are attracted to metals, not repelled, so choice (C) is incorrect. Ethanol is a larger molecule than water, but it does not form more hydrogen bonds than water. So choice (D) is incorrect.

33. **(D)** Starch molecules are too large to pass through the semipermeable membrane, but water and iodine can pass through the membrane. Iodine will move down its concentration gradient from side A to side B and react with the starch molecules, which will turn dark blue. Since the starch molecules cannot move down their concentration gradient and will remain on side B, the water potential on side B will be lower than the water potential on side A. Water will move from the area of higher water potential on side A to the area of lower water potential on side B, raising the water level on side B of the U-tube apparatus.

34. **(C)** Competitive inhibitors bind to the active site of an enzyme, preventing the substrate from binding to the active site, so adding a competitive inhibitor would be an effective way to slow the rate of cell division in cells containing the BCR-ABL kinase. Cofactors and coenzymes increase the rate of enzyme-catalyzed reactions, so choices (A) and (B) are incorrect. Transcription factors increase the rate of transcription of RNA from DNA, so choice (D) is incorrect.

35. **(D)** Water moves from areas of higher water potential to areas of lower water potential, and distilled water has a higher water potential than the inside of a red blood cell (due to the solutes in the red blood cell). So water would move into the red blood cell, and the cell would eventually burst. Choices (A) and (B) are incorrect because red blood cells have a lower water potential than distilled water. Choice (C) is incorrect because water moving into the red blood cells would cause them to swell and lyse, not shrivel.

36. **(B)** The presence of wildebeests has a disproportionately large effect on many other species in the ecosystem, so it is a keystone species. Choice (A) is incorrect because an apex predator consumes

other organisms and has no predators itself, but the wildebeest is a primary consumer because it eats grass. Producers (such as grass) perform photosynthesis, so choice (C) is incorrect. Secondary consumers eat primary consumers, not producers. So the wildebeest is not a secondary consumer, and choice (D) is incorrect.

37. **(C)** Bacteria do not have nuclei but do have cell membranes and cell walls. Choice (A) is incorrect because bacteria do not have nuclei, mitochondria, or chloroplasts. Bacteria do not have nuclei or mitochondria, so choices (B) and (D) are incorrect.

38. **(A)** Plants depend on pollinators for reproduction, and plants have chloroplasts. Choices (B), (C), and (D) are incorrect because they do not have chloroplasts.

39. **(B)** Human muscle tissue would have nuclei and mitochondria but not chloroplasts or cell walls. Choice (A) is incorrect because human cells do not contain chloroplasts. Choice (C) is incorrect because human cells have nuclei, and choice (D) is incorrect because human cells do not have cell walls.

40. **(B)** Hydrolysis uses water to break down polymers into monomers, and dehydration synthesis combines monomers into polymers while removing a water molecule. Choice (A) is incorrect because hydrolysis does not build monomers nor does dehydration synthesis break down monomers. Hydrolysis reactions consume, not release, water, so choice (C) is incorrect. All macromolecules, not just carbohydrates and proteins, can be broken down by hydrolysis reactions, so choice (D) is incorrect.

41. **(D)** Water has a high specific heat because hydrogen bonds form between the oxygen atom on one water molecule and the hydrogen atom on another water molecule. Choices (A) and (C) are incorrect because the high specific heat of water is directly due to hydrogen bonds between water molecules, not covalent bonds between water molecules. Hydrogen bonds do not form between oxygen atoms, so choice (B) is incorrect.

42. **(D)** Alpha helices and beta sheets are forms of secondary structure and are formed by hydrogen bonds. Choices (A) and (B) are incorrect because alpha helices and beta sheets are not forms of primary structure. Secondary structure is formed by hydrogen bonds, not peptide bonds, so choice (C) is incorrect.

43. **(C)** Nucleic acids contain carbon, hydrogen, oxygen, nitrogen, and phosphorus. Choice (A) is incorrect because carbohydrates do not usually contain nitrogen or phosphorus. Lipids usually do not contain nitrogen, so choice (B) is incorrect. Choice (D) is incorrect because proteins usually do not contain phosphorus.

44. **(C)** The R-group (also known as the side chain) of an amino acid is unique to that amino acid and determines whether the amino acid is hydrophobic or hydrophilic. Therefore, it influences how the amino acid would interact with the aqueous environment of the cell. All amino acids contain a carboxyl group. An amino group and a hydrogen atom are attached to the central carbon but are less likely to affect the interaction of an amino acid with the aqueous environment of the cell. For those reasons, choices (A), (B), and (D) are incorrect.

45. **(D)** Signal amplification involves one signaling molecule (also known as a ligand) triggering multiple chemical reactions, magnifying the effect of the chemical signal. Choices (A), (B), and (C) are incorrect because they each involve one reaction triggering just one event.

46. **(C)** DNA replication occurs during the S phase of the cell cycle. Choices (A), (B), and (D) are incorrect because DNA replication does not occur during those phases.

47. **(B)** Positive feedback occurs when the stimulus increases the response and moves the condition further away from homeostasis. Choice (A) is incorrect because it is an example of the formation of a secondary messenger. Excess tryptophan shutting down the operon that produces tryptophan is an example of negative feedback, so choice (C) is incorrect. The release of insulin lowers blood sugar,

returning blood sugar levels to normal levels, so choice (D) is also incorrect.

48. **(C)** Mutations in a splice site could result in errors in intron splicing and the loss of exons. Choice (A) is incorrect because a nonsense mutation (which would form a stop codon) would result in a truncated protein, not the deletion of two exons at different locations in the gene. Translocations would most likely form a completely different protein, so choice (B) is incorrect. A point mutation would not result in the deletion of two different exons, so choice (D) is incorrect.

49. **(D)** The F508del mutation results in a misfolded protein, so a treatment that helps the CFTR protein fold correctly would be an effective treatment for the result of this mutation. Activation of transcription of the *CFTR* gene with the F508del mutation would not be effective. It would just produce more of the misfolded protein, so choice (A) is incorrect. Choice (B) is incorrect because repression of the number of ribosomes would result in less production of all proteins in the cell. The F508del mutation does not code for a premature stop codon, so choice (C) is incorrect.

50. **(A)** Since there would be fewer copies of the CFTR protein, a drug that causes the CFTR channel to stay open and function longer might help alleviate the symptoms of the disease. Choice (B) is incorrect because suppression of the expression of the *CFTR* gene would make the symptoms worse, not better. Similarly, decreasing the number of chloride ions exiting the cell would also make the symptoms worse, so choice (C) is incorrect. Stimulating the Na$^+$/K$^+$ pump would not help with the migration of chloride ions, so choice (D) is incorrect.

51. **(A)** Since cystic fibrosis is a recessive disorder, an individual would have to inherit a copy of the cystic fibrosis allele from both parents. If one of the parents does not have the cystic fibrosis allele, a child with a person who does have cystic fibrosis would have no chance of inheriting the disease, though their child would be a heterozygous carrier of the allele, as shown in the figure.

	Parent with cystic fibrosis	
	C	C
Parent without the cystic fibrosis allele C	Cc	Cc
C	Cc	Cc

52. **(C)** The product formed by the enzymes of a repressible operon acts as a corepressor to help the repressor shut off the operon when an excess of the product is present. Choice (A) is incorrect because arginine would not induce the operon to turn on nor would arginine stimulate transcription, so choice (B) is also incorrect. Arginine does not stimulate the production of cyclic AMP, so choice (D) is incorrect.

53. **(B)** Polymerase chain reaction (PCR) is used to make multiple copies of a specific DNA sequence. Choice (A) is incorrect because gel electrophoresis is used to separate DNA fragments by size and charge. CRISPR-Cas9 is used in gene editing, so choice (C) is incorrect. Choice (D) is incorrect because DNA sequencing is the process of reading the sequence of nucleotide bases in DNA.

54. **(A)** Horizontal gene transfer is the transfer of genetic information between organisms in the same generation; conjugation, transformation, and transduction are all forms of horizontal gene transfer. Choice (B) is incorrect because having an atypical number of chromosomes is an aneuploidy. Horizontal gene transfer can occur in both prokaryotic and eukaryotic cells, so choice (C) is incorrect. Horizontal gene transfer does not reduce genetic diversity, so choice (D) is incorrect.

55. **(D)** If the birth rate exceeds the death rate in a population, the population size will increase. Choice (A) is incorrect because a population's size could increase if the birth rate was less than 1.0 and if the death rate was less than the birth rate. If the death rate is greater than 0 but the birth rate is greater than the death rate, the population size will increase, so choice (B) is incorrect. There is no evidence there is an unlimited food source, so choice (C) is incorrect.

56. **(D)** A natural disaster is a density-independent factor. Choices (A), (B), and (C) are all density-dependent factors and are therefore incorrect.

57. **(B)** Both species are competing for the same food source (nutrient-rich sap from the host plant), so this is an example of interspecies competition. Choice (A) is incorrect because intraspecies competition is when members of the same species compete for a resource. Mutualism is when both species benefit. *F. formicaria* is harmed by the presence of *F. geoica*, so choice (C) is incorrect. Both *Fordinae* species are parasites of the host plants but not of each other, so choice (D) is incorrect.

58. **(D)** Both *Fordinae* species are parasites on the host plant because they obtain nutrition from the host plant but provide no benefit to the host plant. Choices (A), (B), and (C) are incorrect, as explained in the answer explanation to Question 57.

59. **(B)** Catfish declined by over 90% after the introduction of the carp. Bass decreased by about 50%, so choice (A) is incorrect. Steelheads decreased by about 65%, so choice (C) is incorrect. Choice (D) is incorrect because trout decreased by about 33%.

60. **(B)** The Simpson's Diversity Index of the population in 1985 (0.7145) was greater than the Simpson's Diversity Index of the population in 1991 (0.6575). So the 1985 population would be more resilient and more able to recover from an ecological disturbance. Choice (A) is incorrect because even though the population size was the same both years, the population was much less diverse in 1991 than in 1985. Choice (C) is incorrect because a higher Simpson's Diversity Index makes a population more likely to recover from an ecological disturbance. Choice (D) is incorrect because even though there was one more species present in 1991 than in 1985, there was less diversity in 1991 than in 1985.

Self-Analysis Chart for Section I

Use this chart to help identify areas where you need to focus your review. After completing the multiple-choice section and reviewing the answer explanations, circle the questions in this chart that you answered incorrectly. Note the units that you scored well in as well as the units that might require further study. Be sure to also review the answer explanations for Section II as all of those questions cover a wide variety of units and topics.

Unit	Questions
Unit 1: Chemistry of Life	32, 40, 41, 42, 43, 44
Unit 2: Cell Structure and Function	21, 33, 35, 37, 38, 39
Unit 3: Cellular Energetics	13, 14, 15, 16, 17, 18, 19, 22, 34
Unit 4: Cell Communication and Cell Cycle	23, 24, 45, 46, 47
Unit 5: Heredity	20, 25, 27, 28, 29, 51
Unit 6: Gene Expression and Regulation	26, 48, 49, 50, 52, 53, 54
Unit 7: Natural Selection	1, 2, 3, 4, 5, 6, 7, 8, 9, 10, 11, 12
Unit 8: Ecology	30, 31, 36, 55, 56, 57, 58, 59, 60

Section II

1. (A-i) Turnips placed in solutions with a higher water potential than the turnip cells would have a positive percent change in mass because water flows from an area of higher water potential to an area of lower water potential. Water would flow into the turnip cells, and the turnips would gain mass. Turnips placed in solutions with a lower water potential than the turnip cells would have a negative percent change in mass because water would flow out of the turnip cells into the surrounding solution and the turnips would lose mass.

   (B-i and ii) The molarity of the turnip cell would be about 0.45 molar. The line of best fit crosses the x-axis at the coordinates (0.45 molar, 0%), which indicates a 0.45 solution that would be isotonic to a turnip cell and in which a turnip cell would have 0% change in mass.

   (C-i) The line of best fit crosses the x-axis at approximately 0.45 molar, so that would be the best estimate of the concentration of solutes in a turnip cell. Since the turnip is open to the atmosphere, the pressure potential is 0. The water potential of the turnip cell would then depend solely on the solute potential, which is calculated as follows:

   $$\Psi_s = -iCRT$$

   $i = 1$ (because sucrose is a covalent compound and does not form ions)

   $$T = 21 + 273 = 294 \text{ K}$$

   $$\Psi_s = -(1)\left(0.45 \frac{\text{moles}}{\text{liter}}\right)\left(0.0831 \frac{\text{liters-bars}}{\text{mole-Kelvin}}\right)(294 \text{ Kelvin})$$

   $$\Psi_s = -10.99 \text{ bars}$$

   > **TIP**
   > Remember to always use temperatures in Kelvin when calculating water potential!

   (D-i and ii) The turnip in the 0.5 molar solution would probably have a negative percent change in mass because the line of best fit is below 0% change in mass for a molarity of 0.5.

2. (A-i) A plant needs to open its stomata to take in the carbon dioxide it needs for photosynthesis. However, each time a plant opens its stomata, it can lose water to the atmosphere. The plant needs to balance its needs for carbon dioxide and water, and this is called the photosynthesis-transpiration compromise.

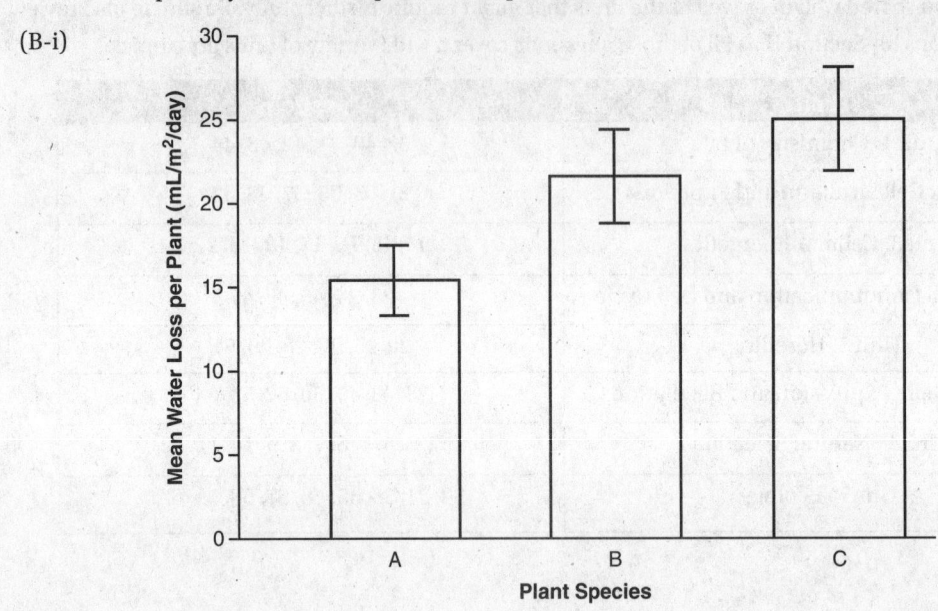

(B-i)

(C-i) There is a statistically significant difference between plant species A and the other two plant species because the 95% confidence interval for species A does not overlap with the 95% confidence interval for either species B or species C. However, the 95% confidence intervals for species B and species C do overlap, so we cannot say there is a statistically significant difference between species B and species C.

(D-i and ii) Species A would be best suited for a dry climate because its rate of water loss due to transpiration is lower than that of species B and species C.

3. (A-i) This is an example of artificial selection because a person is selecting for a particular trait (antibiotic resistance).

(B-i and ii) Since plate 2 contains the antibiotic, only bacteria with resistance to the antibiotic will grow on plate 2. Plate 1 does not contain antibiotic, so bacteria with and without resistance to the antibiotic will grow on plate 1. There will be fewer bacteria on plate 2 than on plate 1.

(C-i) Plate 6 will have bacteria with the greatest resistance to the antibiotic.

(D-i) Plate 6 has the greatest concentration of ampicillin, so the bacteria that survive on that plate will have the greatest resistance to ampicillin.

4. (A-i) Seagulls have no predators in this ecosystem. Penguins are prey for seals. Penguins are primary consumers (because they eat plankton) and tertiary consumers (because they eat squid, squid eat krill, krill eat plankton). Seagulls are secondary consumers (because they eat krill, krill eat producers) and tertiary consumers (because they eat squid, squid eat krill, krill eat plankton).

(B-i and ii) Penguins and krill are primary consumers because they both eat producers (plankton).

(C-i) The number of krill would decrease.

(D-i) A bloom of a phytoplankton *Coccolithophore* (which krill cannot eat) would outcompete the phytoplankton that krill can eat, depleting the food supply for the krill. So krill would die.

5. (A-i) Juxtacrine signaling involves direct contact between cells.

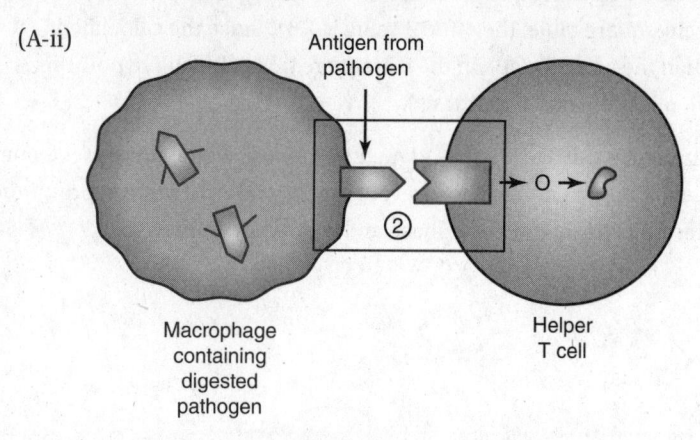

   (A-ii)

   (B-i) Cytokines function as signaling molecules or ligands.

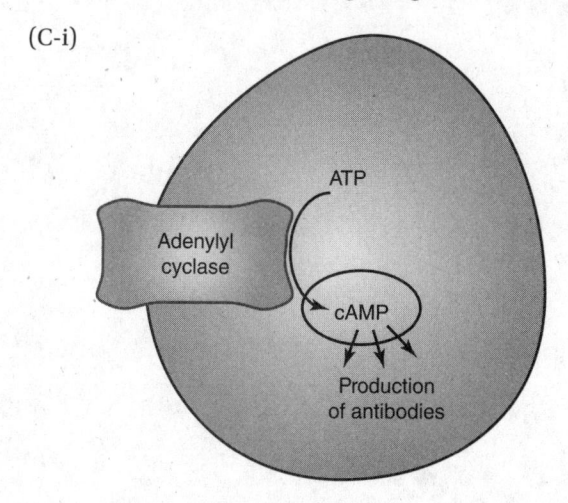

   (C-i)

   (D-i) Secondary messengers amplify a response in a cell by triggering multiple chemical reactions through the process of signal amplification.

6. (A-i) In the distilled water group, $\frac{20}{114} \times 100 = 17.5\%$ of the cells are in mitosis, and $\frac{94}{114} \times 100 = 82.5\%$ of the cells are in interphase. In the caffeine group, $\frac{30}{150} \times 100 = 20.0\%$ of the cells are in mitosis, and $\frac{120}{150} \times 100 = 80.0\%$ of the cells are in interphase.

   (B-i) The onions grown in distilled water are the control group.

   (B-ii) The independent variable is the presence or absence of 2 millimolars of caffeine in the growth medium.

   (B-iii) The dependent variable is the number of cells in mitosis and interphase in each group.

   (C-i) Using the formula for chi-square:

   $$\chi^2 = \sum \frac{(\text{observed} - \text{expected})^2}{\text{expected}}$$

   $$\chi^2 = \frac{(20-22)^2}{22} + \frac{(94-92)^2}{92} + \frac{(30-28)^2}{28} + \frac{(120-122)^2}{122} = 0.4010$$

There are four possible outcomes in the experiment, so the number of degrees of freedom is $4 - 1 = 3$. Using a p-value of 0.05 and the chi-square table, the critical value is 7.81. Since the calculated chi-square value (0.4010) is less than the critical value in the chi-square table, the null hypothesis is supported (or we can say we fail to reject the null hypothesis).

(D-i) A drug that binds irreversibly to the active site of the cyclin-dependent kinase would prevent the binding of cyclin to the cyclin-dependent kinase and therefore prevent the formation of mitosis-promoting factor. The cell cycle would be arrested at the end of interphase and would not enter mitosis.

Appendices

A

Frequently Used Formulas and Equations

Statistics

The proper use of statistics is important in biology. Here are some of the terms and formulas you should be familiar with for the AP Biology exam.

Mode, median, and mean are used to describe the central tendency of a data set.

Mode—the most frequently used value in a data set

Median—the middle value of a data set

Mean—the average of a data set; to find the mean, add all of the data points in a data set, and then divide by the number of members in the data set

Here is another way to define the mean:

$$\bar{x} = \frac{1}{n}\sum_{i=1}^{n} x_i$$

n = the number of members in the data set

Range, standard deviation, and standard error of the mean are used to describe the spread of a data set.

Range—the highest value of the data set minus the lowest value of the data set

Standard Deviation*

$$s = \sqrt{\frac{\sum (x_i - \bar{x})^2}{n-1}}$$

Standard Error of the Mean*

$$SE_{\bar{x}} = \frac{s}{\sqrt{n}}$$

To construct a 95% confidence interval, start at the mean of the data set. Add two times the standard error of the mean to find the upper limit of the 95% confidence interval. To find the lower limit of the 95% confidence interval, start at the mean of the data set, and then subtract two times the standard error of the mean.

Chi-Square

The chi-square formula is appropriate for evaluating a null hypothesis about a data set that involves counts (for example, whether or not the observed numbers of cells in different phases of the cell cycle are significantly different from the expected numbers). The formula for chi-square is:

$$\chi^2 = \sum \frac{(\text{observed} - \text{expected})^2}{\text{expected}}$$

Once the chi-square value for a data set is calculated, it is compared to the critical value found in the chi-square table. Find the critical value at the intersection of the degrees of freedom (df) and the *p*-value.

*Note: While you do need to understand how to apply standard deviation and standard error of the mean, and how to use them to construct 95% confidence intervals, you will not be required to calculate standard deviation or standard error of the mean on the AP Biology exam.

Chi-Square Table

p-value	\multicolumn{8}{c}{Degrees of Freedom}
	1	2	3	4	5	6	7	8
0.05	3.84	5.99	7.81	9.49	11.07	12.59	14.07	15.51
0.01	6.63	9.21	11.34	13.28	15.09	16.81	18.48	20.09

Genetics Problems and Laws of Probability

When working through genetics problems, it is useful to understand the basic **laws of probability**.

If the results of one event do not affect the results of a second event (independent events), the probability of both events happening equals the product of their individual probabilities:

$$P(A \text{ and } B) = P(A) \times P(B)$$

For example, if the probability of a tall plant is $\frac{1}{2}$ and the probability of the plant having purple flowers is $\frac{1}{2}$ (and the gene that determines tallness has no effect—is independent of—the gene that determines flower color), the probability of the plant being tall AND having purple flowers is $\frac{1}{2} \times \frac{1}{2} = \frac{1}{4}$.

If the results of one event do affect the probability of a second event happening (mutually exclusive events), the probability of both events happening equals the sum of their individual probabilities:

$$P(A \text{ or } B) = P(A) + P(B)$$

For example, in a heterozygous cross, the probability of a homozygous recessive offspring is $\frac{1}{4}$ and the probability of a heterozygous offspring is $\frac{1}{2}$, so the probability of an offspring being either homozygous recessive OR heterozygous is $\frac{1}{4} + \frac{1}{2} = \frac{3}{4}$.

Hardy-Weinberg Equations

Use this equation when finding allele frequencies:

$$p + q = 1$$

p is the frequency of the dominant allele.
q is the frequency of the recessive allele in the gene pool.

Use this equation when looking for genotype frequencies:

$$p^2 + 2pq + q^2 = 1$$

p^2 is the frequency of the homozygous dominant genotype.
$2pq$ is the frequency of the heterozygous genotype.
q^2 is the frequency of the homozygous recessive genotype.

Water Equations

The total water potential is the sum of the pressure potential and the solute potential:

$$\Psi = \Psi_{pressure} + \Psi_{solute}$$

In open containers in equilibrium with their environment:

$$\Psi_{pressure} = 0$$

Solute potential can be calculated with the following equation:
$$\Psi_{solute} = -iCRT$$

i = ionization constant for the solute
C = concentration of the solute in moles per liter
R = pressure constant $\left(\dfrac{0.0831 \text{ liters-bars}}{\text{mole-K}}\right)$
T = temperature in Kelvin (Note: Kelvin = °C + 273)

pH balance is vital to living organisms and can be calculated with the following equation:
$$pH = -\log[H^+]$$

Equations Related to Ecology

Greater biodiversity can help ecosystems withstand disruptions (environmental changes). One way to measure the biodiversity of an ecosystem is by using the Simpson's Diversity Index:

$$\text{Simpson's Diversity Index} = 1 - \sum\left(\dfrac{n}{N}\right)^2$$

N = the total number of organisms of all species in an entire ecosystem
n = the total number of organisms of one particular species in an ecosystem

Measuring population growth is important in understanding ecosystems. The following equations can be used to describe population growth.

General rate equation:
$$\text{Rate} = \dfrac{dY}{dt}$$

dY is change in Y.
dt is change in time.

The population growth rate can be measured by comparing the birth rate and the death rate in a population:
$$\dfrac{dN}{dt} = B - D$$

dN is change in population size.
dt is change in time.
B is the birth rate.
D is the death rate.

If a population is growing **exponentially**, you may use the following equation:
$$\dfrac{dN}{dt} = r_{max} N$$

r_{max} is the maximum per capita growth rate of a population.
N is population size.

If a population is growing **logistically**, you may use the following equation:
$$\dfrac{dN}{dt} = r_{max} N\left(\dfrac{K-N}{K}\right)$$

K is the carrying capacity.

Equations for Surface Area and Volume

The surface area to volume ratio of a cell has a big effect on how efficiently it can exchange materials with its environment. The table lists some commonly used surface area formulas and volume formulas for various shapes. In these formulas, note the following:

h is height.
l is length.
r is radius.
s is length of one side of a cube.
w is width.

Shape	Surface Area Formula	Volume Formula
Sphere	$4\pi r^2$	$\frac{4}{3}\pi r^3$
Rectangular Solid	$2lh + 2lw + 2wh$	lwh
Cylinder	$2\pi rh + 2\pi r^2$	$\pi r^2 h$
Cube	$6s^2$	s^3

Metric Prefixes

Here are some commonly used metric prefixes.

Prefix	Symbol	Factor
Giga	G	10^9
Mega	M	10^6
Kilo	k	10^3
Deci	d	10^{-1}
Centi	c	10^{-2}
Milli	m	10^{-3}
Micro	μ	10^{-6}
Nano	n	10^{-9}
Pico	p	10^{-12}

B

Choosing the Right Graph in AP Biology

One of the keys to success on the AP Biology exam is understanding the various types of graphs you are likely to encounter on the exam and the appropriate use for each type of graph. This section will cover the key features of the most commonly used graphs in AP Biology and provide advice for when to use each graph on test day. When constructing each of these graphs, make sure your axes are labeled appropriately with units, use appropriate scaling, plot your data points with error bars, and include a legend, where appropriate.

Line Graph

- Line graphs show the connections between two continuous variables.
- Those continuous variables are on both the *x*-axis and the *y*-axis.
- Line graphs are very useful for showing *trends, patterns,* or *changes* in data over time. For example, a line graph can illustrate the rate of production of a product over time.
- Larger data sets can often be effectively illustrated with a line graph, but line graphs may be less useful than *X-Y* graphs (scatterplots) for showing *individual* data points.

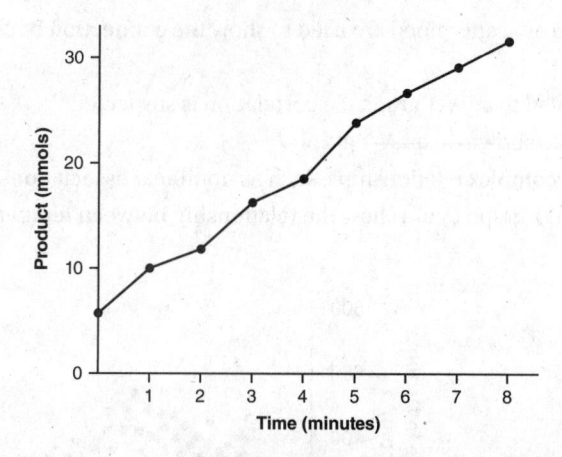

Dual Y Graph

- A dual *Y* graph has two separate *y*-axes and a single *x*-axis.
- These graphs are useful when comparing *two* different data sets that may have *different scales*. For example, a dual *Y* graph might illustrate changes in temperature and rainfall (each of which would be shown on its own *y*-axis) over a given time period (which would be shown on the *x*-axis).

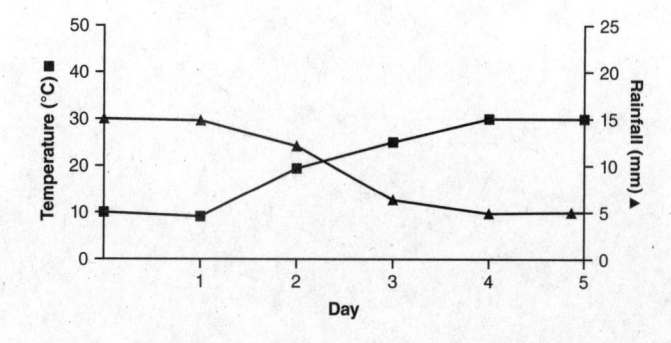

Log Y Graph

- In this type of graph, the y-axis is scaled *logarithmically* so that a larger range of data points may be shown.
- Log Y graphs are very useful when the range of the data shown on the y-axis spans several orders of magnitude. For example, a log Y graph might show the relationship between the number of base pairs in a DNA sequence and the distance the DNA sequence migrated in an electrophoresis gel.

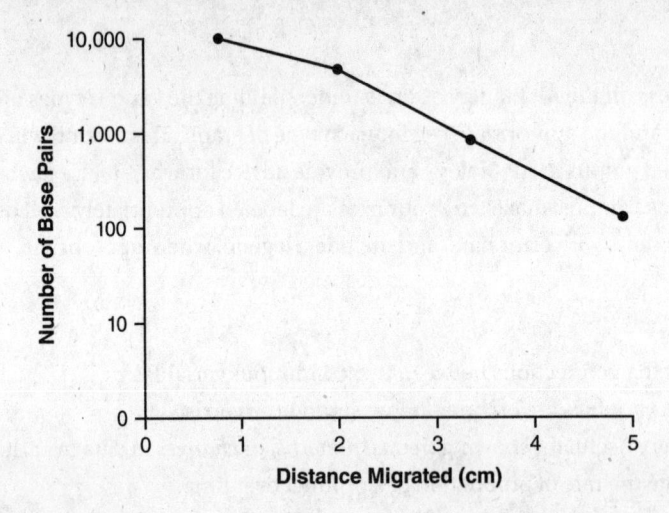

X-Y Graph (Scatterplot)

- *X-Y* graphs (also known as scatterplots) are used to show the connection or correlation between two variables.
- A trend line may be added to an *X-Y* graph if a correlation is suspected.
- Multiple data sets can be shown on an *X-Y* graph.
- *X-Y* graphs can handle complex relationships, such as nonlinear associations, clustering, or outliers in the data. For example, an *X-Y* graph could show the relationship between temperature and heart rate for a sample of *Daphnia*.

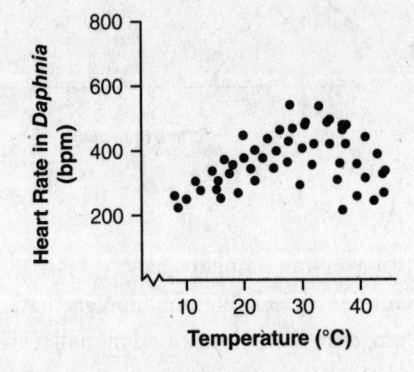

Bar Graph

- Bar graphs usually have a categorical variable on the x-axis and a numerical value on the y-axis.
- There are gaps between the bars to emphasize the categorical nature of the data; this helps distinguish a bar graph from a histogram.
- Bar graphs are great for comparing categorical data side-by-side.
- If your data falls into discrete categories that do not have a specific order, a bar graph would be appropriate. For example, if there are four different plots of land and you need to graph the number of trees in each plot, a bar graph could illustrate that data effectively.

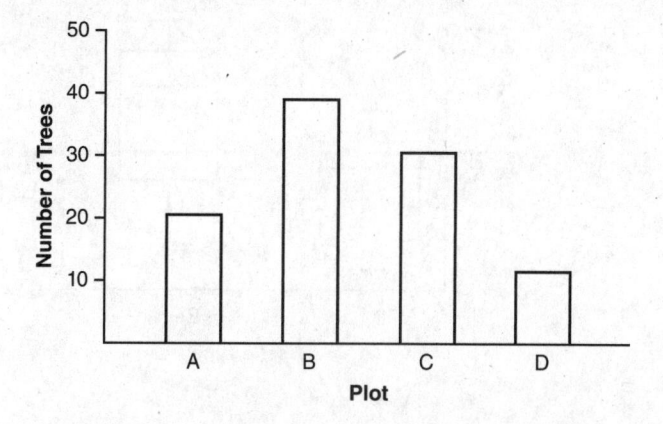

Histogram

- Histograms show the frequency of data falling into specific intervals, or "bins."
- Each bar on the x-axis shows an interval of a continuous numerical variable, and the y-axis shows the frequency, or counts, of data points in each interval.
- The bars in a histogram touch each other because there are no gaps between the intervals and the data on the x-axis is continuous.
- Histograms are used to show the distribution of continuous data. For example, the number of individuals in several age brackets could be illustrated by a histogram.

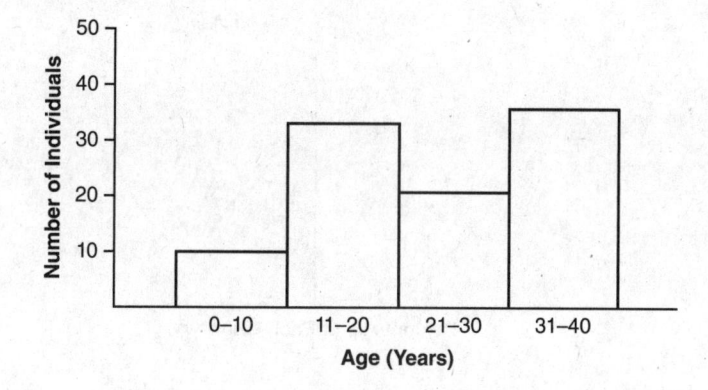

Box and Whisker Plot (Box Plot)

- Box and whisker plots (also known as box plots) show the median and quartiles of a data set. They can also show any outliers in the data set.
- Box and whisker plots are useful for showing a data set's overall distribution and for making comparisons between data set distributions. For example, a box and whisker plot might display the levels of expression of a gene (mRNA levels) in different tissue types.

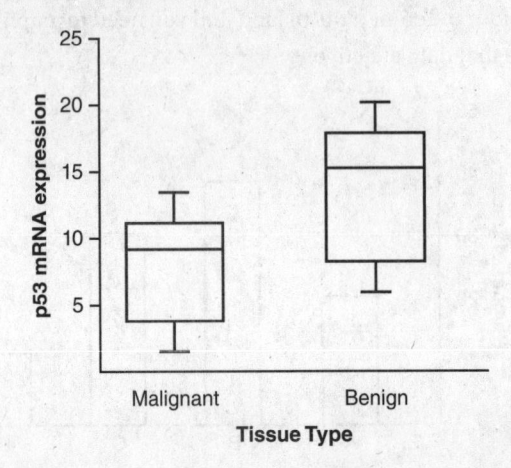

Pie Chart

- A pie chart is a circular diagram divided into slices that represent the proportions of different categories of data within a whole. The size of the slice in a pie chart is proportional to the quantity it represents.
- Pie charts are useful for showing the composition of different parts in relation to the whole. For example, a pie chart could be used to present the percentage of each genotype in a population.

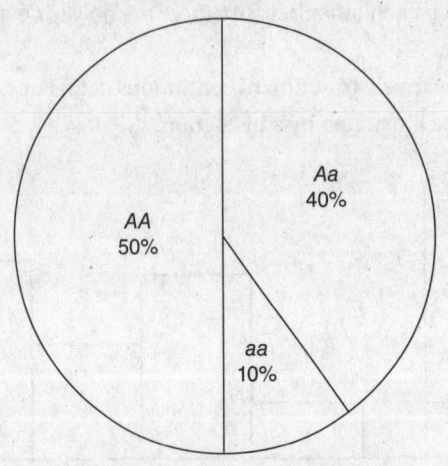

C

Biogeochemical Cycles

Matter and nutrients cycle between the environment and organisms via biogeochemical cycles. Each cycle is necessary for the continuity of life on Earth and each demonstrates the conservation of matter. Biogeochemical cycles include biotic and abiotic reservoirs where matter and nutrients are stored. The four biogeochemical cycles you need to know for the AP Biology exam are the **hydrologic (water)** cycle, the **carbon** cycle, the **nitrogen** cycle, and the **phosphorus** cycle.

The Hydrologic Cycle

The hydrologic (water) cycle involves the movement and storage of water. Water is stored in abiotic reservoirs, including oceans, surface water, groundwater, and the atmosphere. Living organisms are biotic reservoirs and they too store water. Water cycles between these reservoirs through the following processes:

- **Evaporation:** liquid water changes into water vapor in the atmosphere; this process is primarily driven by heat energy from the Sun
- **Condensation:** water vapor in the air is changed into liquid water; this process is essential for cloud formation
- **Precipitation:** condensed water vapor in clouds becomes heavy enough to fall to the earth as rain, snow, sleet, or hail; this process replenishes water on Earth's surface
- **Transpiration:** plants absorb water from the soil through their roots and release it into the air as water vapor through stomata on their leaves

The Carbon Cycle

During the carbon cycle, carbon atoms from carbon dioxide in the atmosphere are repurposed into carbohydrates in living organisms and are eventually re-released back into the atmosphere as carbon dioxide. The major processes that move carbon through Earth's biosphere are as follows:

- **Photosynthesis:** plants use light energy from the Sun to absorb carbon dioxide from the air and convert it into organic compounds (refer to Chapter 8 for more details)
- **Cellular Respiration:** organisms break down organic molecules to release energy, producing carbon dioxide that is released into the atmosphere (refer to Chapter 9 for more details)
- **Decomposition:** bacteria and fungi break down complex organic molecules (in dead organisms, urine, and feces) into simpler carbon compounds; these carbon compounds return carbon to the environment, allowing it to be used by other organisms
- **Combustion:** organic materials are burned in the presence of oxygen, releasing carbon dioxide into the atmosphere

The Nitrogen Cycle

The largest reservoir of nitrogen is the atmosphere, which is approximately 78% nitrogen gas (N_2). However, N_2 needs to be converted into a more usable form of nitrogen for living organisms. Microorganisms in the soil help convert nitrogen into usable forms that support life. The processes of the nitrogen cycle are outlined below:

- **Nitrogen Fixation:** bacteria convert nitrogen gas in the air into biologically available (a.k.a. biologically usable) forms, such as ammonia, nitrites, and nitrates
- **Assimilation:** plants and other organisms absorb inorganic nitrogen compounds (such as ammonia or nitrates) and incorporate them into organic nitrogen compounds (such as amino acids and nucleotides)
- **Ammonification:** organic nitrogen compounds in dead plants and animals are broken down into ammonia, returning usable nitrogen to the soil
- **Nitrification:** bacteria convert ammonia into nitrites and nitrates, which are usable by plants
- **Denitrification:** bacteria in the soil convert nitrates into nitrogen gas, returning it to the atmosphere

The Phosphorus Cycle

This biogeochemical cycle involves the movement of phosphorus through the environment and living organisms. Human activities, such as the use of fertilizers and detergents that contain phosphates, can impact the phosphorus cycle. The processes of the phosphorus cycle are as follows:

- **Weathering and Erosion:** the weathering and erosion of rocks releases phosphates into the soil and groundwater
- **Plant Uptake:** plants absorb inorganic phosphates from the soil through their roots and incorporate phosphorus into phospholipids, nucleotides, and other compounds
- **Animal Consumption:** animals consume plants and other animals, absorbing phosphorus in the process
- **Decomposition:** decomposers break down dead plants and animals and release phosphorus back into the soil
- **Sedimentation:** some phosphorus in the ocean is deposited on the continental margins and on the bottom of the ocean

Index

A

Abiotic factors, 297
Accessory pigments, 102
Acetylation, histone proteins, 216
Acetyl coenzyme A, 115
Acids, pH sensitivity, 28
Acquired characteristics, 238
Activation energy (E_A), 91–92
Activator proteins, 216
Active site (enzyme interaction), 89
Active transport, 64–65
Adaptation
    biodiversity, 297
    convergent evolution, 238
    ecosystems, 280
Adaptive radiation, 267
Adenine, 41–42, 191–192, 203
Adenylyl cyclase, 129
Aerobic organisms, 113–114
Alcohol fermentation, 118
Algae, 59, 61, 111
Alleles. *See also* genes
    Hardy-Weinberg equilibrium, 255–257, 309, 402
    population genetics, 253–255
    segregation, 165
Allopatric speciation, 267
Allosteric inhibitor, 90
Allosteric site, 90–91
Alpha helixes, 40, 63
Alternative splicing, transcription, 202
Amino acids
    peptide bonds, 40
    in proteins, 39–41
    sequencing, 310
Amino group, 39
Amish population, founder effect in, 255
Ammonification, 410
Amphibians
    common ancestors, 266
    ecosystems, 280
Amphipathic lipids, 38
Ampicillin, bacterial transformation and, 318
Amyloplasts, 60
Amylose, 38
Anaerobic organisms
    cellular processes, 113–114
    fermentation, 118
Anaphase
    meiosis I and II, 151–153
    mitosis, 139

Anchorage dependence, 141
Androgen insensitivity syndrome (AIS), 130
Aneuploidy, 155, 217
Animal cells, 59–60
    artificial selection in, 242–243
    mutations, 217
Animal consumption, 410
Anole lizards, 296
Antennapedia, gene coding for, 216
Antibiotic resistance genes, 225
Antibiotic-resistant bacteria, 239, 269, 317
Anticodon, 203–204
Antidiuretic hormone (ADH), 130
Antigen-presenting cells, 128
Antiparallel strands
    DNA, 42, 192–193
    RNA, 201–202
AP Exam
    biology units, 3
    format, 3
    scoring, 7
    sections, 4–6
    study plans, 7–8
Apoptosis, cell cycle and, 140–142
Aposematism, 279–280
Aquaporins, 64
Arabinose operon, 215
Artificial selection, 242–243, 308–309
Assimilation, 410
ATP
    Calvin cycle, 101–104
    glycolysis, 114
    Krebs cycle, 115–116
    light-dependent reactions, 101
    oxidative phosphorylation, 116–118
    photolysis, 102–103
ATP synthase, 103
    oxidative phosphorylation, 115–116
Audible signals, 279
Autocrine signaling, 127
Autosomes, 178
Autotrophs, 101, 281

B

Bacteria
    antibiotic-resistant bacteria, 239
    cells, 55–56
Bacterial transformation, 225, 317–318
Bar graph, 407

Base-pairing rules, 192
Bases, pH sensitivity, 28
B cells, 202
Behavioral isolation, 268
Behavioral responses, 279–280
Beta-pleated sheets, 40
Bicarbonate ions, 28
Big Ideas, xii
Bilayer, plasma membrane, 62–63
Binary fission, 61
Biodiversity, 297, 403
Biogeochemical cycles, 409–410
Biogeography, evolution and, 238
Bioinformatics, 310
Biological macromolecules, 37–39
Biomagnification, 297
Biotechnology
    bacterial transformation, 225, 317–318
    CRISPR-Cas9, 226–227
    gel electrophoresis, 226
    polymerase chain reaction (PCR), 226
Biotic factors, 297
Birds
    behavioral response, 279–280
    reptiles and, 266
    sexual selection in, 243
Birth rate, 293, 403
BLAST (Basic Local Alignment Search Tool), 310
Blood plasma, 28
Blood sugar, 130
Blue-footed booby bird (*Sula nebouxii*), 243
Bottleneck effect, 253–254
Bottom-up ecosystem regulation, 282
Box and whisker plot (box plot), 408
Brassica oleracea, 242–243
Brassica rapa, 308–309, 320
BRCA gene, 142
Breakdown, hybrid, 268
Breeding
    artificial selection, 242–243
    seasons, 268
    speciation, 267–268
Buffers, water and, 28

C

C3 photosynthesis. *See* Light-independent reactions
C4 photosynthesis, 104
Cabbage butterfly (*Pieris rapae*), 320–321

411

Calvin cycle, 101–104
CAM photosynthesis, 104
Cancer
    autocrine signaling, 127
    cell cycle and, 139–142
    centrosome dysfunction, 59–60
    enzymes, 91
Capillary action, water, 26
Carbohydrates, 38
Carbon
    cellular respiration, 116
    cycle, 409
    DNA/RNA structure, 192
    fixation, 103
    glycolysis, 114
    macromolecules, 37
Carbon dioxide
    Calvin cycle, 103–104
    cellular respiration, 116, 314–315
    fermentation, 118
    Krebs cycle, 113
    pH and, 27–28
    photosynthesis, 101, 313
    plasma membrane transport, 62–63
    pyruvate oxidation, 113
    transpiration, 321
Carbonic acid, 28
Carboxylic acid group, 39
Carnivores, 280
Carrying capacity, 294, 403
Catabolism, inducible operons, 214–215
Catabolite activator protein (CAP), 214–215
Catalysts, enzymes, 89, 323
Cell cycle
    lab experiments, 316
    meiosis, 151–155
    phases, 139–140
    regulation, 140–142
Cell division. See meiosis I and II; mitosis
Cell membrane. See plasma membrane
Cell membrane receptors, 129
Cells, gene expression and specialization of, 216
Cell(s). See also organelles
    animal, 60–61
    communication and signaling, 127–130
    compartmentalization, 61–62
    endosymbiosis hypothesis, 61
    eukaryotic and prokaryotic, 55–56
    homeostasis, 55
    movement in water, 73–76
    plant, 60–61
    recognition, 63
    structure and function, 55–61
    surface to volume ratio, 62
Cellular respiration
    carbon cycle, 409
    fermentation, 118
    glycolysis, 113
    Krebs cycle, 115–116
    lab experiments, 314–315
    oxidative phosphorylation, 116–118
    processes, 113–114
    pyruvate oxidation, 115

Cellulose, 38, 64
Cell wall, 64
Celsius, solution temperature, 75
Center of data
    mean, 11
    median, 11–12
    mode, 401
Central dogma, genetic information, 201
Centromeres, meiosis, 151
Centrosome, 59–60
Channel proteins, 64
Chaperonins, 41
Checkpoints, cell cycle, 140
Chemical reactions
    activation energy, 91–92
    coupled reactions, 92–93
    free energy, 91–92
Chemical signals, 279
Chemiosmosis
    oxidative phosphorylation, 116–118
    photosynthesis, 103
Chemoautotrophs, 281
Chi-square test, 9–11, 401–402
    cell cycle, 316
    fruit fly experiment, 322
Chlorophyll, 102
Chloroplast DNA (cpDNA), 59, 179
Chloroplasts
    endosymbiosis hypothesis, 61
    nonnuclear inheritance, 179
    photosynthesis, 59, 102
Cholesterol, 39
Chromatids
    sister, 152
    tetrads, 153–154
Chromosomes
    cell cycle, 140
    DNA in, 55
    epigenetic changes, 216
    lab experiments, 318
    linked genes, 154, 177–178
    meiosis I and II, 151–155
    polyploidy, 267–268
    sex chromosomes, 178
Cicadas, speciation in, 268
Cisternae, Golgi complex, 57
Citric acid cycle. See Krebs cycle
Clade and cladograms, 265–267
Climate, water effects, 27
Clutch size, stabilizing selection, 241
Codominance, 178
Codons, 201, 203–205
Coenzyme A, pyruvate oxidation, 115
Coenzymes, 91
Cofactors, 91
Cohesion, water, 26
Collagen, 41
Color blindness, 178
Combustion, 409
Commensalism, 297
Common ancestry, 265–266, 310
Community ecology, 295–297
Compartmentalization, 61
Competition, community ecology, 296
Competitive inhibitors, 90–91

Concentration gradient
    active transport, 64–65
    passive transport, 64
Concentration of solute, 74
Condensation, 409
Conjugation, 217
Consumers
    ecosystems, 280–283, 297
    energy dynamics, 320–321
Continuing evolution, 269
Contractile vacuole, 76
Convergent evolution, 238
Cooperative behaviors, 280
Coral bleaching, 280
Corepressor, 215
Cortisol, 39
Coupled reactions, 92–93
Covalent bonds, 25–26
CRISPR-Cas9, 226–227
Cristae, 62
Crosses, 166–168
Crossing-over, 151–154
CUtTheP (cytosine, uracil and thymine), 192
Cyanobacteria, 102
Cyclic AMP (cAMP), 129, 214–215
Cyclin-dependent kinase, 140–141, 316
Cyclin protein, 140–141
Cyclins, 140–141, 316
Cytokinesis, 139–140, 152
Cytoplasmic ribosomes, 203
Cytosine
    denaturation and, 41–42
    DNA structure, 191–192
Cytoskeleton, 60
Cytosol, 55, 129

D
Darwin, Charles, 238–239
Darwin's finches, 267
Databank of sequences, 310
Data set, spread of, 11–12
Daughter cells, cell cycle, 140
Death rate, 293, 403
Decomposers, 281
Decomposition, 409, 410
Degrees of freedom (df), 10
Dehydration, 130
Dehydration synthesis, 37–38, 40
Denaturation
    DNA, 226
    enzymes, 90
Denitrification, 410
Density-dependent factors, population growth, 294
Density-dependent inhibition, cell cycle, 141
Density-independent factors, population growth, 294
Deoxyribose, 192
Descriptive statistics, 11–15
    mean, 11, 401
    median, 11–12, 401
    95% confidence interval, 13–15
    standard deviation, 12
    standard error of the mean, 12–13, 401

Detritivores, 281
Dialysis tubing, 311
Dietary partitioning, 296
Differential gene expression, 216
Differential reproductive success, 238
Diffusion, 64
    lab experiments, 310-312
Dihybrid cross, 167-168
Diploid (2n) parent cells, 151-153
    lab experiments, 316
Directionality, DNA molecules, 193-194
Directional selection, 240-242
Disruptive selection, 241-242
DNA. *See also* RNA
    bacterial transformation,
        225, 317-318
    cell cycle, 139
    cell organelles, 55
    CRISPR-Cas9 sequencing, 226-227
    denaturation of, 226
    epigenetic changes, 216
    extranuclear, 191
    formation of, 42
    gel electrophoresis, 226
    as information carrier, 205-206
    meiosis and replication of, 152
    Mendelian genetics, 165-166
    replication, 193-194, 202
    restriction enzyme analysis, 318-320
    sequencing, 310
    structure, 191-192
DNA ligases, 225
DNA polymerase, 193-194, 201-202
Dominant traits, 166
    Hardy-Weinberg equilibrium,
        255-257, 309, 402
Double helix, DNA structure, 192-194
Double-stranded DNA (dsDNA), 42, 216
Double-stranded RNA (dsRNA), 216
Down syndrome, 155, 217
Drosophila
    gene expression, 216
    lab experiments on, 322
Dual *Y* graph, 405

E
Ecology
    community ecology, 295-297
    environmental change and, 279-283
    equations, 403
    population, 293-294
Ecosystems
    consumers in, 297
    energy flow in, 280-283
Ectotherms, 280
Electrical signals, 279
Electronegativity, water, 25
Electrons
    hydrogen bonds, 25
    macromolecules, 37
Electron transport chain (ETC)
    fermentation, 118
    oxidative phosphorylation, 116-118
    photosynthesis, 102-103
Elephant seal, population
    bottleneck, 254

Elisia crispata, 281
Elongation, of translation, 204-205
Endergonic reaction, 91-93
Endocrine signaling, 128
Endocytosis, 65
Endomembrane system, 61
Endoplasmic reticulum, 56-57
Endosymbiosis hypothesis, 61
Endotherms, 280
Energetics, 2
Energy
    activation energy (E_A), 91-92
    carbohydrates and, 38
    coupled reactions, 92-93
    dynamics, lab experiments,
        320-321
    ecosystems and flow of, 280-283
    in mitochondria, 58-59
Enhancers, 216
Environment
    ecology and, 279-280
    enzyme function in, 90-91
    gene expression and phenotypic
        plasticity, 179
    natural selection and, 239-242
Enzymes. *See also* specific enzymes
    activity, 89-90, 323
    environmental factors, 90-91
    inhibitors, 90-91
    structure and function, 89
Enzyme-substrate complex, 90
Epigenetics, 216, 238
Erosion, 410
Estradiol, 39
Euchromatin, 216
Eukaryotes
    cell structure, 55-56
    common ancestor for, 267
    gene expression, 216
    information flow in, 205-206
    transcription, 201-203
    translation, 203-205
Eutrophication, 297
Evaporation, 409
Evolution, 2
    artificial selection, 242-243, 308-309
    evidence of, 237-238
    modern examples of, 269
    natural selection, 238-242
    phylogeny, 265-267
    population genetics, 253-257
    sexual selection, 243
Exergonic reaction, 91-93
Exocytosis, 65
Exons, 202
Expected results, 9-10
Exponential population growth,
    293-294, 403
Extranuclear DNA, 191

F
Facilitated diffusion, 64
FAD^+
    glycolysis, 114
    Krebs cycle, 115-116
    oxidative phosphorylation, 117

$FADH_2$
    Krebs cycle, 115-116
    oxidative phosphorylation, 117
Fast Plants, 308-309, 320-321
Fatty acids, 38-39
Feedback mechanisms, 130
Females
    nonnuclear inheritance in, 179
    sex-linked genes, 179
Fermentation, 118
Fertility, hybrids, 268
Fertilization
    meiosis, 151, 315-316
    mitosis, 315-316
    segregation, 165
First Law of Thermodynamics, 92
Fish
    common ancestors, 266
    ecosystems, 280
Five-carbon sugars, 191
Fixation, carbon, 103
Fleas, 296
Fluid mosaic model, 63-64
Food chains, 280-281
Food webs, 281
Fossils, 237, 265
Founder effect, 254-255
Frameshift mutation, 217
Free energy, chemical reactions, 91-92
Fruit flies. *See Drosophila*
Fungi, 64
Fur color, 241-242

G
G0 phase, cell cycle, 140
G1 phase, cell cycle, 140
G2 phase, cell cycle, 140
Galápagos Islands, 267
Gametes
    haploid, 151, 153
    meiosis I and II, 151-153
    Gametic isolation, 268
GAPDH (glyceraldehyde-3-phosphate
    dehydrogenase) gene, 237
Gel electrophoresis, 226
    restriction enzyme analysis, 318-320
Gene expression
    cell specialization, 216
    eukaryotes, 216
    prokaryotes, 213-215
Gene flow, 253, 255
Genes
    antibiotic resistance, 225
    bacterial transformation, 317-318
    linked genes, 154, 177-178
    mutations, 130
    sex-linked, 178
    structural genes, 213
Genetic drift, 253-257, 309
Genetics
    central dogma, 201
    independent assortment, 165
    information flow, 205-206
    law of segregation, 165
    meiosis and, 151-155
    Mendelian genetics, 165-168
    mutations, 217

non-Mendelian, 177–179
pedigree, 166
population genetics, 253–257
probability in, 166–168, 402
recombination, 151–154
Genome
cell cycle, 140
evolution, 269
Genotype, 166–167
environment and, 179
Hardy-Weinberg equilibrium, 255–257, 309, 402
test cross, 168
Glucagon, endocrine signaling, 128
Glucose
cellular respiration, 116
glycolysis, 116
Glyceraldehyde-3-phosphate (G3P), 103–104
Glycogen, 38
Glycolipids, 63
Glycolysis, 113–114, 116
Glycoproteins, 63
Golgi complex, 57, 206
Gonadal tissue, androgen insensitivity syndrome (AIS), 130
Gradualism, 267
Grana, 59
Graphs, 405–408
5' GTP cap, 203
Guanine, 41–42, 191–192

H
Habitat destruction, 297
Habitat isolation, 268
Habitat partitioning, 296
Haploid (n) gametes, 151, 153
lab experiments, 318
nonnuclear inheritance, 179
Hardy, Godfrey, 255
Hardy-Weinberg equilibrium, 255–257, 309, 402
Heat shock technique, 317
Helicase, DNA replication, 193
Hemoglobin, 40, 41
Hemophilia, 178
Herbivores, 280
Heterochromatin, 216
Heterotrophs, 101, 281
Heterozygosity, 166–167, 259
Histogram, 407
Histone proteins, 216
Homeostasis
cell, 55
feedback mechanisms, 130
Homologous chromosomes, 152, 237
Homology, 310
Homozygosity, 166, 259
Horizontal transmission, 217
Hormones, endocrine signaling, 128
Hox genes, 216
Hybrid breakdown, 268
Hybrid fertility and viability, 268
Hydrogen
ions, 28
in macromolecules, 37

oxidation and reduction, 101
pH and, 27–28
in water, 25–27
Hydrogen bonds
in DNA, 42
DNA/RNA structure, 192
protein structure, 40–41
in water, 25–27
Hydrologic cycle, 409
Hydrolysis, 37–38
Hydrolytic enzymes, 57
Hydrophilic ligands, 129
Hydrophilic lipids, 63–64
Hydrophobic ligands, 129
Hydrophobic lipids, 63–64
Hypertonic solution, 73
Hypotheses, 9, 401
Hypotonic solution, 73–74

I
Incomplete dominance, 178
Independent assortment
lab experiments, 316
meiosis metaphase I, 154–155
Mendelian genetics and, 165–166
Inducer molecules, 214
Inducible operons, 214–215
Information storage and transmission, 3, 205–206
Inheritance patterns. *See also* genetics
nonnuclear inheritance, 179
pedigree, 166
Inhibitors, enzyme, 90–91
Initiation of translation, 203–204
Insulin
endocrine signaling, 130
recombinant, 227
Interphase (cell cycle), 139–142
Intersexual selection, 243
Intestine, surface area to volume ratio, 62
Intracellular receptors, 129
Intrasexual selection, 243
Introns, 202–203
Invasive species, 297
Ionization constant, 74, 403
Ions
cell membrane transport, 62–63
water and, 27–28
Isotonic solution, 73–74

J
Jumping genes, transposition and, 217
Juxtacrine signaling, 127–128

K
Kelvin temperature scale, 75, 403
Keystone species, 297
Kinases, 129
Kleptoplasty, 61, 281
Klinefelter syndrome, 217
Knockout genes, 227
Krebs cycle
oxidative phosphorylation, 116–118
processes, 113, 115–116
K-selected populations, 294–295

L
Lab experiments
artificial selection, 308–309
bacterial transformation, 317–318
BLAST, 310
cellular respiration, 314–315
core practices and skills, 309–310
diffusion and osmosis, 310–312
energy dynamics, 320–321
enzyme activity, 323
fruit fly behavior, 322
Hardy-Weinberg equilibrium, 309
mitosis and meiosis, 315–316
photosynthesis, 313–314
restriction enzyme analysis, DNA, 318–320
transpiration, 321
Lac operon, 214–215
Lactic acid fermentation, 118
Lactose, 214–215
Lagging strand replication, 194
Lag phase, population growth, 294
Lamarck, Jean-Baptiste, 238
Last universal common ancestor (LUCA), 267
Law of independent assortment, 154–155, 165, 316
Law of segregation, 165, 316
Laws of probability
genetics, 166–168, 402
Hardy-Weinberg equilibrium, 256, 402
Leading strand replication, 194
Ligands, cell signaling and, 127–130
Ligase, DNA replication, 194
Light-dependent reactions, 101–103
photosynthesis, 313–314
Light-independent reactions, 101, 103–104
Line graphs, 405
Linked genes, 154, 177–178
Lipids, 38–39
Local regulators, paracrine signaling, 128
Logarithmic graphs, 406
Logistic population growth, 403
Log phase, population growth, 294
Log *Y* graph, 406
LUCA (last universal common ancestor), 267
Lumen, Golgi complex, 57
Lysosomes, 57–58

M
Macromolecules
carbohydrates, 38
elements, 37–39
lipids, 38–39
proteins, 39–41
Magicicada septendecim, 268
Magicicada tredecim, 268
Males
nonnuclear inheritance in, 179
sex-linked genes, 178
Mammals, common ancestry, 265–267
Map units, 177–178
Mass, solution molarity, 312
Maternal inheritance, 179

Matrix, mitochondrial, 58
Mature RNA, 203
Mean, 11, 14, 401
Mechanical isolation, 268
Median, 11-12, 401
Mediators, 216
Meerkats, 280
Meiosis I and II
    cell division, 151-153
    genetic diversity, 153-155
    lab experiments, 315-316
    linked genes, 177-178
    mitosis *vs.*, 153
    mutations, 217
Melting temperature (Tm), 46
Membrane. *See* Plasma membrane
Membrane potential, 65
Membrane proteins, active and passive transport, 64-65
Mendelian genetics, 165-168
    lab experiments, 316
Messenger RNA (mRNA), 192, 201-203
    transcription, 205-206
    translation, 203-205
Metabolic rate, 280
Metabolism
    aerobic and anaerobic, 113-114
    coupled reactions, 92-93
    ecosystem, 280
Metaphase
    meiosis I and II, 151-155
    mitosis, 140
Methylation, 216
Metric prefixes, 404
MicroRNA, 192
Microtubules, 59
Miller-Urey experiment, 267
Minerals, as coenzymes and cofactors, 91
Mitochondria
    endosymbiosis hypothesis, 61
    energy production and, 58-59
    Krebs cycle, 115-116
    nonnuclear inheritance, 179
Mitochondrial DNA (mtDNA), 58, 179
Mitosis
    lab experiments, 315-316
    meiosis and, 151-155
    mutations, 217
    phases of, 139-142
Mitosis-promoting factor (MPF), 141
Mode (statistics), 401
Molarity, of solution, 312
Molecular clocks, 265-267
Molecules
    evolution and, 237-238
    ligands, 127
Monohybrid cross, 166-167
Morphology, 265
    evolution and, 237
M phase, cell cycle, 139-142
Mutations, 217
    cancer and, 141
    Hardy-Weinberg equilibrium, 255-257, 309, 402
    signal transduction, 129
Mutualism, 297

N

NAD^+ electron carrier
    enzyme substrate, 91
    fermentation, 118
    glycolysis, 114
    Krebs cycle, 115-116
    oxidative phosphorylation, 117
    pyruvate oxidation, 115
NADH
    fermentation, 116
    glycolysis, 114
    Krebs cycle, 115-116
    oxidative phosphorylation, 117
$NADP^+$, 103
NADPH
    Calvin cycle, 101-104
    oxidative phosphorylation, 117
Natural selection, 238-242, 308-309
NCHOPS (nitrogen, carbon, hydrogen, oxygen, phosphorus, sulfur), 37
Negative feedback, 130
Nephrogenic diabetes insipidus (NDI), 130
Neurotransmitters, 128
Niche partitioning, 296
95% confidence interval, 13-15, 308-309, 401
Nitrification, 410
Nitrogen
    cycle, 410
    fixation, 410
    macromolecules, 37
    membrane transport, 63-64
Nitrogenous bases, 191-192
    nucleic acids, 41-42
Nodes, phylogenetic trees, 265-267
Noncompetitive inhibitors, 90-91
Nondisjunction, 155
Nondividing cells, 140
Non-Mendelian genetics, 177-179
    linked genes, 154, 177-178
Nonnuclear inheritance, 179
Nonsense codons, 205
Nonsense mutation, 217
Nucleic acids, 39, 41-42
Nucleoid, 55
Nucleolus, 60
Nucleotides
    DNA and RNA structure, 191-192
    sequences, 310
    structure, 41
    transcription, 201
Nucleus
    information flow from, 205-206
    structure, 55
Null hypothesis (H_0), 9-11, 398

O

Observed results, 10
OILRIG (Oxidation Is Losing hydrogen atoms; Reduction is Gaining hydrogen atoms), 101
Okazaki fragments, 194
Oncogenes, 141
Operators, 213
Operons, 213-215

Organelles
    endosymbiosis hypothesis, 61
    photosynthesis, 102
Origin of replication (ori) site, 193
Osmolarity, 75-76
Osmosis, 64
    lab experiments, 310-312
Outgroup, 265
Oxidation
    cellular respiration, 116-118
    photosynthesis, 101
    pyruvate, 113-116
Oxidative phosphorylation, 113, 116-118
Oxygen
    Calvin cycle, 101-104
    cellular respiration and, 314-315
    macromolecules, 37
    oxidative phosphorylation, 116-118
    photosynthesis and, 313-314

P

Paracrine signaling, 128
Paramecium, osmolarity, 75-76
Parasitism, 296
Passive transport, 64
Pathogen evolution, 269
Peacock (*Pavo cristatus*), 243
Pedigree, 166
    sex-linked traits, 178
Peppered moths, evolution of, 239-241
Peptide bonds, 40
Permeability, plasma membrane, 63-64
Peroxidase, 323
Peroxisome, 60
Pesticide resistance, 269
PH (power of hydrogen)
    hydrogen, 28
    protein interactions, 40
    water, 27-28
Phenotype, 166
    mutations, 217
    natural selection and, 239-240
    plasticity, 179
Pheromones, 128, 279
Phosphatases, 129
Phosphate groups
    cell cycle, 140
    coupled reactions, 92-93
    DNA structure, 42, 191-192
    gel electrophoresis, 226
    nucleotides, 41
    phospholipids, 38
    RNA structure, 191-192
    substrate-level phosphorylation, 115
    transduction, 129
Phospholipids, 38-39
    bilayer, plasma membrane, 62-64
Phosphorus
    cycle, 410
    macromolecules, 37
Phosphorylation
    oxidative, 113, 116-118
    substrate-level, 115-116
Photoautotrophs, 101
Photolysis, 102-103
Photophosphorylation, 102

Photosynthesis
  carbon cycle, 409
  cellular respiration, 314–315
  energy dynamics, lab experiments, 320–321
  lab experiments, 313–314
  light-dependent reactions, 102–103
  light-independent reactions, 103–104
  overview, 101–102
Photosystems (PSI, PSII), 102–103
Phylogenetic trees, 265–267, 310
Phylogeny, 265–267
Physiological response, 279
Pie chart, 408
Pigments, light-dependent reactions, 102
Pilus, conjugation and, 217
Pleiotropy, 178
Plant cells
  mutations, 217
  structure, 60–61
Plants
  artificial selection in, 242–243
  cell wall, 64
  juxtacrine signaling in, 127–128
  photosynthesis in, 103–104
  polyploidy, 267–268
  transpiration experiment, 321
  uptake, 410
Plasma membrane
  cell organelles, 55
  information flow to, 205–206
  permeability experiments, 311–312
  receptors, 129–130
  structure, 62–63
  transport, 63–64
Plasmids, 55, 225, 317–318
Plasmodesmata, 127–128
Plasmolysis, 312
Point mutations, 217
Polar covalent bond, 25–26
Polarity, water, 25–27
3' poly-A tail, 203
Poly-A polymerase, 203
Polymerase chain reaction (PCR), 226
Polyploidy, 267–268
Polyribosomes, 203–205
Population ecology, 293–294, 403
  K-selected vs. r-selected populations, 294–295
Population genetics, 253–257
Population growth and size, 255, 293–294
Positive feedback, 130
Postzygotic barriers, 268
Potential
  membrane, 65
  pressure, 74, 402
  solute, 74, 402
Power of hydrogen. See pH (power of hydrogen)
Precipitation, 409
Predator/prey interactions, community ecology, 296
Pre-MRNA, 202–203

Pressure constant, 74, 403
Pressure potential, 74, 310–312, 402
Prezygotic barriers, 268
Primary protein structure, 40
Primers, annealing of, 226
Probability equations, dihybrid cross, 167–168
Probability in genetics, 166–168
Programmed cell death, 142
Prokaryotes
  cell structure, 55–56
  endosymbiosis hypothesis, 61
  gene expression regulation, 213–215
  photosynthesis, 102
  transcription, 201–203
  translation, 203–205
Promoters, 201, 213–216
Prophase
  cell cycle, 140
  meiosis I and II, 151–153, 177
Proteins. *See also* membrane proteins
  histone, 216
  membrane, 64–65
  mutations, 130
  regulatory proteins, 216
  structure, 39–41
Proton gradient
  chemiosmosis, 102–103, 117
  mitochondria, 58
Protons, oxidative phosphorylation, 117
Proto-oncogenes, 141
Punctuated equilibrium, 267
Punnett square, 166–167
PureAsGold (purines, adenine and guanine), 192
Purines, 42
P-value, 10
  cell cycle, 315–316
  fruit fly experiment, 322
Pyrimidines, 42, 192
Pyruvate oxidation, 113–116

Q
Quaternary protein structure, 40

R
Random mating, 255
Reaction profiles, enzymes, 92
Receptors, signal transduction, 129–130
Recessive traits, 166–167
  Hardy-Weinberg equilibrium, 256–257, 309, 402
Recombinant DNA, 225
Recombination
  genetic, 151–154
  linked genes, 177–178
Redox (reduction-oxidation) reactions, 102
Reduction
  Calvin cycle, 101–104
  photosynthesis, 101
Regeneration, of RuBP, 104
Regulatory switches, 216
Respirometer, 314–315

Replication, DNA, 193–194
  lab experiments, 316
  meiosis and, 152
Repressible operons, 213, 215
Repressor proteins, 216
Reproduction
  mitochondria, 58–59
  strategies, 238, 257
Reptiles
  common ancestors, 266
  ecosystems, 280
Resistance
  bacterial transformation, 317–318
  evolution and, 239, 269
Response
  environmental change, 279
  signal transduction, 129
Restriction endonucleases (restriction enzymes), 225–226
Restriction enzyme analysis, 318–320
Retroviruses, 206
Reverse transcriptase, 206
R-group, 39
Ribose, 192
Ribosomal RNA (rRNA)
  structure, 56, 192
  translation, 203–204
  transcription, 201
Ribosomes
  cell structure, 55–56
  chloroplast, 59
  transcription, 205–206
  translation, 203–205
Ribozymes, 89, 201
Ribulose-bisphosphate (RuBP), 103–104
Ribulose-bisphosphate-carboxylase (Rubisco), 103
RNA
  DNA *vs.*, 42
  as information carrier, 205–206
  Mendelian genetics, 165–166
  structure, 191–192
RNA polymerase
  operon transcription, 214–215
  promoters, 215
  replication, 193
  transcription, 201–202
RNA primer, 193
Root, phylogenetic tree, 265–266
Rough endoplasmic reticulum (RER), 56–57, 206
R-selected populations, 294–295

S
Saturated fatty acids, 38–39
Scatterplot, 406
Science practice and skills, 307
Sea otters, 296
Second Law of Thermodynamics, 92
Secondary messengers, signal transduction, 129–130
Secondary protein structure, 40
Sedimentation, 410
Segregation, Mendelian law of, 165
  lab experiments, 316
Semiconservative replication, 193–194

INDEX

Semipermeable membrane, 311
Sex chromosomes, 178
    mutations, 217
Sex-linked dominant traits, 166
Sex-linked genes, 178
Sex-linked inheritance, 179
Sex-linked recessive traits, 166
Sexual selection, 243, 268
Shared derived characteristics, 265
Signal amplification, 129
Signals
    cell signaling, 127–130
    environmental change, 279–280
Signal transduction, 129
Signal transduction pathways, 129–130
Silencers, 216
Silent mutation, 217
Simple diffusion, 64
Simpson's Diversity Index, 295–297, 403
Single-stranded RNA structure, 192
Sister chromatids, 152
Small interfering RNA (siRNA), 216
Small nuclear ribonucleoproteins (snRNPs), 202
Small nuclear RNAs (snRNAs), 202
Smooth endoplasmic reticulum, 56–57
Sodium-potassium (Na+/K+ pump), 64–65
Solute, concentration of, 74
Solute potential, 74, 311, 402–403
Solutions, 73–74
Solvent, water as, 27
Somatic cells, cell cycle, 141
Sordaria fimicola, 316
Speciation, 266–268
Species
    defined, 267
    diversity, 269, 297
    evolution in, 238
    invasive, 297
    keystone, 297
Specific heat, water, 26
S phase, cell cycle, 140
Sphere, surface-to-volume ratio, 62
Spindle fibers, 59
Spliceosomes, 202
Stabilizing selection, 241–242
Standard deviation, 12, 401
Standard error of the mean, 12–14, 401
Starch, amyloplast storage, 60
Start codons, translation, 203–205
Statistics
    chi-square test, 9–11
    descriptive, 11–15
    mean, 11, 401
    median, 11–12, 401
    95% confidence interval, 13–15
    null hypothesis, 9
    standard deviation, 12
    standard error of the mean, 12–13, 401
Steroids, 39
Stimulus response, 279

Stomata, 321
Stop codons, 204–205
Stroma, chloroplast, 59, 102
Structural genes, 213
Substrate, enzymes, 89, 90
Substrate-level phosphorylation, 115–116
Sucrose, mass *vs.* molarity of, 312
Sugars
    carbohydrates, 38
    cellular respiration, 113–114
    DNA/RNA structure, 192
    nucleotides, 41
    photosynthesis, 101–104
Sulfur, macromolecules, 37
Supercoiling, 193
Surface area to volume ratios, 62, 404
Surface tension, water molecules, 26
Sympatric speciation, 267
Synapsis, 153–154
Systems interactions, 3

T
Tactile signals, 279
Target cells, 129
TATA box, 201
Telophase
    meiosis I and II, 152–155
    mitosis, 140
Temperature
    enzymes and, 90
    of solution, 75
    water potential, 73–74
Template strand, DNA, 202
Temporal isolation, 268
Termination, of translation, 205
Tertiary protein structure, 40, 41
Test cross, 168
Testosterone, 39
Tetrads, 153–154
Thylakoids, 59, 103
Thymine, 41–42, 191–192, 201
Top-down ecosystem regulation, 282–283
Topoisomerase, 193
Total water potential, 402–403
Transcription, prokaryotes *vs.* eukaryotes, 201–203
Transcription factors, 201, 216
Transduction
    mutations, 217
    signal transduction, 129
Transfer RNA (tRNA), 201
    translation, 203–205
Transformation, mutations, 217
Transition, 91–92
Translation
    codons, 318
    double-stranded RNA, 216
    prokaryotes and eukaryotes, 203–205
    ribosomes, 56

Translocations, 154, 204–205
Transpiration, 409
    lab experiment, 321
Transpirational pull, 321
Transpiration-photosynthesis compromise, 321
Transport, active and passive, 64–65
Transposition, 217
Trichomes, 308–309
Triglycerides, 39
Trisomy 21, 155, 217
Trophic cascades, 296
Trophic levels, 280–282
Trp operon, 215
Tryptophan, 215
Tumor cells, 269
Tumor suppressor genes, 141
Turgor pressure, 312

U
Unsaturated fatty acids, 38–39
Upregulation, gene expression, 214–215
Uracil
    classification, 42
    DNA structure, 191–192

V
Vacuoles, 58, 76
Vesicles, 57, 61, 65
Vestigial structures, 237–238
Viability, hybrids, 268
Villi, 62
Visual signals, 279–280
Vitamins, as coenzymes and cofactors, 91
Volume
    cell, 58
    equations, 403–404

W
Water
    cell movement in, 73–76
    cohesive and adhesive molecules, 26
    equations, 402–403
    freezing and expansion, 27
    hydrogen bonds, 25–27
    pH, 27–28
    photolysis, 102–103
    polarity, 25–26
    potential, 73–75, 310–312, 402–403
    as solvent, 27
Weathering, 410
Weinberg, Wilhelm, 255

X
X chromosomes, 178
X-Y graph, 406

Y
Y chromosome, 178
Yeast, 118

Z
Zygotes, meiosis I and II, 153